Cholesterol in Membrane Models

Editor
Leonard Finegold
Department of Physics
Drexel University
Philadelphia, Pennsylvania

CRC Press
Boca Raton Ann Arbor London Tokyo

Library of Congress Cataloging-in-Publication Data

Cholesterol in membrane models / edited by Leonard Finegold.
p. cm.
Includes bibliographical references and index.
ISBN 0-8493-4207-4
1. Cholesterol—Physiological effect. 2. Lipid membranes. 3. Bilayer lipid membranes. I. Finegold, Leonard.
[DNLM: 1. Cell Membrane—metabolism. 2. Cholesterol—metabolism. 3. Models, Biological. QU 95 C5465]
QP752.C5C52 1993
599′.087′5—dc20
DNLM/DLC
for Library of Congress

92-49640
CIP

Direct all inquiries to CRC Press, Inc., 2000 Corporate Blvd., N. W., Boca Raton, Florida, 33431.

International Standard Book Number 0-8493-4207-4

Library of Congress Card Number 92-49640
Printed in the United States 1 2 3 4 5 6 7 8 9 0
Printed on acid-free paper

To Mary, Alan,
Geoffrey, and Gina

From the Editor

INTRODUCTION

We explain the rationale and timeliness of this volume. Cholesterol is an essential component of real membranes, and its "purpose" has yet to be explained. In the last few years, basic molecular modeling and experiments on cholesterol in membranes have approached one another, so this is an appropriate time for a volume. The reasons for this approach lie in (a) the establishment of "liposome" or "multibilayer" membrane systems as good models for biological membranes, (b) the refinement of many physical techniques so they can be powerfully applied to the model systems, (c) the advancement of high-speed computers and concomitant theoretical models, and (d) the interest of researchers in the general problems of membranes and cholesterol. We shall concentrate on physical models. An introductory biological chapter makes the connection with mammalian cells.

Model membranes are those composed of perhaps only one or two kinds of simple phospholipids, typically of the same chain length, whereas biological membranes contain many varieties of phospholipids, of many chain lengths. A measure of the present interest in model membranes is given by Figure 1, which shows the increase in the number of published papers effectively treating lipids—which are the main architectural components of membranes—over the years. Caffrey et al. (1991a, 1991b) have established a database *Lipidat* for lipid phase transitions—which occur in model membranes—covering the years since 1966, and have kindly provided the data from which Figure 1 is made. A similar recent rise appears in the number of *Lipidat* papers in which cholesterol is a component.

The topic of cholesterol in model membranes will appeal to a wide range of readers, from medical and cell researchers who require a simple system on which to test ideas derived from *in vivo* observations, to experimenters and theorists who are expert in various techniques, and wish to apply themselves to cholesterol and membranes. This volume will be useful to people already working in the area, and we trust that it will become a source book.

Presentation. Each chapter begins with an overview, deliberately written at a level to make the chapter accessible to a wide audience. (Readers who wish to refresh themselves on membranes are referred to the attractively presented chapters in recent cell biology or biochemistry texts, e.g., the latest edition of the university level texts by Alberts et al. 1989 and Darnell et al. 1990.) Chapters then proceed to the state-of-the-art. Authors were encouraged to finish with an adventurous statement of future evolution of their topic. So that the book will continue to be a valuable reference in the years to come, each chapter incorporates useful references, complete with title. Detailed author and subject indexes are provided.

A historical note to explain the genesis of the volume. During a Biophysical Society meeting in Baltimore in 1990, several people who were working in membranes encountered each other randomly, and refreshed themselves on what the others were doing. They discovered that they were all now working on

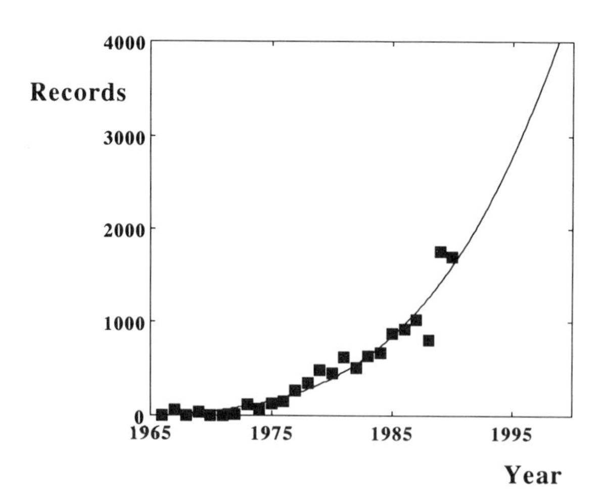

FIGURE 1. The number of published references on lipid phase transition properties increases with time: the number of records in the *Lipidat* database vs. year (Caffrey et al., 1991a, 1991b), replotted from data generously provided by Drs. J. Hogan and M. Caffrey. The line is to aid the eye. (Figure composed by Camu Zdat, who also did the word processing for this editorial.)

aspects of cholesterol in membrane models, and for all of them this was a new direction. After one of the symposia, some of this group happened to meet and, during the usual suggestions of improving the state of the world, also noted (with some force) the absence of biophysics cholesterol-in-membranes at the meeting. One of us (modesty forbids identification) suggested that this topic be considered as a workshop topic for the next Biophysical Society meeting; a proposal was duly made to the Society, duly accepted and—after many a phone and fax communication—a workshop resulted in San Francisco (*Biophysical Journal* 59:2 1991). The format of the workshop was that speakers presented "opposing" viewpoints and indicated future directions. The success of the workshop (which effectively ended after midnight at a local bar/pizza parlor) showed that the present is a good time for a complete volume on the topic.

THE CONTRIBUTIONS

How does one give an overview of the setting of the chapters? The first chapter, "The Biophysics and Cell Biology of Cholesterol: An Hypothesis for the Essential Role of Cholesterol in Mammalian Cells" by P. L. Yeagle, stands alone, in that this chapter provides an essential link between the physically minded reader and the biologically minded one, to explain to the former the biological context in which one is modeling, and to explain to the latter why they should know about membranes. This first chapter is also an overview of "why" cholesterol exists in mammalian cell membranes, and why it is so important for cell biology. Why did Nature choose *cholesterol*, out of "all" possible sterically similar molecules? The answer involves the evolution of enzymatic pathways. Cholesterol is the sterol in animal membranes; plants use other sterols. What is it about cholesterol which makes it unique? A way of looking at this is to consider variations of cholesterol. Subtle variations (e.g., on which side of the molecule an OH group appears) are extremely important biochemically, but have to be ignored completely at the present stage of modeling.[1] (One overall physical viewpoint is mentioned in the chapter by M. J. Zuckermann et al.; an accessible description is Bloom's, 1992.)

Research into membranes demands the application of many biochemical and physical techniques, which probe different aspects of time and of space. Although each researcher tends to feel that her or his approach is the royal road, in practice various techniques complement one another. The order of chapters presented in this book reflects this diversity of approach, in that after the first chapter, the remaining chapters are simply ordered alphabetically by authors' names. (A thematic experimental classification night have been the time scale of the molecular motions being probed, as in Singer, 1991.) One arbitrary way to classify the chapters is along the theory-experiment continuum.

The two theoretical contributions (of H. L. Scott and M. J. Zuckermann et al.) illustrate a fundamental problem faced by theorists: to balance the conflicting demands of a (perhaps too-) simple model system which can be solved mathematically (or numerically by computer) to one's mathematical satisfaction, and a more realistic system which is too complex to permit mathematical solution, and in which the mathematics must be simplified. Two time-tested approaches in condensed-matter (i.e., including liquid- and solid-like state) physics present different viewpoints: one can model a very few (say, a hundred) cholesterol and lipid molecules, by making them on one's desk from a biochemist's kit (e.g., CPK Corey-Pauling-Koltun) of space-filling scaled atoms, and shaking them to imitate thermal agitation. All this, with interactions, is effectively done by computer using the "Monte Carlo" method[2] described by H. L. Scott in his chapter "Lipid-Cholesterol Phase Diagrams: Theoretical and Numerical Aspects." The power of the Monte Carlo method is that it is transparent: it is easy to see what approximations are introduced. This clarity is obtained at the large cost of computation. The jiggling of even a hundred molecules requires the fastest supercomputers available, which are exceedingly expensive. The other time-honored approach, taken by M. J. Zuckermann, J. H. Ipsen, and O. Mouritsen in their chapter "Theoretical Studies of the Phase Behavior of Lipid Bilayers Containing Cholesterol," is immediately to average over all the molecules by the "mean-field" method. The advantage of the mean-field method is that initial averaging is free of extensive computer use. This gave the correct phase diagram, but it becomes hard to see just how sensitive are the final results to the initial assumptions. This advantage can to some extent be tested by also using Monte Carlo methods for the simple models.

---

[1] Cholesterol is, of course, an essential and necessary component of mammalian membranes (P. L. Yeagle, this volume). Not all cholesterol is good to all mammals: One is often asked about cholesterol in one particular mammal, where it has received a bad press. Anthropocentric questioners can be confidently referred to an excellently written popular – yet rigorous – small book (Yeagle, 1991).

[2] A sampling technique, named after gambling activities in that city.

In nature, cholesterol is not manufactured in the membrane, so the question immediately arises as to how cholesterol becomes inserted into membranes. Synthetic membranes, as topologically closed sphere-like "vesicles" or liposomes, provide models for cells. R. Bittman's chapter "Kinetics of Cholesterol Movement Between Donor and Acceptor Bilayer Membranes" describes the spontaneous movement of cholesterol from one vesicle to another, via diffusion in the surrounding medium or even direct contact from a cholesterol-donating membrane to a cholesterol-receiving membrane.

Many interactions of cholesterol with the lipid bilayer are conveniently measured thermodynamically by a "phase diagram." A homely example of a phase transition is that between the two phases of solid ice and liquid water. Phase diagrams (a plot of the temperature of a phase transition vs. the fraction of the component—here the cholesterol concentration) use phase transitions to delineate the many phases which are found in lipid systems—see for example the last figure in J. H. Davis' chapter. Even more phases are seen in all systems as the pressure is increased above atmospheric pressure (e.g., Driscoll et al., 1991). Much of the current interest in the physical modeling of cholesterol in membranes stems from the publication by M. Vist and J. H. Davis of a comprehensive phase diagram for cholesterol in lipid membranes (see Chapter 4 by J. H. Davis). Theoretical phase diagrams, to compare with experiment, are given in the chapters by H. L. Scott and by M. J. Zuckermann et al.; H. L. Scott also poses and answers the question of the least number of parameters needed to describe cholesterol-lipid phase diagrams. In his chapter, "The Molecular Dynamics, Order and the Thermodynamic Phase Equilibria of Cholesterol/Phosphatidylcholine Mixtures," J. H. Davis gives the detailed fundamentals of nuclear magnetic resonance involving deuterium, and its applications to the effect of cholesterol on phospholipid membranes. This extensive discussion leads to the pivotal phase diagram studies, and a connection of the deuterium nuclear magnetic resonance results with those of freeze fracture and neutron scattering.

The spatial distribution of cholesterol in model membranes, as it is distributed in the plane of the membrane in different lateral phases, is discussed by S. W. Hui in "Visualization of Cholesterol Domains in Model Membranes." The microscopy uses light and electrons, probing the geometric areas defined by the light and electron beams; visualization of cholesterol employs fluorescence light microscopy, freeze-fracture electron microscopy, and diffraction contrast electron microscopy. The limitations of present results, and their extrapolation to biological membranes, are carefully discussed.

A powerful technique for studying condensed matter is that of infrared and Raman spectroscopy, which probe the vibrations of molecules. Of recent years, instrumental advances have enabled us to examine tiny quantities of lipids, aided by sophisticated computer averaging. T. J. O'Leary's chapter "Vibrational Spectroscopy of Cholesterol-Lipid Interactions" shows one how to interpret the relevant vibrational peaks in the spectra of phospholipid bilayers, and then proceeds to study the same systems with the addition of cholesterol. The effects of cholesterol on model systems are complex; a brief discussion is given of the effects of cholesterol upon enzymes in model membranes, and upon erythrocyte membranes (which are themselves regarded as models for the membranes of nucleated cells).

The previously discussed experimental techniques in this volume are "table-top" in that the apparatus resides on a laboratory bench, or at most occupies a single laboratory room at one's home institution. Neutron diffraction demands a high flow of neutrons, which are supplied by nuclear fission in a controlled chain reaction, so the "apparatus" is the size of a large building, and is generally a national resource. T. M. Bayerl and E. Sackmann in "Neutron Scattering as a Tool to Probe the Effect of Cholesterol on Phase Separation, Structure Formation, and Lipid/Protein Interaction in Artificial Membranes" discuss a series of sophisticated and cunning methods using neutrons on model membranes. It is the nature of neutrons to probe matter differently from X-rays and light; there is an especially large difference between hydrogen and the biologically practically-identical deuterium, which gives a very useful differential contrast.

Biological membranes contain phospholipids of a variety of acyl chain lengths. A technique for directly exploring the thermodynamics of cholesterol–lipid interactions, which is free of exogeneous probes, is that of differential scanning calorimetry. L. Finegold and M. A. Singer, in "Cholesterol/Phospholipid Interactions Studied by Differential Scanning Calorimetry: Effect of Acyl Chain Length and Role of the C(17) Sterol Side Group," use this approach, with a systematic variation of acyl chain length and of cholesterol analogs, to find which part (mainly the ring structure) is of prime importance in cholesterol–lipid interactions. The results enable predictions to be made of cholesterol–lipid phase diagrams for lipids of chain lengths 10 through 20, and a dynamic model is presented.

THE REVIEWERS

Each manuscript was read by at least one carefully chosen reviewer in the appropriate field, as well as by me. Comments and suggestions were incorporated, and the reviewers apprised of the changes. The reviewers were critical, thorough and thoughtful, and their effort is an important part of the book. We thank them. They are, in alphabetical order:

Martin Zuckermann, Department of Physics, McGill University, Montreal
Philip Yeagle, Department of Biochemistry, State University of New York, Buffalo
David Siminovitch, Department of Physics, University of Guelph, Guelph
Xi Shan, Department of Physics, University of Guelph, Guelph
Larry Scott, Department of Physics, Oklahoma State University, Stillwater
Erich Sackmann, Department of Physics, Technical University, Munich
Scott Prosser, Department of Physics, University of Guelph, Guelph
Michael Phillips, Department of Physiology and Biochemistry, Medical College of Pennsylvania, Philadelphia

Dianne O'Leary, Department of Cellular Pathology, Armed Forces Institute of Pathology, District of Columbia
John Nagle, Department of Physics, Carnegie-Mellon University, Pittsburgh
Ole Mouritsen, Department of Physical Chemistry, Technical University of Denmark, Lyngby
Michael Morrow, Department of Physics, University of Guelph, Guelph
W. Scott McCullough, Department of Physics, Oklahoma State University, Stillwater
Jeffrey Mason, Department of Cellular Pathology, Armed Forces Institute of Pathology, District of Columbia
John Ipsen, Department of Physical Chemistry, Technical University of Denmark, Lyngby
Myer Bloom, Department of Physics, University of British Columbia, Vancouver
Michele Auger, Department of Physics, University of Guelph, Guelph.

Research is a living entity. The contributors give our present knowledge; in most cases the body of informed opinion is in agreement. There exist cases where people of good will may hold divergent views on certain aspects, and this is so in an instance in the present volume. These authors have reviewed each other's chapter, have referenced each other, and still beg to differ. It has been uplifting to me to observe this friendly difference, in the highest traditions of science.

THE VISUALIZATION OF MEMBRANES

One finds it useful to see ideas of membranes expressed as pictures ("and what is the use of a book ... without pictures ... ?" (Carroll, 1990)), even though the pictures perforce are cartoons. An example is the seminal drawing of the fluid mosaic model (Singer and Nicolson, 1972). Excellent scale molecular models of membranes, drawn by S. Smith, are shown in Robertson's 1983 book. It is arresting to see membranes drawn to scale in the context of the cytoplasm, in all its complexity (Goodsell, 1991). Good drawings—helpful for giving colloquia—are also to be found in the pages of *Scientific American* and *Trends in Biochemical Sciences*. Printed stereo drawings are becoming common; most of us can see them without any optical aids, such as stereoscopic viewers.

INFORMATION TOOLS

To extend the future usefulness of this book, this paragraph is offered. Of the making of books, there *is* an end, for eventually they must be printed in "hard copy." The following suggestions are made to answer the questions "What are good background references?", "What has been done in the field?" and "What is the current thinking?" Current answers to all three will be found in this book, with the help of the index. A useful book is that of Small, 1986, which gives much data in good graphical or pictorial form, with extensive description and discussion. Marsh (1990) gives mainly tabular data.

Further tools are databases such as the Index Medicus of the National Library of Medicine (U.S.A.) and Chemical Abstracts, via print media, optical compact disc storage (and whatever storage forms the future brings) or through an on-line computer service. Still, these resources have limitations at present in that they are accessed by index words (instead of by a full text "wild-card" search) and lipid data (e.g., nomenclature or phase transitions) may be referred to in many ways, or perhaps missed by the indexers. Three databases which specialize in lipids and membranes will now be discussed; they differ in approach, frequency of renewal, and cost. The Caffrey and colleagues computer compendium (Caffrey et al., 1991a, 1991b, 1993) treats the problems of very large databases, will include pictorial phase diagrams, and is to be widely distributed. It is intended that the printed version (Caffrey et al., 1993) will be periodically updated. This database *Lipidat* (LIPid DATa) is somewhat unusual in that whole papers are carefully screened by workers who are active in the field, since quite often lipid data will be missed by a computer search of only title and abstract. Also, contributors to journals (*Biochimica et Biophysica Acta, Biophysical Journal* and *Chemistry and Physics of Lipids*) are now asked to directly submit lipid phase transition data from their papers directly to *Lipidat*.[3] (Contact: Martin Caffrey, *Lipidat*, Department of Chemistry, The Ohio State University, 120 West 18th Avenue, Columbus, Ohio 43210, phone: (614) 292-8437, fax: (614) 292-1532, E-mail: caffrey+@ohstmail, E-mail: caffrey+@osu.edu; available in an IBM-PC oriented version as National Institute of Standards and Technology Standard Reference Database 34, phone (301) 975-2208; fax (301) 926-0416). The University of Sheffield has long issued periodical printed summaries (author, reference, title) of papers, books and symposia in many areas, produced by human searching. Of interest here is the useful "Cell Membranes", "Cholesterol and Lipoproteins" and "Liposomes" bulletins, of which they send a copy to authors whose publication they listed. They now issue also bulletins (with authors and their addresses, reference and abbreviated titles) on diskette as *Subidex*. For IBM-PC DOS-compatible users, free searching and database (*Subibase*) software is provided; Apple Macintosh users will have to do a little work. (Contact: Sheffield University Biomedical Information Service, Sheffield S10 2TN, U.K., phone: (+44) (0)742-768555 x6234, fax: (+44) (0)742-7797066.) Another database is *Lifo* (LIposome InFOrmation Electronic Liposome Library), which is distributed by diskette, and is provided with searching software. Authors, titles, references and keywords are continually updated; some Boolean searching is possible. This database is available only in IBM-PC DOS-compatible format.[4] (Contact Mark Ruzbie, Avestin Inc., 60 MacLaren St., Suite #404, Ottawa, Canada, K2P 0K7, phone: (613) 234-7709, fax: (613) 563-8214.) These data banks are the only ones of which I'm aware in the field. The data banks try to distribute the data so that it can be read by any software; fortunately there are now computers which can read both IBM-PC DOS-compatible and Apple

---

[3] I found a preliminary Macintosh-oriented version to be quite convenient.

[4] I was not able to test Lifo, since I do not have an IBM-PC DOS-compatible machine: The limitation is mine.

Macintosh formats, so reducing the Tower-of-Babel effect. I am grateful to Martin Caffrey (*Lipidat*), Teresa Hourihan (*Subidex*) and Mark Ruzbie (*Lifo*) for information.

It has been said that one half of human information is to be found in libraries, and the other half is to be found in humans. Researchers are part of "invisible colleges" (Ziman, 1976); often the fastest and best way of finding information is by contacting someone, be it in person, by telephone, fax, electronic mail or by letter. To this end, contact information for the authors of this volume is given with their addresses. The abstracts of the annual meetings of societies such as the Biophysical Society of one's country are helpful, and some abstracts are issued on floppy disc with the facility to search for any word. There exist also local biochemists' and lipids clubs.

Technology of text preparation. In a decade's time perhaps, authors will dictate manuscripts into their portable computers, and then electronically mail the text and figures to the publisher, using completely compatible software and protocols. In the interim, the present volume was prepared with a variety of software, some of which is very sophisticated, and transmitted by floppy disc (with paper copy) to the publishers. This editor now has an enhanced appreciation of the complexities—all of which are mercifully unseen by the reader—of producing a book, and thanks Publisher Jeff Holtmeier, Carol Campbell, Chris Richardson and their colleagues at CRC Press for their friendly cooperation.

ACKNOWLEDGMENTS

Eons ago, the American Cancer Society and the National Institutes of Health gave me a Research Scholar Award and a Special Fellowship to educate me in membranes, with S. Jon Singer at the Department of Biology, University of California at San Diego, in La Jolla. I like to think that this book repays, in some measure, the debt that I owe all three.

Michael Singer, of Queen's University, over many years has patiently listened to me, and reined in my often all-too-exuberant imagination. I thank him for this free exchange, which has been one of the delights of research.[5]

REFERENCES

Alberts, B., D. Bray, J. Lewis, M. Raff, K. Roberts, and J. D. Watson. 1989. *Molecular Biology of the Cell,* 2nd Ed. Garland Publishing, New York.

Bloom, M. 1992. The physics of soft, natural materials, *Physics in Canada/ La Physique au Canada,* 48:7–16.

Caffrey, M., D. Moynihan and J. Hogan. 1991a. A database of lipid phase transition temperatures and enthalpy changes, *Chem. Phys. Lipids,* 57:275–291.

[5] Coincidence of names: Michael is no relation to Jon.

Caffrey, M., D. Moynihan and J. Hogan. 1991b. A database of lipid phase transition temperatures and enthalpy changes, *J. Chem. Inf. Comput. Sci*, 31:275–284.
Caffrey, M. 1993. LIPIDAT: A Centralized Database for Thermodynamic Data on Lipid Mesomorphic and Polymorphic Phase Transitions, CRC Press, Boca Raton, FL, in press.
Carroll, L. 1990. Alice's Adventures in Wonderland, in *The Annotated Alice,* M. Gardner, Ed., Random House, New York, chap. 1.
Darnell, J., H. Lodish and D. Baltimore. 1990. *Molecular Cell Biology*, Freeman and Company, New York.
Driscoll, D. A., J. Jonas and A. Jonas. 1991. High pressure ^{2}H nuclear magnetic resonance study of the gel phases of dipalmitoylphosphatidylcholine, *Chem. Phys. Lipids,* 58:97–104.
Goodsell, D. S. 1991. Inside a living cell, *Trends Biochem. Sci,* 16:203–206.
Marsh, D. 1990. *CRC Handbook of Lipid Bilayers,* CRC Press, Boca Raton, FL.
Robertson, R. N. 1983. *The Lively Membranes,* Cambridge University Press, New York.
Singer, S. J. and Nicolson, G. L. 1972. The fluid mosaic model model of the structure of cell membranes, *Science,* 175:720–731.
Singer, M. A. 1991. Time and the Membrane. (A Symposium.) 34th Annual Meeting, Canadian Federation of Biological Societies/Federation Canadienne des Societes de Biologie, Kingston, Ontario.
Small, D. M. 1986. *The Physical Chemistry of Lipids,* Plenum Press, New York.
Yeagle, P. L. 1991. *Understanding Your Cholesterol,* Academic Press, New York.
Ziman, J. M. 1976. *The Force of Knowledge: the Scientific Dimension of Society,* Cambridge University Press, New York.

AUTHOR ADDRESSES

Thomas M. Bayerl

Neutron Scattering as a Tool to Probe the Effect of Cholesterol on Phase Separation, Structure Formation, and Lipid/Protein Interaction in Artificial Membranes

Physik Department E22
Technische Universitat Munchen
D-8046 Garching
Munich, Germany
89-3209-2480
Fax 89-3209-2469
E-mail thomas.bayerl@bl.physik.tu-muenchen.de

Robert Bittman

Kinetics of Cholesterol Movement between Donor and Acceptor Bilayer Membranes

Department of Chemistry and Biochemistry
Queens College of The City University of New York
Flushing, NY 11367, USA
(718) 997-3279
Fax (718) 997-3349
E-mail bittman@qcvax.acc.qc.edu

James H. Davis

The Molecular Dynamics, Orientational Order and the Thermodynamic Phase Equilibria of Cholesterol/Phosphatidylcholine Mixtures

Department of Physics
University of Guelph
Guelph, Ontario, Canada N1G 2W1
office (519) 824-4120 x 2659
lab x 8541
Fax (519) 836-9967
E-mail phydavis@vmucguelph.ca

Leonard Finegold

Cholesterol/Phospholipid Interactions Studied by Differential Scanning Calorimetry: Effect of Acyl Chain Length and Role of the C(17) Sterol Side Group

Department of Physics
Drexel University
Philadelphia, PA 19104, USA
(215) 895-2740
Fax (215) 895-5934
E-mail lxf@coasmail.physics.drexel.edu

Sek Wen Hui

Visualization of Cholesterol Domains in Model Membranes

Membrane Biophysics Laboratory
Roswell Park Cancer Institute
Buffalo, NY 14263, USA
(716) 845-8595
Fax (716) 845-3545
E-mail roswhui@ubvms.cc.buffalo.edu

John H. Ipsen

Theoretical Studies of the Phase Behavior of Lipid Bilayers Containing Cholesterol

Department of Physical Chemistry
The Technical University of Denmark
2800-Lyngby, Denmark
45-931222 –2467
Fax 45-934808
E-mail ipsen@lipid.fki.dth.dk

Ole G. Mouritsen

Theoretical Studies of the Phase Behavior of Lipid Bilayers Containing Cholesterol

Department of Physical Chemistry
The Technical University of Denmark
2800-Lyngby, Denmark
45-931222 –2462
Fax 45-934808
E-mail fymearn@vm.uni-c.dk

Timothy J. O'Leary

Vibrational Spectroscopy of Cholesterol-Lipid Interactions

Department of Cellular Pathology
Armed Forces Institute of Pathology
Washington, DC 20306, USA
(202) 576-2915
Fax (202) 576-2164

Erich Sackmann

Neutron Scattering as a Tool to Probe the Effect of Cholesterol on Phase Separation, Structure Formation, and Lipid/Protein Interaction in Artificial Membranes

Physik Department E22
Technische Universitat Munchen
D-8046 Garching
Munich, Germany
89-3209-2471
Fax 89-3209-2469

H. Larry Scott

Lipid-Cholesterol Phase Diagrams: Theoretical and Numerical Aspects

Department of Physics
Oklahoma State University
Stillwater, OK 74078, USA
(405) 744-5796
Fax (405) 744-6811
e-mail u3029ab@unx.ucc.okstate.edu

Michael A. Singer

Cholesterol/Phospholipid Interactions Studied by Differential Scanning Calorimetry: Effect of Acyl Chain Length and Role of the C(17) Sterol Side Group

Department of Medicine
Queen's University
146 Stuart Street
Kingston, Ontario, Canada K7L 4N6
(613) 545-6197
Fax (613) 548-0686

Philip L. Yeagle

The Biophysics and Cell Biology of Cholesterol: An Hypothesis for the Essential Role of Cholesterol in Mammalian Cells

Department of Biochemistry
School of Medicine and Biomedical Sciences
140 Farber Hall
University at Buffalo (SUNY)
Buffalo, NY 14214, USA
(716) 829-2700
Fax (716) 829-2725
E-mail bchphil@ubvms.cc.buffalo.edu

Martin J. Zuckermann

Theoretical Studies of the Phase Behavior of Lipid Bilayers Containing Cholesterol

Department of Physics and Centre for the Physics of Materials
McGill University
Montreal, Quebec, Canada H3A 2T8
(514) 398-6524
Fax (514) 398-8434
E-mail martin@physics.mcgill.ca

CONTRIBUTORS

Thomas Bayerl
Department of Physics
Technical University of Munich
Munich, Germany

Robert Bittman
Department of Chemistry and
Biochemisty
Queens College of The City
University of New York
Flushing, New York, USA

James H. Davis
Department of Physics
University of Guelph
Guelph, Ontario, Canada

Leonard Finegold
Department of Physics
Drexel University
Philadelphia, Pennsylvania, USA

Sek Wen Hui
Department of Biophysics
Roswell Park Cancer Institute
Buffalo, New York, USA

John Hjort Ipsen
Department of Physical Chemistry
Technical University of Denmark
Lyngby, Denmark

Ole G. Mouritsen
Department of Physical Chemistry
Technical University of Denmark
Lyngby, Denmark

Timothy J. O'Leary
Department of Cellular Pathology
Armed Forces Institute of
Pathology
Washington, DC, USA

Erich Sackmann
Department of Physics
Technical University of Munich
Munich, Germany

H. Larry Scott
Department of Physics
Oklahoma State University
Stillwater, Oklahoma, USA

Michael A. Singer
Department of Medicine
Queen's University
Kingston, Ontario, Canada

Philip L. Yeagle
Department of Biochemistry
School of Medicine and
Biomedical Sciences
University at Buffalo (SUNY)
Buffalo, New York, USA

Martin J. Zuckermann
Department of Physics
McGill University
Montreal, Quebec, Canada

To
Curiosity
and to
Hope

"Curiouser and curiouser!" cried Alice. . .

Lewis Carroll, *Alice in Wonderland* (chapter 2, line 1)

True hope is swift, and flies with swallow's wings;
Kings it makes gods, and meaner creatures kings.

William Shakespeare, *Richard III* (act V, scene II, line 23)

TABLE OF CONTENTS

Chapter 1

THE BIOPHYSICS AND CELL BIOLOGY OF CHOLESTEROL: AN HYPOTHESIS FOR THE ESSENTIAL ROLE OF CHOLESTEROL IN MAMMALIAN CELLS

Philip L. Yeagle

TABLE OF CONTENTS

0-8493-4207-4/93/$0.00+$.50

ABSTRACT

An understanding of the essential role of cholesterol in mammalian (and other cholesterol-requiring) cells has long been the object of intense interest. Cholesterol can stimulate or inhibit the function of membrane proteins critical to cell viability. The specific sterol required varies from one cell type to another and is unrelated to the ability of that sterol to affect the bulk properties of the membrane. An encompassing hypothesis can now be described that includes two mechanisms for this modulation. In one mechanism, the sterol modulates membrane protein function through direct sterol-protein interactions. This mechanism provides an explanation for the *stimulation* of the activity of important membrane proteins and for the essential requirement of a structurally specific sterol for cell viability. In the other mechanism, the requirement of "free volume" by integral membrane proteins for conformational changes as part of their functional cycle is antagonized by the presence of high levels of cholesterol in the membrane.

I. INTRODUCTION

Cholesterol, the topic of this book, is one of a family of neutral lipids with a common chemical structure: the steroid ring system. Sterols are found in most plant and animal cells. However, all plants and animals do not utilize the same sterols. Each sterol-requiring species requires its own specific sterol. In the animal kingdom, including in humans, cholesterol is the most common essential sterol.

The role of sterols in the functioning of cells is one of the great unanswered questions of cell biology and biochemistry. Sterols are widely found to be a requirement for cell growth and cell differentiation. Sterols are synthesized by complicated pathways independent of the pathways for the synthesis of the other common lipids. Thus, considerable cellular energy is expended in producing specific sterol structures. This would lead to the deduction that sterols are important to cell survival and, further, that only a particular sterol structure is maximally useful to a particular cell type.

Yet the accumulation of the mammalian sterol, cholesterol, in the form of cholesterol ester and free cholesterol in atherosclerotic plaques is deadly to the human organism. Thus, one has an apparent dichotomy between the essential and lethal natures of one natural compound. In mammalian systems both the requirement for cholesterol and the lethal nature of cholesterol have attracted an intense level of research (for reviews, see Yeagle, 1985; Yeagle, 1988). Part of this research effort has uncovered fascinating capabilities of cholesterol to modify the physical properties of lipid bilayers and biological membranes. The results of some of those efforts will be evident in other chapters in this book.

Some of the fundamental properties of cholesterol should be kept in mind at all times when considering the role cholesterol may play in cells. For example, cholesterol is a largely hydrophobic compound. The only polar functionality on cholesterol is the hydroxyl. The remainder of the molecule is hydrophobic. Thus, as is typical for amphipathic compounds, cholesterol must be found in hydrophobic environments with a polar interface, such as is offered by lipid bilayers: both simple bilayers and lipid bilayers that are part of cell membranes.

Within a cell, unesterified cholesterol is found exclusively in the membranes of the cell. Interestingly, there is a disproportionation of cholesterol among the membranes of typical mammalian cells. Apparently as much as 90% of the total cellular cholesterol is found in the plasma membrane of the cell (Lange and Ramos, 1983). Therefore, when searching for the role of cholesterol in mammalian cells, the focus of such investigations would properly be on the influence of cholesterol, most fundamentally on membrane properties and functions, and most specifically on the structure and function of the plasma membrane of cholesterol-requiring cells.

The other major repository of unesterified cholesterol in mammalian systems is in the circulating serum lipoproteins. For example, low-density lipoprotein (LDL) is the major serum carrier of cholesterol in humans. The LDL offers the amphipathic cholesterol an environment in the surface lipid monolayer that is similar to the environment offered by a cell membrane. Esterified cholesterol, in which the polarity of the hydroxyl is largely neutralized by involvement in the ester bond, is found in the hydrophobic core of the LDL.

This chapter will concern itself with using the described behavior of cholesterol in membrane systems to understand the essential role of cholesterol in mammalian cells. The goal will be to attempt to synthesize what has been learned about cholesterol into a unifying hypothesis for the role of cholesterol in cell biology. This chapter will end with a look at the future of this large field: both the future to be explored by the membrane biochemist and cell biologist and the future to be explored by the biophysicist.

II. STEROL SPECIFICITY FOUND IN NATURE

A variety of sterol structures are found in cells, each species emphasizing a particular sterol structure. As is well known, the dominant and the required sterol in mammalian cells is cholesterol. Mammalian cells do not grow in the absence of cholesterol (i.e., when endogenous cholesterol biosynthesis is inhibited and the cells are in a medium depleted of cholesterol). However, in yeast the dominant and required sterol is ergosterol. The differences in chemical structure between ergosterol and cholesterol lead to significantly different physical properties of ergosterol in bilayers. Yet ergosterol is the required sterol for yeast growth and is the sterol synthesized by yeast. This observation provides one of the important clues to understanding the role of sterols in biology and will be discussed in detail later.

Although there are many sterol structures found in nature, one more example should illustrate the point under consideration here. Sitosterol and stigmasterol are common plant sterols. These plant sterols are not normally found in abundance in mammalian systems and cannot substitute for cholesterol in the mammalian systems. In humans, dietary cholesterol is effectively absorbed, while dietary sitosterol, campesterol, and stigmasterol are not effectively absorbed in most cases (less than 10% of the absorption efficiency of cholesterol). There is obviously a selectivity for sterols based on their chemical structure, even though they are all largely hydrophobic and will partition into membranes. It has been suggested that the differences in chemical structure of these sterols leads directly to differences in their absorption efficiency (Kan and Bittman, 1991).

These observations lead to the deduction that most cells require a sterol of a specific chemical structure for proper cell function. Cholesterol cannot substitute for ergosterol in an ergosterol-requiring cell (for example, yeast), and ergosterol cannot substitute for cholesterol in a cholesterol-requiring cell (for example, mammalian cells). It is reasonable to suggest, therefore, that sterols must play important roles in cell biology and that while the roles may be analogous from one cell type to another, finely tuned recognition systems must play a role so that subtle changes in the chemical structure of the sterol structure (from the required structure) render it incompetent for its essential role in a foreign cell type. As will be seen below, at least some of the roles sterols must play in cells then become deconvoluted from the physical chemical effects of the sterols on phospholipid bilayers. Therefore, this discussion will eventually be split into two separate parts: one in which the physical chemical properties of the sterols are related to biological effects, and a second in which the specificity of biological effects with respect to the chemical structures of the sterols are understood in terms of recognition systems for sterol structure that are specific to each cell type.

III. "COST" OF PRODUCING CHOLESTEROL

Sterols are synthesized according to long, complicated, and energetically expensive biosynthetic pathways. To synthesize cholesterol from acetyl-CoA requires about 30 enzymatically catalyzed steps, an extensive use of reducing equivalents (NADH and NADPH), a number of enzymes and cofactors (all of which must be synthesized by the cell), and molecular oxygen. To quote from a recent review: "the process of synthesizing a cholesterol molecule . . . is a laborious task costing the cell a great deal of reduced pyridine nucleotide consumption presumably due to the importance of the involvement of the end product in overall cellular function" (Faust et al., 1988).

Along the biosynthetic pathway of cholesterol, a large number of sterols are synthesized as intermediates. Starting with lanosterol as the first sterol intermediate synthesized in the cholesterol biosynthetic pathway, 19 different sterols are made on the way to cholesterol. Extensive cellular energy in the form of, for example, NADPH is utilized. Molecular oxygen is also extensively utilized, which was only possible, in an evolutionary sense, in higher organisms after oxygen became available.

The question that is inexorably raised by these facts is why expend so much cellular energy to make cholesterol? Or any other cell-specific sterol? As indicated in the quote above, it is reasonable to propose that (for cholesterol-requiring cells) cholesterol fulfills important functions not fulfilled by other sterol structures. As will be seen below, cholesterol affects essential cellular functions through the modulation of particular membrane enzymes. This is in addition to the role of cholesterol as the starting material for the synthesis of a family of hormones and for the synthesis of bile salts, important to function on an organismal level.

IV. DISTRIBUTION OF CHOLESTEROL AMONG CELL MEMBRANES

Cholesterol is nonhomogeneously distributed among the membranes of mammalian cells. As mentioned above, greater than 90% of the cholesterol may be located in the plasma membrane of the cell. Only small amounts of cholesterol are found in intracellular membranes. The question has arisen how such a strong disproportionation of cholesterol is achieved and maintained within the cell (Clejan and Bittman, 1984a). Certain model studies have suggested that cholesterol transfers between membranes by collision of the donor and acceptor membranes (Steck et al., 1988). In the cell, sterol carrier proteins could facilitate intermembrane transfer. However, it has been suggested that cholesterol can even transfer through the aqueous phase without contact of the donor and acceptor membranes (Backer and Dawidowicz, 1981; McLean and Phillips, 1981; McLean and Phillips, 1982). Therefore, the disproportionation question raised above becomes even more interesting.

Available studies have suggested that several factors contribute to the disproportionation of cholesterol among mammalian cell membranes. The membrane content, both with respect to phospholipids and with respect to proteins, plays a role. Both in experiments of cholesterol distribution between lipid bilayers of different phospholipid content (Yeagle and Young, 1986) and in *in vitro* experiments on cholesterol distribution between membranes of cellular organelles (Wattenberg and Silbert, 1983), a role for membrane composition appeared. A good example of this can be found in the membranes of the retinal rod outer segment (Boesze-Battaglia et al., 1990). New disk membranes are made from the plasma membrane and are initially high in cholesterol content, like the plasma membrane. However, upon separation of the new disks from the plasma membrane, a significant increase in phosphatidylethanolamine content occurs, which has been shown to create a thermodynamically unfavorable environment for membrane cholesterol. Cholesterol then moves out of the disks as they age.

However, available evidence does not support an exclusive role for membrane composition in the distribution of cholesterol among cell membranes. Experiments have shown that unique vesicular transport pathways for movement of cholesterol from site of synthesis to the plasma membrane exist, and these may be involved in the concentration of the cellular cholesterol in the plasma membrane (Echevarria et al., 1990; Kaplan and Simoni, 1985; Lange and

Steck, 1985). There may also be kinetic or physical barriers to the movement of cholesterol from one "site" to another within the cell (Clejan and Bittman, 1984b). Much remains to be understood.

The end result is the accumulation of the majority of the cellular cholesterol in the plasma membrane of the cell. Thus, a logical place to explore the specific roles of cholesterol on cellular function would be in the plasma membrane. As a background to that, it is necessary to briefly review the kinds of properties cholesterol can confer upon a membrane and, subsequently, how those might modulate membrane function and, as a consequence, cellular function.

V. FEATURES AND PROPERTIES OF CHOLESTEROL

Among the important physical features of the cholesterol molecule is the planar steroid ring that is relatively conformationally inflexible. These features of its chemical structure govern much of the interaction between cholesterol and the lipid bilayer. For example, when cholesterol is adjacent to a lipid hydrocarbon chain of a phospholipid in a liquid crystalline state, the flat steroid ring system restricts the conformational flexibility of the lipid hydrocarbon chain or, in other words, orders the chain (Stockton and Smith, 1976). The increase in motional order decreases packing defects in the membrane, decreasing passive permeability. No other component of biological membranes affects the membrane in quite this manner (Yeagle, 1992).

The more highly ordered lipid hydrocarbon chains reflect the adoption of a more extensive *trans* configuration of the carbon-carbon single bonds. This reduces the cross-sectional area occupied on average by a hydrocarbon chain. This physical effect of cholesterol has been termed condensation. It can be readily measured as a reduction in surface area per phospholipid, due to the introduction of cholesterol into a phospholipid monolayer (Demel et al., 1972b).

Cholesterol can also act as an impurity in a physical chemical sense and inhibit the formation of the gel state of a lipid bilayer by inhibiting the packing of one chain next to another (Estep et al., 1978). However, inhibition of the formation of a gel state of lipid bilayers by cholesterol is not generally important to cell membranes since the dominant lipid species in cellular membranes contain unsaturated hydrocarbon chains. A single double bond in the middle of a chain can do more to inhibit the formation of a gel state than can high levels of membrane cholesterol (Yeagle, 1987). Most biological membranes containing cholesterol will remain in the liquid crystalline state at 37°C in the absence of cholesterol. Furthermore, in the above biosynthetic context the formation of a double bond is bioenergetically much less expensive than the synthesis of cholesterol. Therefore, this potential "role" of cholesterol will not be discussed further here.

VI. CHOLESTEROL AS A MODULATOR OF THE LIPID BILAYER AND BIOLOGICAL CONSEQUENCES

The physical effects of cholesterol on the ordering of the phospholipid hydrocarbon chains in a bilayer offer one possible mode for the modulation of biological membrane function by cholesterol. Litman has developed the concept of fractional volume to describe the effect of motional ordering by cholesterol on the properties of a lipid bilayer, in terms that can explain some of the effects of cholesterol on membrane function (Straume and Litman, 1987). Conceptually, fractional volume refers to "free volume" elements in the bilayer created by the formation of kinks in the lipid hydrocarbon chains. Kinks form from coupled *trans-gauche* isomerizations such that two bends in the hydrocarbon chain partially compensate, forming a configuration with the dominant direction of most parts of the chain perpendicular to the membrane surface, but with a kink in the chain. These kinks can form and disappear, but at any given instant of time they define packing defects in the lipid bilayer (Seelig and Seelig, 1974). The "excess" volume occupied by these kinks, relative to the volume of the lipid chains with an all-*trans* configuration, is free volume that could alternatively be used for other purposes if the chains were more ordered (i.e., more *trans* configurations about the carbon-carbon single bonds).

A companion concept that is important to this discussion is that most functional proteins, including enzymes, undergo conformational changes during their active cycle. If the relevant protein is a membrane protein, then a requisite conformational change may involve a change in the volume occupied by the protein in the lipid bilayer (Straume and Litman, 1988). If the change in protein conformation involves an increase in the volume occupied by the protein within the membrane, where is that volume to come from? It could be scavenged from the volume elements inherent in packing defects in the membrane.

If cholesterol is added to a membrane, a motional ordering of the lipid hydrocarbon chains would be expected, as has been measured by a variety of physical techniques. The result is a significant decrease in the number of gauche bonds per hydrocarbon chain, due to the presence of cholesterol in the membrane (Mendelsohn et al., 1991). This would decrease the packing defects in a membrane (thus reducing passive permeability to small solutes, as is observed). If this membrane contained membrane proteins that must undergo conformational changes during their activity cycles, one would expect an inhibition of these proteins by cholesterol at high membrane cholesterol levels, due to an inability to recruit sufficient volume elements for the necessary protein conformational changes. The photopigment rhodopsin provides an example of these effects (Mitchell et al., 1990).

Therefore, one means by which cholesterol might inhibit membrane enzymes is to inhibit required changes in membrane protein conformation. This effect is directly related to the ability of cholesterol to condense or order lipid hydrocarbon chains. As has been shown previously, such an effect is best realized by cholesterol itself. Required for an effective sterol under this definition is a 3-β-hydroxyl, a planar steroid ring system, and a hydrophobic "tail" (where the cholesterol "tail" proved to be the most effective) (Demel et al., 1972).

As important as this concept appears to be, this idea does not adequately explain important biological observations, including the results on yeast, where ergosterol is the most effective sterol at promoting cell growth and membrane function (Dahl and Dahl, 1988; Rodriguez et al., 1985), not cholesterol, even though cholesterol is more effective than ergosterol at ordering membranes (Demel et al., 1972a). Therefore, the structurally specific role of sterols in cell biology is not adequately explained by these purely physical effects of cholesterol on lipid bilayer properties.

VII. CHOLESTEROL AS A DIRECT MODULATOR OF MEMBRANE ENZYMES AND BIOLOGICAL CONSEQUENCES

In vitro experiments have shown that lanosterol cannot fully substitute for cholesterol as the essential sterol for mycoplasma cell function (Dahl and Dahl, 1988). In particular, *Mycoplasma mycoides* can be adapted to grow on low-cholesterol media. However, they cannot grow in the total absence of cholesterol in the medium, since they do not make their own cholesterol, and cholesterol is required for cell growth and function. Supplementation in the medium of lanosterol will not support cell growth in the absence of cholesterol. However, cell growth will occur at nearly the same rate in cells fed low-cholesterol levels supplemented with high (relatively) lanosterol, as in cells fed high (relatively) cholesterol levels (Dahl et al., 1980). Thus, cholesterol appears both adequate and necessary for minimal cellular function in these mycoplasma, while for optimum cellular function higher membrane sterol content is required, but without the structural specificity associated with the requirement for minimal cellular function.

The requirement of cholesterol for normal function of mycoplasma is mirrored in yeast by an analogous requirement for ergosterol (Dahl and Dahl, 1988). Yet one sterol cannot substitute for the other; cholesterol cannot fully substitute for ergosterol in yeast, and ergosterol cannot support normal mammalian cellular function. Anaerobic growth of *Saccharomyces cerevisiae* requires ergosterol supplement to the culture medium. Supplementation only with cholesterol will not support normal growth (Andreason and Stier, 1953).

How can these observations be explained? In addition to modulating the physical properties of lipid bilayers, cholesterol could directly interact with membrane proteins in membranes and thus cholesterol could act as an effector of membrane protein function. By interacting directly with membrane proteins, specific chemical structures would be required to modulate protein activity.

For example, recent experiments have revealed that activation of the Na^+K^+ATPase by sterol at low-to-moderate sterol levels in the membrane is highly structurally specific (Vemuri and Philipson, 1989; Yeagle et al., 1988). In the absence of cholesterol the enzyme has little or no activity. Addition of small amounts of cholesterol leads to activation of the enzyme. Ergosterol was found ineffective at activating the mammalian Na^+K^+ATPase. Recently the list has been widened such that the necessity for the cholesterol structure to support the activity of this enzyme is even better defined. Analysis has revealed that the ability of the set of sterols so far studied to activate this enzyme is not related to the ability of the same sterols to "condense" lipid bilayers (Yeagle, 1991).

Therefore, another explanation for the activation of the enzyme is required. The structural specificity in the activation of the Na^+K^+ATPase can be best explained by a direct interaction of sterol with membrane protein. The shape of the activation curve of the Na^+K^+ATPase is similar to a binding isotherm. A sterol-protein interaction would provide a basis for the structural specificity observed.

At high membrane cholesterol levels, increasing cholesterol leads to an inhibition of Na^+K^+ATPase activity. Such a biphasic effect of cholesterol on membrane enzyme activity (stimulation at low membrane cholesterol levels and inhibition at high membrane cholesterol levels) has been observed for other enzymes (Shouffani and Kanner, 1990; Vemuri and Philipson, 1989).

The combined data on cholesterol modulation of the Na^+K^+ATPase can be explained by the competition of two effects. One is stimulation of the enzyme by cholesterol at low-to-moderate cholesterol levels. Stimulation in this model would result from a direct interaction of the sterol with the protein. The other is inhibition of the enzyme by cholesterol at high membrane cholesterol levels. Inhibition in this model would result from a restriction of conformational changes by the enzyme required for function, due to a reduction in free volume available within the lipid bilayer by the presence of the cholesterol (see previous section).

VIII. HYPOTHESIS: THE ROLE OF CHOLESTEROL IN CHOLESTEROL-REQUIRING CELLS

This discussion has led to the hypothesis that cholesterol modulates the function of biological membranes by more than one mechanism.

1. Cholesterol alters the bulk biophysical properties of membranes. Cholesterol increases the orientational order of the lipid hydrocarbon chains of membranes and reduces the "free volume" available to membrane proteins for conformation changes that may be required for membrane protein function. In this role cholesterol likely inhibits membrane protein function, at high membrane cholesterol levels.
2. Cholesterol may bind directly to membrane proteins and regulate their function. In this role as an effector or modulator, cholesterol may stimulate or may inhibit membrane function. Cholesterol would operate from within the membrane, as a membrane component interacting directly with the transmembrane portion of the membrane protein. Even at low levels of cholesterol in a membrane the sterol could act in this manner.

IX. QUESTIONS FOR THE BIOPHYSICIST

This analysis and the subsequent development of the above hypothesis leads to a number of important questions that need to be addressed from the point of view of biophysics. Among them are the following:

1. The interesting concept of volume elements within the membrane that could be recruited for protein conformational changes deserves more attention.
2. To what extent do membrane proteins undergo small conformational fluctuations in membranes (i.e., "breathe")? To what extent are changes in conformation required for expression of membrane enzyme activity? How does cholesterol affect these phenomena?
3. Does cholesterol bind to specific sites on membrane enzymes such that membrane enzyme conformation and activity are modified? What is the "affinity" of cholesterol for those sites?

X. QUESTIONS FOR THE CELL BIOLOGIST/BIOCHEMIST

Important questions for the cell biologist/biochemist are raised by this hypothesis. Among them are the following:

1. Will a mammalian enzyme exhibit full activity when expressed in an organism that requires and expresses a sterol other than cholesterol?
2. Are there other examples of specific activation by cholesterol (or other sterols) of membrane enzymes?
3. By what vesicular pathway is newly synthesized cholesterol transported to the cell plasma membrane? How is the high cholesterol content of the plasma membrane maintained in the face of vesicular movement of plasma membrane components from the surface to sites within the cell?
4. At any time does a variable level of cholesterol become regulatory?

Only through combined advances on both the physical and biological side will the mystery of cholesterol finally be solved.

REFERENCES

Andreason, A. A. and T. J. B. Stier. 1953. Anaerobic nutrition of Saccharomyces cerevisiae. Ergosterol requirements for growth in a defined medium. *J. Cell. Comp. Physiol.*, 41:23.

Backer, J. M. and E. A. Dawidowicz. 1981. Mechanism of cholesterol exchange between phospholipid vesicles. *Biochemistry*, 20:3805–3810.

Boesze-Battaglia, K., S. J. Fliesler, and A. D. Albert. 1990. Relationship of cholesterol content to spatial distribution and age of disk membranes in retinal rod outer segments. *J. Biol. Chem.*, 265:18867–18870.

Clejan, S. and R. Bittman. 1984a. Kinetics of cholesterol and phospholipid exchange between Mycoplasma gallisepticum cells and lipid vesicles. *J. Biol. Chem.*, 259:441–448.

Clejan, S. and R. Bittman. 1984b. Decreases in rates of lipid exchange between Mycoplasma gallisepticum cells and unilamellar vesicles by incorporation of sphingomyelin. *J. Biol. Chem.*, 259:10823–10826.

Dahl, C. and J. Dahl. 1988. Cholesterol and cell function. In *Biology of Cholesterol*, Yeagle, P. L., Ed., CRC Press, Boca Raton, FL, pp. 147–172.

Dahl, J. S., C. E. Dahl, and K. Bloch. 1980. Sterol in membranes: growth characteristics and membrane properties of Mycoplasma capricolum cultured on cholesterol and lanosterol. *Biochemistry*, 19:1467–1472.

Demel, R. A., K. R. Bruckdorfer, and L. L. M. van Deenen. 1972a. The effect of sterol structure on the permeability of liposomes to glucose, glycerol and Rb+. *Biochim. Biophys. Acta*, 255:321–330.

Demel, R. A., W. S. M. G. V. Kessel, and L. L. M. van Deenen. 1972b. The properties of polyunsaturated lecithins in monolayers and liposomes and the interactions of these lecithins with cholesterol. *Biochim. Biophys. Acta*, 266:26–40.

Echevarria, F., R. A. Norton, W. D. Ness, and Y. Lange. 1990. Zymosterol is located in the plasma membrane of cultured human fibroblasts. *J. Biol. Chem.*, 265:8484–8489.

Estep, T. N., D. B. Mountcastle, R. L. Biltonen, and T. E. Thompson. 1978. Studies on the anomalous thermotropic behavior of aqueous dispersions of dipalmitoylphosphatidylcholine-cholesterol mixtures. *Biochemistry*, 17:1984–1989.

Faust, J. R., J. M. Trzaskos, and J. L. Gaylor. 1988. In *Biology of Cholesterol*, Yeagle, P. L., Ed., CRC Press, Boca Raton, FL, pp. 19–38.

Kan, C.-C. and R. Bittman. 1991. Spontaneous rates of sitosterol and cholesterol exchange between phospholipid vesicles and between lysophospholipid dispersions: evidence that desorption rate is impeded by the 24 α-ethyl group of sitosterol. *J. Am. Chem. Soc.*, 113:6650–6656.

Kaplan, M. R. and R. D. Simoni. 1985. Transport of cholesterol from the endoplasmic reticulum to the plasma membrane. *J. Cell Biol.*, 101:446–453.

Lange, Y. and B. V. Ramos. 1983. Analysis of the distribution of cholesterol in the intact cell. *J. Biol. Chem.*, 258:15130–15134.

Lange, Y. and T. L. Steck. 1985. Cholesterol-rich intracellular membranes: a precursor to the plasma membrane. *J. Biol. Chem.*, 260:15592–15597.

McLean, L. R. and M. C. Phillips. 1981. Mechanism of cholesterol and phosphatidylcholine exchange or transfer between unilamellar vesicles. *Biochemistry*, 20:2893–2900.

McLean, L. R. and M. C. Phillips. 1982. Cholesterol desorption from clusters of phosphatidylcholine and cholesterol in unilamellar vesicle bilayers during lipid transfer or exchange. *Biochemistry*, 21:4053–4059.

Mendelsohn, R., M. A. Davies, H. F. Schuster, Z. Xu, and R. Bittman. 1991. CD_2 rocking modes as quantitative infrared probes of one-, two-, and three-bond conformational disorder in dipalmitoylphosphatidylcholine and dipalmitoylphosphatidylcholine/cholesterol mixtures. *Biochemistry*, 30:8558–8563.

Mitchell, D., M. Straume, J. Miller, and B. J. Litman. 1990. Modulation of metarhodopsin formation by cholesterol-induced ordering of bilayers. *Biochemistry*, 29:9143–9149.

Rodriguez, R. J., C. Low, C. D. K. Bottema, and L. W. Parks. 1985. Multiple functions for sterols in Saccharomyces cerevisiae. *Biochim. Biophys. Acta*, 837:336–343.

Seelig, A. and J. Seelig. 1974. The dynamic structure of fatty acyl chains in a phosholipid bilayer measured by deuterium magnetic resonance. *Biochemistry*, 13:4839–4845.

Shouffani, A. and B. I. Kanner. 1990. Cholesterol is required for reconstitution of the sodium- and chloride-coupled GABA transporter from rat brain. *J. Biol. Chem.*, 265:6002–6008.

Steck, T., F. Kezdy, and Y. Lange. 1988. An activation-collision mechanism for cholesterol transfer between membranes. *J. Biol. Chem.*, 263:13023–13031.

Stockton, B. W. and I. C. P. Smith. 1976. A deuterium NMR study of the condensing effect of cholesterol on egg phosphatidylcholine bilayer membranes. *Chem. Phys. Lipids*, 17:251.

Straume, M. and B. J. Litman. 1987. Influence of cholesterol on equilibrium and dynamic bilayer structure of unsaturated acyl chain phosphatidylcholine vesicles as determined from higher order analysis of fluorescence anisotropy decay. *Biochemistry*, 26:5121–5126.

Straume, M. and B. J. Litman. 1988. Equilibrium and dynamic bilayer structural properties of unsaturated acyl chain phosphatidylcholine-cholesterol-rhodopsin recombinant vesicles and rod outer segment disk membranes as determined from higher order analysis of fluorescence anisotropy decay. *Biochemistry*, 27:7723–7733.

Vemuri, R. and K. D. Philipson. 1989. Influence of sterols and phospholipids on sarcolemmal and sarcoplasmic reticular cation transporters. *J. Biol. Chem.*, 264:8680–8685.

Wattenberg, B. W. and D. F. Silbert. 1983. Sterol partitioning among intracellular membranes. *J. Biol. Chem.*, 258:2284–2289.

Yeagle, P. L. 1985. Cholesterol and the cell membrane. *Biochim. Biophys. Acta*, 822:267–287.

Yeagle, P. L. 1987. *The Membranes of Cells*, Academic Press, Orlando, FL.

Yeagle, P. L. 1988. *The Biology of Cholesterol*, CRC Press, Boca Raton, FL.

Yeagle, P. L., D. Rice, and J. Young. 1988. Cholesterol effects on bovine kidney Na+K+ATPase hydrolyzing activity. *Biochemistry*, 27:6449–6452.

Yeagle, P. L. and J. Young. 1986. Factors contributing to the distribution of cholesterol among phospholipid vesicles. *J. Biol. Chem.*, 261:8175–8181.

Yeagle, P. L. 1991. Modulation of membrane function by cholesterol. *Biochemie*, 73:1303–1310.

Yeagle, P.L. 1992. *The Structure of Biological Membranes*, CRC Press, Boca Raton, FL.

Chapter 2

NEUTRON SCATTERING AS A TOOL TO PROBE THE EFFECT OF CHOLESTEROL ON PHASE SEPARATION, STRUCTURE FORMATION, AND LIPID/PROTEIN INTERACTION IN ARTIFICIAL MEMBRANES

Thomas M. Bayerl and Erich Sackmann

TABLE OF CONTENTS

0-8493-4207-4/93/$0.00+$.50

ABSTRACT

Neutron scattering represents a unique tool for studying the structure and the phase behavior of lipid membranes. This chapter introduces applications of small-angle neutron scattering in combination with contrast variation techniques for the elucidation of phase diagrams of binary lipid mixtures (e.g., lipid/cholesterol mixtures) as well as for the determination of the excess volume of the membrane constituents. The advantages of the inverse contrast variation technique for demixing studies in membranes and the determination of phase boundaries are discussed for binary mixtures of dimyristoyl-phosphatidylcholine with cholesterol. An interesting feature of neutron scattering is its extremely short experimental timescale, which enables the observation of critical fluctuation processes in membranes. Among the latest applications of neutron scattering in lipid membranes is the specular reflection of neutrons. This method proves as very powerful for the study of the interfacial structure of supported lipid membranes and its changes due to the presence of cholesterol.

I. INTRODUCTION

Neutron scattering offers a unique opportunity for the exploration of three fundamental aspects of membrane research:

1. By small-angle neutron scattering (SANS) in combination with contrast variation, phase diagrams of lipid mixtures can be determined in thermodynamic equilibrium, and information on the lateral organization can be obtained, even in cases where the components exhibit phases of the same symmetry (e.g., fluid-fluid demixing).
2. Incoherent and coherent quasielastic neutron scattering (QENS) allows one to distinguish between various molecular and collective motional processes in membranes and to determine both time constants (correlation times) and geometric constraints of the membrane dynamics.
3. Specular reflection of neutrons (SRN) offers unique advantages for structural studies of lipid monolayers and supported bilayers, as well as for the determination of the penetration depth of extrinsic proteins into the monolayer/bilayer and the associated changes of the lipid head group hydration.

Cold neutrons with wavelengths λ in the range 0.4 to 3 nm, which covers the size range of molecules like membrane phospholipids, are used for the scattering experiments. Neutrons interact with atomic nuclei, unlike X-rays, which interact primarily with atomic electrons, and so neutrons give a different, complementary picture to that given by X-rays. This interaction, which results in the scattering of part of the incident neutron beam from its original direction in space, is remarkably different for hydrogen and deuterium, and so provides a unique tool that is not available for X-rays. This tool, the contrast variation method, provides the basis for all three above-cited applications of neutrons in membrane research. It allows one to screen out some constituents or segregated phases in SANS studies of phase separation processes. In QENS experiments the motion of the constituents (e.g., cholesterol) or of parts of molecules (e.g., the lipid head groups) can be selectively evaluated by deuteration of the other constituents (e.g., phospholipids) or of the residual parts of the molecule (e.g., the lipid hydrocarbon chains). In SRN studies of supported membranes (or of monolayers at the air water interface), the total reflectivity of the membrane can be modified by partial deuteration of the lipids as well as by variation of the D_2O content of the bulk water. In the case of protein-adsorption studies, the scattering of the latter may be suppressed as well. The key point is, however, that a high degree of resolution and reliability of the structure determination may be achieved by continuous variation of the contrast.

A limiting factor in the use of neutron scattering is the availability of high-flux neutron beams. Typically, cold neutrons come from a large nuclear reactor or from a spallation source (which is driven by a powerful particle accelerator) that is a national resource shared by many users, so "beam time" is scarce. Since the neutron flux is low compared to synchrotron radiation sources used for X-ray experiments, experiments can be long and tedious. However, sample damage as a result of the exposure of the membrane sample to the cold neutron beam, mainly due to inelastic neutron scattering, is usually negligible.

A. LIPID PHASE DIAGRAMS

Since the early days of model membrane research, these central questions have been addressed: (1) whether the large number of lipid molecules in biomembranes are a present necessity or an evolutionary left-over, and (2) whether this complex composition is essential for the formation of functional enzyme complexes and the maintenance of lateral heterogeneity and of the asymmetry of the membranes. Such an active role would certainly be associated with processes of lateral and transverse phase separation of lipids and selective lipid/protein interactions. In order to elucidate this complex problem, phase separation in mixed lipid model membranes has been extensively studied by calorimetric and various spectroscopic techniques (cf. Marsh and Cevc, 1987, for references). Since the working state of biomembranes is the liquid crystalline (L_α) phase, phase separation in this phase is in the center of interest. SANS in combination with contrast variation is a very powerful — albeit time consuming and expensive — technique to determine phase diagrams in thermodynamic equilibrium states and to evaluate simultaneously the detailed structure of the membrane undergoing phase instabilities.

As in conventional mixtures, phase separation is expected if the components exhibit phases of different symmetry (e.g., tilted and nontilted) or differ strongly in their molecular structure (e.g., phospholipids of different chain lengths; phospholipids and cholesterol). There exist, however, two remarkable new features in membranes:

1. Owing to the two dimensionality of membranes, phase separation may be induced isothermally, that is, induced in the presence of charged lipids by the adsorption of two-valent ions or extrinsic proteins. The adsorption process may even lead to the local formation of inverted micelles (de Kruijff et al., 1980). One interesting aspect of the charge- (e.g., Ca^{2+}) induced phase separation is its pronounced hysteresis (Sackmann 1984).
2. Phase separation in vesicles is, in general, associated with domain formation that is again a consequence of the two dimensionality and of the unique elastic curvature properties of bilayers (Sackmann et al., 1980). The initial process may be described in terms of spinodal decomposition, but the lateral growth (coarsening) of the segregated phases is impeded by the coupling between phase separation and local variations in curvature, leading to an interfacial energy that increases with the size of the segregated domains (for review, cf. Albrecht et al., 1981; Leibler and Andelmann, 1987). The domain formation is essential for the stability of vesicles undergoing lateral phase separation. Thus, unlimited growth of the domains leads, in general, to the detachment of segregated vesicles.

SANS is a powerful technique to explore both of the above features. The thermodynamics of the phase behavior of lipid/lipid mixtures can be very well described in terms of the regular solution theory, that is, the free energy of mixing is determined by the excess enthalpy of lipid-lipid interaction (Lee, 1977; Maksymiw et al., 1987). In contrast, the behavior of protein solutions in bilayers is largely determined by the nonideality of the mixing entropy.

The phase diagrams of binary lipid mixtures have the general shape shown in Figure 1. The miscibility gap ($\beta_1 + \beta_2$) in the crystalline state may extend into the fluid/solid miscibility gap, leading to either peritectic or eutectic behavior (depending on the differences of the chain melting transition temperatures). Furthermore, the fluid/fluid miscibility gap may be hidden below the liquidus line.

Both the ($\beta + \beta$)- and ($\alpha + \alpha$)-miscibility gaps may have very intriguing consequences for the structure formation in membranes. As will be shown (for the case of a DMPC/DSPC mixture, where DXPC is di-X-phosphatidylcholine, X=M=myristoyl ([14 carbons long], X=S= stearoyl [18 carbons long]), pronounced critical concentration fluctuations may be observed some 15°C above the critical point (which may be hidden in the ($\alpha + \beta$)-coexistence region). Insertion of proteins exhibiting a slight preference for one of the lipid components (e.g., owing to a better thickness matching of the hydrophobic cores of the protein and the lipid) (Sackmann, 1990) would stabilize the concentration fluctuations. Since the correlation lengths in the two dimensional mixtures are quite large (some 10 nm), this stabilization would provide a mechanism for transient structure formation in fluid membranes as well as for long-range protein-protein interactions (algebraic decay laws) (Sackmann, 1983).

The chain length of the most abundant lipids in biomembranes varies from 16 to 24 carbon atoms. Moreover, a fluid-fluid miscibility gap has been demonstrated recently for lecithin-cholesterol mixtures (Ipsen et al., 1987, 1989, Vist and Davis, 1990). It is therefore quite likely that critical demixing occurs also in

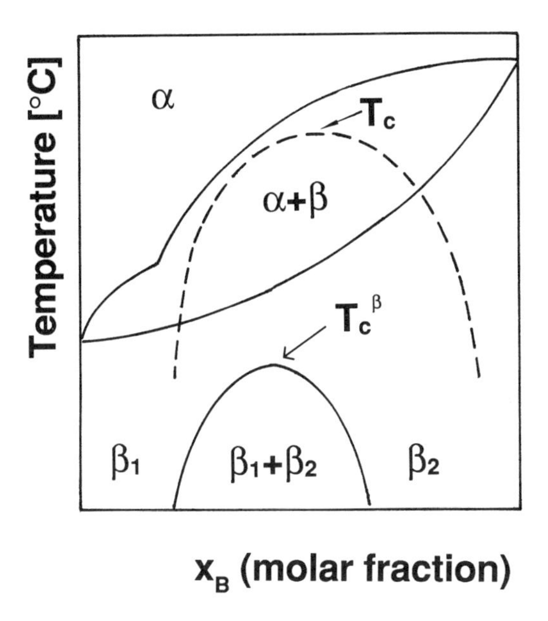

FIGURE 1. Hypothetical phase diagram of binary mixture of lipids A and B exhibiting a fluid (L_α) and a solid (L_{β}, $P_{\beta'}$) phase (at transition temperatures T_A and T_B). The curve defines the solid-solid miscibility gap, and the broken line a fluid-fluid miscibility gap.

biological membranes. Erythrocytes, for instance, exhibit a liquidus line near 20°C, and critical behavior could well prevail at physiological temperatures.

B. MEMBRANE STRUCTURE MODULATION BY PROTEIN ADSORPTION

Various model membrane studies showed that water-soluble proteins (such as chymotrypsin, insulin, or spectrin), or large hydrophilic head groups of receptors, may penetrate into the semipolar lipid/water interface. In the presence of small amounts of charged lipids, adsorption is strongly enhanced by electrostatic forces. This is associated with local lateral phase separation and domain formation (cf. Maksymiw et al., 1987, for references). The intriguing aspect of these findings is that they demonstrate that extrinsic proteins may be much more essential for the local organization of biomembranes than anticipated hitherto. Such adsorptive processes could, for instance, play an important role for the coupling of the cell cytoskeleton to the plasma membrane. Note that for this process the charged phosphatidylinositol is essential. The case of the spectrin-bilayer interaction will be discussed below.

C. MEMBRANE STRUCTURE STUDIED BY NEUTRON REFLECTIVITY

In membrane biophysics it is of great importance to be able to characterize not only the mixing or segregation of lipids in the plane of the bilayer, but also the behavior of lipids at surfaces and interfaces. While the former can be obtained by small-angle neutron scattering (SANS) methods, the latter requires a method that is particularly sensitive to the interfacial structure of a lipid monolayer or bilayer. Specular reflection of neutrons (SRN) is then the method of choice.

Specular reflection of neutrons (or neutron reflectivity) is a relatively new technique for measuring the interfacial structure of membranes. The methodological approach is essentially similar to well-established X-ray reflectivity measurements. However, neutrons provide a distinct advantage over X-rays in that they interact markedly differently with hydrogen and deuterium. Thus, by selective deuteration of a lipid, one can probe different interfaces in a lipid bilayer. Similarly, variation of the bulk water contrast around the membrane by mixing normal water and heavy water or by substituting one species by the other provides information about the hydration properties of the membrane head groups and the penetration depth of water into the membrane interior. Moreover, the relatively high neutron transparency of crystals like silicon enables the use of such solids as a planar support for the membrane in SRN studies (see below). SRN provides excellent spatial resolution down to less than 1 nm, with a penetration depth over hundreds of nanometers.

In general, SRN gives information on the profile of the neutron refractive index normal to the membrane plane. The refractive index is simply related to the scattering length density and can be modified by isotopic substitution. All interfaces of the membrane, at which changes of the neutron refractive index occur, contribute to the total reflectivity in a well-understood manner analogous to the optics of light waves. Therefore, classical methods designated for the analysis of the reflectivity of light, such as the optical matrix method, can be applied in the evaluation of the neutron refractive index profile.

D. MEMBRANE DYNAMICS

Membrane function is closely related to the lipid bilayer dynamics that is determined by a whole hierarchy of motional processes. It involves (1) conformational transitions of the hydrocarbon chains, (2) locally restricted rotational diffusion, as well as in-plane and out-of-plane translational diffusion, (3) long-range lateral diffusion of the lipid molecules, and (4) collective undulatory excitations of the bilayer (membrane flickering). The chain dynamics provide the fluctuating thermal forces for the excitation of the conformational states of the membrane proteins, which is essential for enzymatic function. The lateral diffusion controls the formation of functional enzyme complexes (e.g., in the electron transfer chain of mitochondria or during the association of the receptor-adenylatecylase-G-protein complex). For these processes the two dimensionality of the membranes is essential. For small molecules (diameter a < bilayer thickness) the bilayer behaves as a three-dimensional liquid ($D = k_B T / 6\pi\eta a$, where k_B is the Boltzmann constant and η is the viscosity), whereas a crossover to a logarithmic size dependence ($D \propto \ln a^{-1}$) occurs at $a \approx$ bilayer thickness (Galla et al., 1979). The logarithmic size dependence encourages the rapid formation of enzyme complexes, since very large proteins may diffuse only slightly slower than the lipid molecules unless they are coupled to the cell cytoskeleton. Finally, the undulatory excitations may provide an essential contribution to the repulsive forces between membranes (Helfrich, 1978).

Despite this importance of all aspects of lipid dynamics for biomembrane processes, our knowledge in this field is still rudimentary and contradictory. Quasielastic neutron scattering (QENS) offers outstanding advantages for the exploration of the membrane dynamics. Its particular features are

1. The molecular motions may be studied by the incoherent QENS. Therefore, the motion of parts of the lipid molecules (e.g., of the head groups) may be filtered out by deuteration of the residual parts (e.g., chain and glycerol backbone).
2. By selection of the instrument resolution, motional processes at different time scales may be distinguished (e.g., the chain dynamics from the lateral diffusion).
3. Collective and molecular motional processes may be evaluated by simultaneous measurements with backscattering and/or time-of-flight instruments and neutron spin-echo instruments.
4. Most importantly, in-plane and out-of-plane motional process may be evaluated by variation of the scattering angle with respect to the membrane normal.

Besides the spin echo technique, scattering with ultracold neutrons is a promising technique to study the dynamics in the slow time (ca. 10^{-7} s) regime (Pfeiffer et al., 1988). Some preliminary QENS experiments on DPPC membrane dynamics are presented in the work of Pfeiffer et al. (1989). Since the technique has not been applied yet to cholesterol-containing membranes, we will refrain from discussing QENS in this review and refer to the above-mentioned work and a forthcoming paper (König et al., 1992).

II. EVALUATION OF PHASE DIAGRAMS AND CRITICAL PHENOMENA IN LIPID MIXTURES BY SANS-CV TECHNIQUE

A. THE CONTRAST VARIATION TECHNIQUE

The low-angle scattering of thin-shelled vesicles of sizes larger than 50 nm (as obtained by slight sonication) can be analyzed in terms of the Kratky-Porod model of two-dimensional scatterers (Knoll et al., 1981a,b). The low-angle scattering density is

$$I_{(q)}q^2 = I_o \exp\left\{-R_d^2 q^2\right\} \tag{1}$$

where R_d is the thickness of gyration of the lipid lamellae and $q = (4\pi/\lambda) \sin(\theta/2)$ is the momentum transfer at the scattering angle 2θ and the neutron wave-

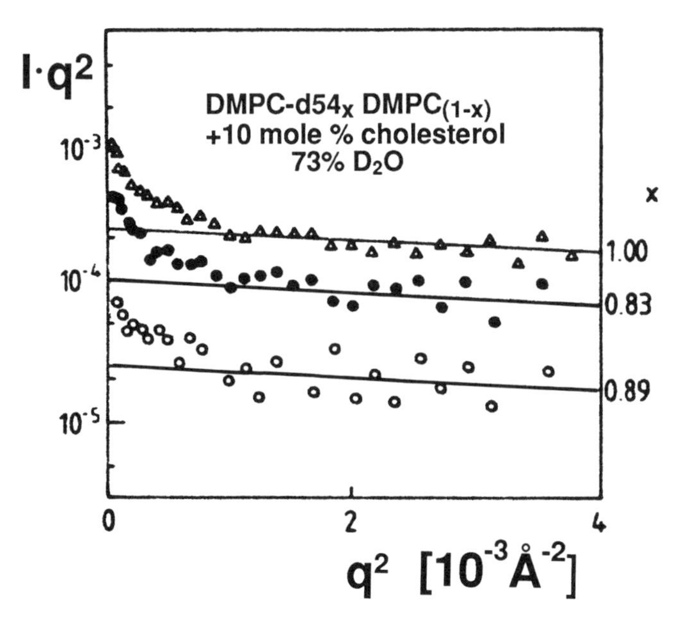

FIGURE 2. Typical Kratky-Porod plots of a slightly sonicated vesicle dispersion in 73:27 D_2O/H_2O solution of a 9:1 DMPC-cholesterol mixture for three different contents of deuterated DMPC-d_{54} (T = 7°C). Extrapolation of straight lines to $q \rightarrow 0$ yields I_0. Their slope is a measure for the bilayer thickness (cf. Knoll et al., 1981).

length λ. It is related to the bilayer thickness d, as $d = (\sqrt{12})R_{d,}$ which holds also for curved shells. The scattering intensity I_0 at zero scattering angle contains all the information on the phase separation. For thin shelled vesicles Equation 1 is only well fullfilled for $q > 0.01 Å^{-1}$, but I_0 can be obtained, to a high degree of accuracy, by extrapolation of the $I_0{}^2$-vs-q^2 plots towards $q \rightarrow 0$ (for a detailed discussion of the reliability of this approach cf. Knoll et al., 1981). Examples of such curves are shown in Figure 2 for the case of an inverse contrast variation experiment (cf. below). From the slopes of the straight lines, one obtains a value for the bilayer thickness of $d = 41 \pm 1$ Å.

The scattering intensity I_0 is determined by the (position dependent) scattering length density of the lipid bilayer (denoted as ρ(r)). It can be expressed as a linear superposition of contributions of the various scatterers of the solution as (Stuhrmann and Duee, 1975; Jacrot 1976)

$$\rho(\mathbf{r}) = (\langle\rho\rangle - \rho_i)\, S(\mathbf{r}) + \rho_f(\mathbf{r}) \tag{2}$$

where $(\langle\rho\rangle - \rho_i)$ is the difference of the average scattering length densities of the solvent $\langle\rho\rangle$ and the solute ρ_i. $S(r)$ is a shape function of the solute. It can be well approximated by a hard core function: S = 1 across the diameter of molecule; S = 0 in solvent. The fluctuation term $\rho_f(r)$ accounts for the internal structure of the solute and is usually ignored. For spherical particles in ordinary solutions, the zero scattering angle intensity is

$$I_0 = [(<\rho>-\rho_i)V_c]^2 \quad (3)$$

where V_c is the particle volume. For lipid lamellae of thickness d

$$I_0 = [(<\rho>-\rho_i)d]^2 \quad 4)$$

Clearly, for an aqueous dispersion of homogeneous scatterers (proteins or vesicles) $I_0 = 0$ for $\rho_i = <\rho>$, which can be achieved by variation of the D_2O/H_2O ratio. It is also obvious that for vesicles exhibiting (lateral) phase separation, contrast matching is not possible.

B. DETERMINATION OF PHASE BOUNDARIES

Consider now a binary mixed membrane containing molar fractions $x_A = x_0$ of component A and $x_B (= 1 - x_0)$ of B. If this mixture decomposes into two phases (α and β) of composition $x_\alpha (= x_{A,\alpha})$ and $x_\beta (= x_{A,\beta})$ (exhibiting average scattering length densities ρ_α and ρ_β), the zero angle scattering intensity becomes (Knoll et al., 1981)

$$I_0 = \frac{x_\beta - x_0}{x_\beta - x_\alpha}\left(\rho_\alpha - <\rho>\right)^2 + \frac{x_0 - x_\alpha}{x_\beta - x_\alpha}\left(\rho_\beta - <\rho>\right)^2 \quad (5)$$

Owing to the linear superposition law of the scattering intensities, ρ_α of the phase α is given by

$$\rho_\alpha = x_\alpha \, \rho_{A,\alpha} + (1 - x_\alpha)\rho_{B,\alpha} \quad (6)$$

where $\rho_{A,\alpha}$ and $\rho_{B,\alpha}$ are the average scattering densities of the components A and B in the phase α. An identical equation holds for ρ_β.

Since I_0 is a quadratic function of ρ, plots of I_0 vs. the solvent scattering length density $<\rho>$ exhibit parabolic curves with minima at a value $<\rho> = <\rho>_{min}$ (cf. Equation 3). A simple consideration (cf. Knoll et al., 1981b; Appendix) yields the following relationship between $<\rho>_{min}$ and the square root of the minimum value of the scattering intensity at $q = 0$, $\sqrt{I_{0,min}}$:

$$\sqrt{I_{0,\min}} = \left(<\rho>_{\min} - \rho_{\beta\cdot}\right)\left(\rho_\alpha - <\rho>_{\min}\right) \quad (7)$$

$<\rho>_{min}$ and $I_{0,min}$ are measurable quantities. In situations of phase separation, the phase boundaries x_α and x_β and the average scattering length densities $<\rho_\alpha>$ and $<\rho_{\beta\cdot}>$ of the phases α and β (4 unknowns) can be determined from the above equations by performing contrast variation experiments for two initial compositions $(x_{0,1}, x_{0,2})$ lying within the phase coexistence region. $<\rho_\alpha>$ and $<\rho_{\beta\cdot}>$ are obtained from Equation 7, while insertion of these values into Equation 6 yields the phase boundaries x_α and x_β.

C. MEASUREMENTS OF EXCESS VOLUMES

The excess molar volumes of homogeneous but nonideal mixtures may be determined from contrast variation measurements. For that purpose, the average scattering length density of the mixed bilayer is determined by searching for the matching H_2O/D_2O composition $<\rho^*>$. The molar (excess) volume V_{ex} is then obtained from the relationship

$$V_{ex} = \frac{\rho_A V_A x + \rho_B V_B (1-x)}{<\rho^*>} - V_{id} \tag{8}$$

V_{id} $(= xV_A + (1-x)V_B)$ is the molar volume of the ideal solution, while V_A and V_B are the molar volumes of the pure components with the scattering length densities ρ_A and ρ_B. A simple example of a mixture (DMPC + DPPC) exhibiting nearly ideal behavior in the fluid and solid phase is shown in Figure 3. The DMPC is chain deuterated. Contrast matching is clearly possible above the liquidus and below the solidus line. The mixture is, however, nonideal both in the solid and fluid phase: a positive excess volume of $V_{ex} = +86$ Å^3 is found for the L_β and of $V_{ex} = +49$ Å^3 for the L_α phase. The free volume is most likely formed at the interface between the opposing monolayers. Since the molar volumes of the two lipids differ by 120 Å^3, a value of $V_{ex} = +60$ Å^3 is expected in good agreement with the SANS data.

A completely different behavior is found for mixtures of diacyl-PCs differing in chain length by more than two methylene groups. Figure 4 shows the phase diagram of a mixture of protonated DSPC with chain-deuterated DMPC. The liquidus and solidus lines were determined by various techniques (Knoll et al., 1981a). The existence of a liquidus line characteristic for peritectic mixtures suggests strongly that the critical point of the (solid phase) miscibility gap is close to the fluid-solid coexistence region. The existence of the miscibility gap was demonstrated by contrast-variation experiments, and the position of its boundaries determined for one temperature.

The most intriguing finding was, however, that the L_α phase of the 1:1 mixture cannot be matched up to 10°C above the liquidus line. As will be shown below, this is the consequence of critical demixing. Its origin is not clear yet. It could be an effect (1) of the solid phase miscibility gap, the critical point of which is hidden within the fluid-solid coexistence, or (2) of a hidden fluid-fluid miscibility gap. Critical fluctuations caused by hidden critical points are a well-known phenomenon in metallurgy (cf. Knoll et al., 1981b).

As an example of chemically induced phase separation, the effect of Ca^{2+} on a mixture of DMPC-d_{54} and a phosphatidylglycerol with 15 C-atoms per chain (di-C15 PG) was also studied by SANS (cf. Knoll et al., 1991). This mixture exhibits a cigar-like fluid-solid coexistence with complete miscibility above the liquidus line. Addition of Ca^{2+} (3.3 m*M*) to vesicles of a 3:1 DMPC-d_{54}:di-C15 PG mixture triggered lateral phase separation. Conductance measurements through Gramicidine channels incorporated in these mixed bilayers provided strong evidence for Ca^{2+}-induced fluid-fluid demixing associated with critical

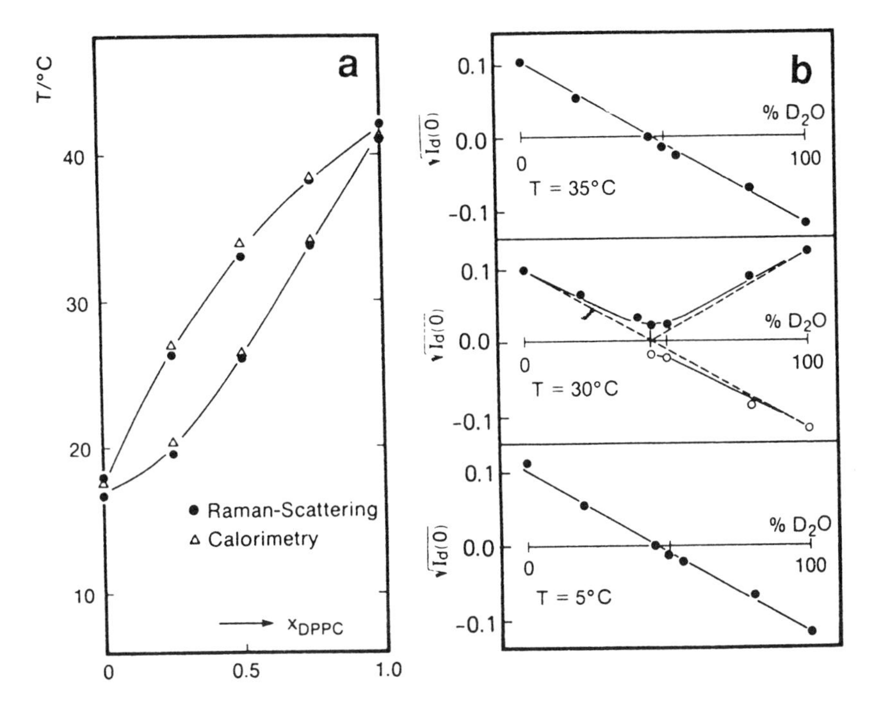

FIGURE 3. (a) Phase diagram of a mixture of deuterated DMPC-d_{54} and protonated DPPC (from calorimetry and Raman spectroscopy). (b) Contrast variation plots of $\sqrt{I_0}$ vs. D_2O content of aqueous phase for a 1:1 DMPC-d_{54}/DPPC mixture at a temperature above (35°C), below (5°C), and within (30°C) the cigar-like fluid-solid coexistence regime. The drawn parabolic lines have been calculated using Equations 4 and 5 for the phase boundaries $x_a = 0.37$ and $x_b = 0.63$.

phenomena (cf. Knoll et al., 1991). This is an example of the modulation of membrane function by lipid phase separation.

D. EVALUATION OF CRITICAL CONCENTRATION FLUCTUATIONS AND OF DOMAIN STRUCTURE

SANS offers unique advantages for studies of the domain structure and critical concentration fluctuations in binary mixtures above the critical point. As is well known from classical fluids, the coherent differential scattering cross section (at zero scattering angle) of binary mixtures is composed of two parts:

$$\frac{1}{N}\left(\frac{d\sigma}{d\Omega}\right)_{q=0}^{coh} = A\kappa + \left[N^2 \frac{v_1^2 v_2^2}{V^2}\left(<\rho_1> - <\rho_2>\right)^2\right] S_{cf}(0) \qquad (9)$$

Here $<\rho_1>$, $<\rho_2>$ are the average scattering length densities of phases 1 and 2; v_1, v_2 the partial molar volumes of the two phases; N, V the total number of molecules and the total volume; κ is the isothermal lateral compressibility, and A is the area of the membrane. $S_{cf}(0)$ is the static structure factor of the lateral

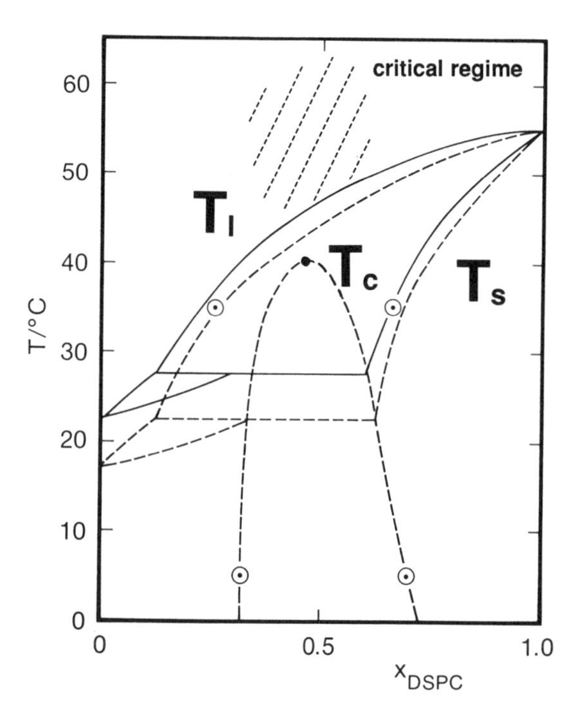

FIGURE 4. Phase diagram of a DMPC/DSPC mixture. (Full line: mixture of protonated lipids; broken line: mixture of DSPC and chain deuterated DMPC-d_{54}.) The liquidus and solidus lines were obtained by various techniques (cf. Knoll et al., 1981, for references). The dip of the former at $x_{DSPC} \approx 0.15$ indicates peritectic behavior. The chain melting transition of DMPC-d_{54} is 6°C below the value for DMPC. The data obtained by the SANS experiment are indicated by open circles, and the full circle denotes the critical temperature T_c.

concentration distribution. It is determined by the 2nd derivative of the Gibbs free energy of the mixture with respect to x_1 or x_2.

$$S_{cf}(0) = Nk_BT \left.\frac{\partial^2 G}{\partial x^2}\right|_{\rho, T, N} \tag{10}$$

Since we deal with demixing in condensed phases, the difference of the compressibility in the two phases is very small and the scattering cross-section at $q \rightarrow 0$ is solely determined by the divergence of the concentration fluctuations.

In this situation the Ornstein-Zernicke approach can be applied, and the structure factor (or the scattering intensity) at small scattering angles is given by the celebrated equation:

$$I(q) \propto S_{cf}(q) \propto \frac{\xi^2}{1+q^2\xi^2} \tag{11}$$

In this equation ξ is the correlation length of the concentration fluctuations ($\approx$ average size of critical domains).

Plots of $I(q)^{-1}$ vs. q^2 should yield straight lines. Figure 5a shows two examples of scattering curves for the DMPC/DSPC mixture, and in Figure 5b the corresponding Ornstein-Zernicke-[$I(q)^{-1}$ -vs.- q^2] plots are presented. Straight lines are clearly found, which provides further evidence that the critical point hidden within the fluid/solid coexistence is located well above the liquidus line.

From intersection of the straight lines with the abscissa, one obtains for the correlation lengths: $\xi = 500$ Å at 45°C, and $\xi = 300$ Å for 48.5°C. In an additional study of the pressure dependence of the critical behavior it was shown that the temperature and pressure dependencies of the correlation length ξ obey the scaling laws predicted by the Ising model for two-dimensional systems, i.e.,

$$\xi \propto (T - T_c)^{-0.92}, \ \xi \propto (p - p_c)^{-1/2} \tag{12}$$

This demonstrates again the power of the neutron scattering techniques for the exploration of subtle structural features of membranes, and it illustrates the usefulness of lipid bilayers as models of two-dimensional systems.

III. THE DMPC-CHOLESTEROL MIXTURE

This biologically important mixture has been extensively studied by calorimetry (Mabrey et al., 1978; Hatta and Imaizumi, 1985), electron microscopy (Hicks et al., 1987), and various spectroscopic methods such as ESR and NMR (Brown and Seelig, 1978; Rubenstein et al., 1979; Recktenwald and McConnell, 1981; Sankaram and Thompson, 1990a,b; Vist and Davis, 1990). The phase behavior is very complex and only partially understood. SANS has been applied (1) to study phase segregation of DMPC-d_{54}-cholesterol at low temperature (7°C) by the contrast matching approach (Knoll et al., 1985a,b) and (2) to explore the modification of the phosphatidylcholine bilayer structure (e.g., bilayer repeat distance, ripple superstructure) by cholesterol (Mortensen et al., 1988). This latter study yielded the phase boundaries at cholesterol concentrations $x_c < 24\%$ and below the chain melting transition of the phospholipid ($L_{\beta'}$, $P_{\beta'}$ phase) as well as below the subtransition (in the L_c phase).

A. EVALUATION OF PHASE BOUNDARIES BY SANS STRUCTURAL STUDY

The bilayer repeat distance was determined by neutrons of wavelength $\lambda = 3.5$ Å, and the ripple structure by neutrons of $\lambda = 8$, 15, and 24 Å. The water content of the DMPC sample in these studies was 17% by weight, corresponding to the following transition temperatures for DMPC-d_{54}: subtransition $L_c - L_{\beta'}$: T_c = 10°C; pretransition $L_{\beta'} - P_{\beta'}$: T_p = 16°C; main transition $P_{\beta'} - L_{\alpha}$: T_m = 22°C. These temperatures are about 3°C higher than those observed in fully hydrated DMPC-d_{54} bilayers.

A first example of this work (Figure 6a) shows a series of intensity-vs.-wavevector (q) plots of DMPC-d_{54} containing 2 mol% cholesterol taken between

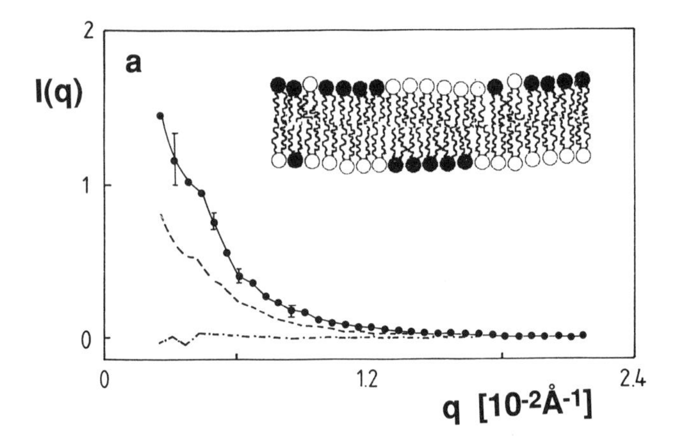

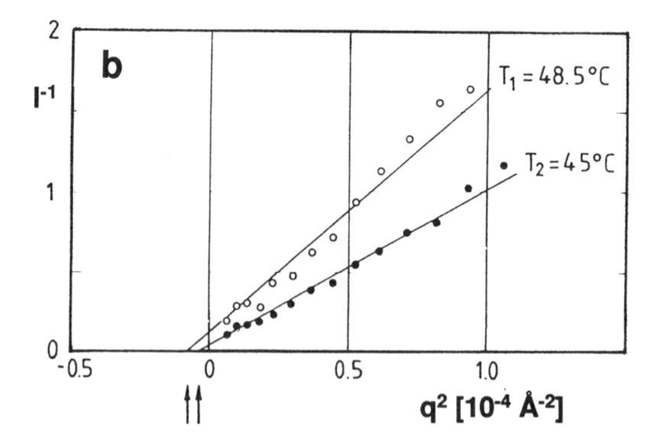

FIGURE 5. (a) Plot of $I(q)$ (defined by Equation 11) vs. q of 1:1 DMPC-d_{54}-DSPC mixture at a H_2O/D_2O ratio of 59/41 for three temperatures 58°C (-------), 48.5°C (- - - -), and 45°C (–●–●–●–). Note that the liquidus line is located at 44°C. (Insert: schematic view of critical concentration fluctuations.) (b) Ornstein-Zernicke plots $(I(q))^{-1}$ for two temperatures (T = 45° and 48.5°C). The arrows point to the intercepts.

4°C and 23°C. Clearly two series of pronounced peaks are observed. One is centered at q ≈ 0.02 $Å^{-1}$ and is observed between 4.2 and 12.9°C (where DMPC is in the $L_{\beta'}$ phase). The second series is centered at q ≈ 0.04 $Å^{-1}$ and starts to appear above 11°C, where DMPC is in the $P_{\beta'}$ phase. Coexistence of the two bands prevails between 11° and 13°C and defines the phase boundaries of a coexistence between a modified $L_{\beta'}$ and $P_{\beta'}$-phase. Similar spectra were taken for other cholesterol concentrations, and from these data other phase boundaries were obtained (cf. Figure 8).

In the second group of experiments, the bilayer repeat distance was measured as a function of temperature and cholesterol concentrations. Figure 6b shows the temperature variation of the bilayer repeat distance for pure DMPC-d_{54} and for mixtures containing 2 mol%, 14 mol%, and 24 mol% cholesterol. Starting from the (metastable) $L_{\beta'}$ phase, the $L_{\beta'} \rightarrow L_{\alpha}$-transitions are clearly indicated at 0, 2, and 14 mol% of cholesterol, but no transition is visible at $x_c = 0.24$. Evaluation of these types of experiments yielded again another set of phase boundaries in the solid phase regime of the DMPC. The phase boundaries obtained from these two types of measurements are also summarized in Figure 8.

B. ON ANOMALOUS THERMAL EXPANSIVITY IN NORMAL DIRECTION

From the temperature dependence of the repeat distance, the thermal expansivity $\alpha_\perp$ in the normal direction can be obtained. This yields for pure DMPC $\alpha_\perp = -3 \cdot 10^{-3}/°C$ in the L_α phase. The corresponding lateral expansivity is $a_{II} = 6.81 \cdot 10^{-3}/°C$ (Evans and Needham, 1986) which yields an isotropic coefficient $\alpha \approx +4 \cdot 10^{-3}/°C$. According to Figure 6b, the out-of-plane expansivity in the fluid phase is much larger at $x_c = 0.14$ (namely $\alpha_\perp = 0.01/°C$) than at $x_c = 0$ and $x_c = 0.02$, but is nearly zero at $x_c = 0.24$. This anomaly at 14 mol% of cholesterol is consistent with the postulate of a fluid-fluid demixing above the chain melting transition and corresponds to the broad endotherm observed by calorimetry (cf. Chapters 3 and 5 in this volume).

C. EVALUATION OF DEMIXING BY INVERSE CONTRAST VARIATION TECHNIQUE

In a detailed contrast variation study, the phase boundaries of the DMPC-d_{54}-cholesterol mixture were determined at 7°C for cholesterol concentrations up to 43%. In order to improve the reliability of this technique, the inverse contrast variation procedure (ICV) was developed. This version of the SANS technique is much more sensitive than the conventional contrast variation, albeit even more time consuming. It avoids, in particular, errors introduced by the incoherent scattering background.

In the inverse contrast variation, the H_2O/D_2O ratio (or $<\rho>$ in Equation 2) is kept constant while the average scattering length densities of the two (segregated) phases $<\rho_\alpha>$ and $<\rho_\beta>$ are varied by replacing the deuterated lipid component (say B) by a mixture of the deuterated (molar fraction x) and the nondeuterated (molar fraction 1 – x) one. The scattering intensity at $q = 0$ can be expressed in terms of the molar fraction x as

$$I_0 \propto \frac{x_\beta - x_0}{x_\beta - x_\alpha}\left(\frac{x_\alpha b_A + x(1-x_\alpha)b_B^d + (1-x)(1-x_\alpha)b_B^p}{V_\alpha} - <p>\right)^2 + \quad (13a)$$

$$+\frac{x_0 - x_\alpha}{x_\beta - x_\alpha}\left(\frac{x_\beta b_A + x(1-x_\beta)b_B^d + (1-x)(1-x_\beta)b_B^p}{V_\beta} - <p>\right)^2 \quad (13b)$$

where b_A, $b_B{}^p$, $b_B{}^d$ are the respective scattering lengths of the (protonated) compound A and of the deuterated and protonated compound B; V_α and V_β are

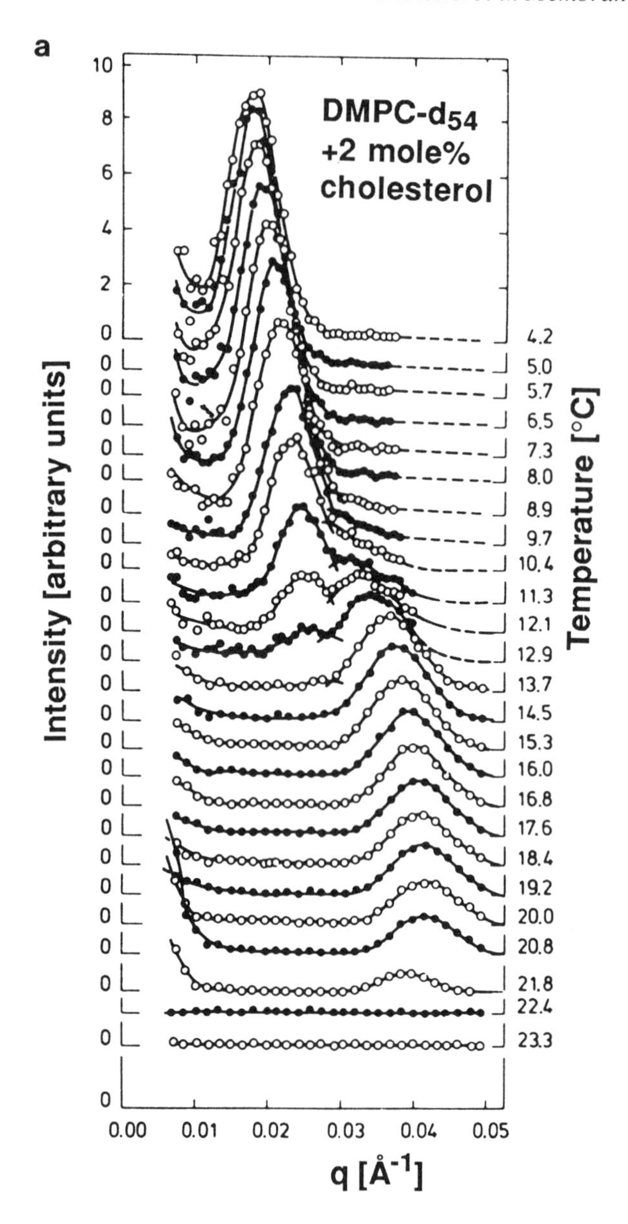

FIGURE 6. (a) q-Dependence of neutron scattering intensity of the reciprocal ripple repeat distance of DMPC-d_{54} vs. wave vector containing 2 mol% of cholesterol at a hydration of 17 wt-% water. The spectra were recorded at increasing temperatures immediately after cooling from the L_α-phase. (b) Temperature dependence of the bilayer repeat distance of DMPC-d_{54} multibilayers at various amounts of cholesterol (0, 2, 14, and 24 mol%). The distances were obtained by Gaussian fits to the diffraction data. The closed symbols correspond to the heating of the system from the $L_{\beta'}$ phase, and the open symbols were obtained by heating from the L_c phase.

b

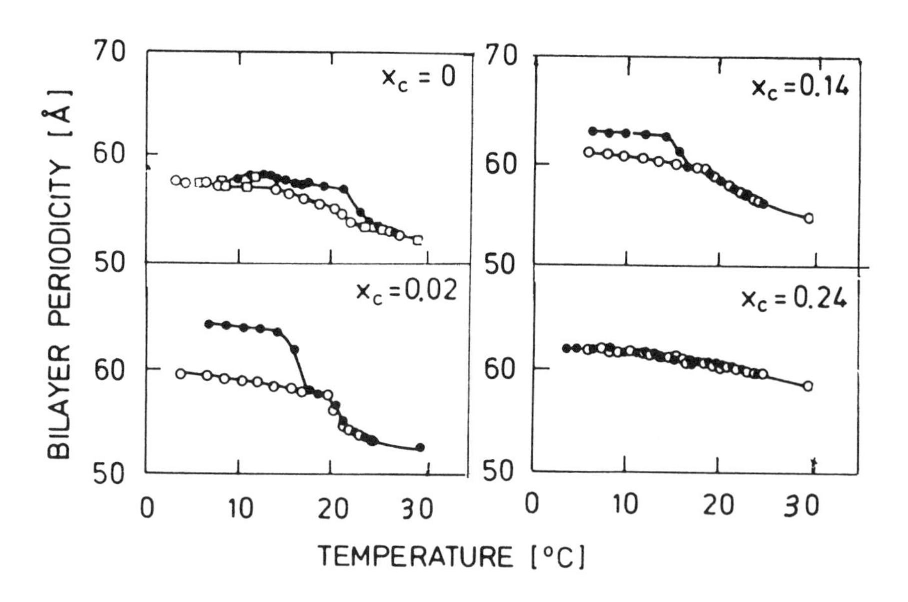

FIGURE 6. (Continued).

again the molar volumes of the two phases, and <ρ> is the average scattering length density of water D_2O/H_2O mixture).

The phase boundaries are obtained by an iterative procedure (Knoll et al., 1985a). First, their approximate positions are determined from the depth of the minima of the plots of $\sqrt{I_0}$ vs. the molar fraction $x_B{}^d$ of the deuterated isomer of DMPC. Then the measured $\sqrt{I_0}$ vs. $x_B{}^d$ curves are simulated with Equation 13 by varying the phase boundaries x_α, x_β until the best fit is achieved. An example is shown in Figure 7b which demonstrates that the phase boundaries may be determined to an accuracy of ±0.5 mol%. The phase boundaries of the DMPC/cholesterol mixture obtained by this procedure for T = 7°C are (x_c molar fraction of cholesterol): $x_{c\alpha} = 0.08$, $x_{c\beta} = 0.24$, $x_{c\gamma} = 0.43$ (Figure 7b). The two former values of x_c fit well to the results obtained by the SANS structural study.

Figure 8 shows the phase diagram of the DMPC-cholesterol mixture as revealed by the contrast variation study of fully hydrated vesicles (Knoll et al., 1985b) and the SANS structural study at 17 wt-% of water (Mortensen et al., 1988). A similar phase diagram was determined by starting from the low temperature, L_c, phase formed after storage for 5 weeks. Further valuable information was obtained by freeze-fracture electron microscopy. Some perti-

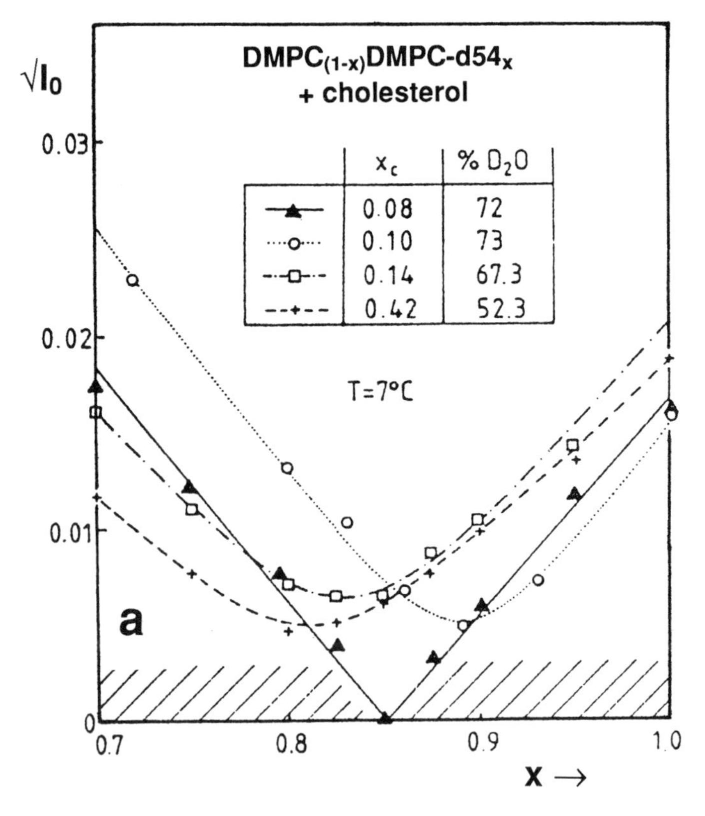

FIGURE 7. (a) Inverse contrast variation (ICV) experiment of a DMPC/DMPC-d_{54}-cholesterol mixture at various D_2O/H_2O ratios (given in the insert). $\sqrt{I_0}$ (as given by Equation 13) is plotted as function of the mole fraction x of the deuterated lipid (DMPC-d_{54}). The dashed horizontal band indicates the limit of detectability of demixing. (b) Calculation of the ICV curve of the DMPC-cholesterol mixture containing 79% DMPC for various positions of the phase boundary curve (- - -): $x_{c\alpha} = 0.08$, $x_{c\beta} = 0.24$; curve (- - -): $x_{c\alpha} = 0.08$, $x_{c\beta} = 0.23$ curve (-----) $x_{c\alpha} = 0.08$, $x_{c\beta} = 0.25$.

nent images are given in Figure 9.

The salient features of the phase diagram are

1. According to the contrast variation experiments, two adjacent miscibility gaps exist at T = 7°C in the domains $0.08 \leq x_c \leq 0.24$ and $0.24 \leq x_c \leq 0.43$. In order to satisfy the phase rule this requires that the mixture containing 24% cholesterol is stoichiometric. This corresponds to the formation of 1:4 cholesterol/PC complexes, as suggested by previous works (Cadenhead and Müller-Landau, 1984). The $x_c = 0.43$ phase line corresponds to the saturation of DMPC with cholesterol leading to the formation of cholesterol precipitates at $x_c > 0.5$. A further verification of a miscibility gap at $x_c \geq 0.24$ (and at low temperature) comes from the finding of a domain structure by freeze-fracture electron micrographs (Figure 9c). Moreover a possible phase separation at $x_c > 0.23$ is suggested by a recent NMR and FTIR- study (Reinl et al., 1991).

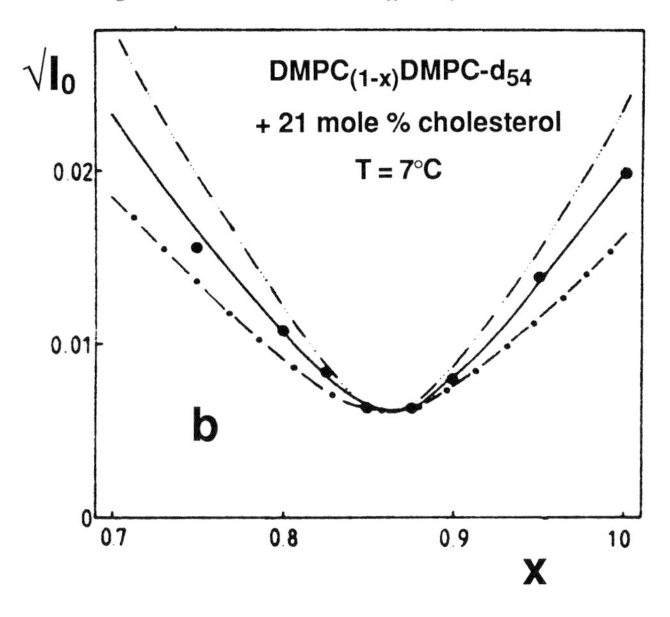

FIGURE 7. (Continued).

2. In the region of the $P_{\beta'}$ phase of DMPC, two types of ripple phases with well-defined repeat distances Δ_r are formed: a phase $P_{\beta'}$ with $\Delta_r = 160$ nm (at 10°C) and a phase $P_{\beta''}$ with $\Delta_r \approx 300$ nm. The former is found at $x_c < 0.08$ and above the pretransition of DMPC, while the latter prevails at $x_c \geq 0.08$, both above and below the pretransition. Both superstructures are also found by freeze-fracture electron microscopy. These show that the former corresponds to the $\Lambda/2$ phase of DMPC-containing cholesterol. The latter is interpreted as a domain-like organization of the former phase (with tilted chains) and the 1:4 cholesterol/DMPC mixture possibly with nontilted chains. The ripple distance increases with increasing cholesterol content (Mortensen et al., 1988).
3. The $L_{\beta'}$_$P_{\beta'}$-transition splits up, leading to the coexistence of the $P_{\beta'}$ ripple phase and a solid solution of (<4 mol%) cholesterol in the $L_{\beta'}$ phase. The transition vanishes at $x_c \geq 0.05$, as is verified by freeze-fracture electron microscopy. The latter reveals clearly a corrugated superstructure with a repeat distance of $\Delta_r \approx 40$ nm. This corresponds to the position of the band at $q = 0.04$ A^{-1} in Figure 6. It is a very impressive demonstration of an ordered defect structure in membranes (Rüppel and Sackmann, 1983).
4. Numerous spectroscopic studies of DPPC bilayers (Rubenstein et al., 1979; Vist and Davis, 1990; Sankaram and Thompson, 1990b) have established a miscibility gap in the L_α phase (at $0.08 \leq x_c \leq 0.24$), with a critical point at $T_c \approx T_m + 10$°C and $x_c = 0.24$. A similar miscibility gap is expected for the DMPC-cholesterol mixture. Evidence is provided by the anomalously large thermal expansion coefficient ($\alpha_\perp = 0.1$/°C) observed at $x_c = 0.14$. Judged from the SANS experiments, this miscibility gap

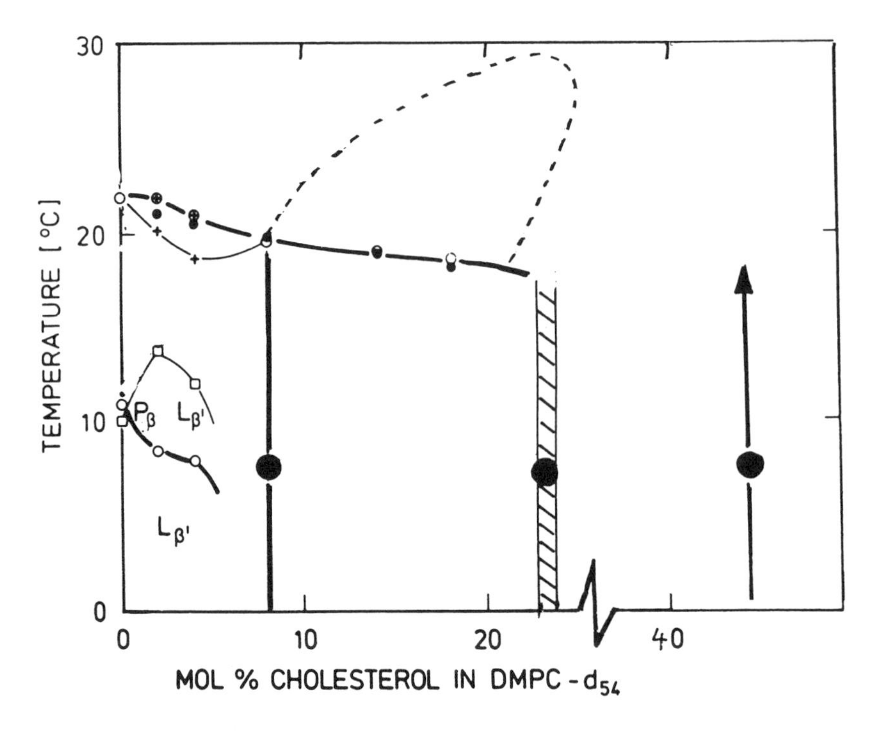

FIGURE 8. Phase diagram of a DMPC-cholesterol mixture as determined by SANS-structural study and contrast variation experiments. The phase boundaries obtained by the latter method are indicated by three large full circles and vertical bars. The points (•, o, +, ⊕) are obtained from measurements of the ripple repeat distance. The vertical cross-hatched bar indicates the stoichiometric 4:1 DMPC-cholesterol mixture. The miscibility gap in the L_α-phase is indicated by anomalous out-of-plane thermal expansivity. At $x_c > 43$ mol% decomposition into a saturated bilayer, and cholesterol occurs both below and above the chain melting transition.

corresponds to the coexistence of the stoichiometric mixture (with $x_c = 0.24$) with a solution of cholesterol in the L_α phase. It is important to note that fluid-fluid demixing is also observed in mixed monolayers of DMPC and cholesterol.

IV. SPECULAR REFLECTION OF NEUTRONS ON PHOSPHOLIPID BILAYERS: THE EFFECT OF CHOLESTEROL

In this part we intend to present some results obtained by specular reflection of neutrons (SRN), a static neutron scattering method that is rapidly gaining recognition as a powerful technique for elucidating interfacial structure. Although it is beyond the scope of this review to treat the method in depth, the most pertinent features of this technique will be described. After a brief outline of the theoretical basics, we will present some examples of its uniqueness in the study of phospholipid membranes without and with cholesterol.

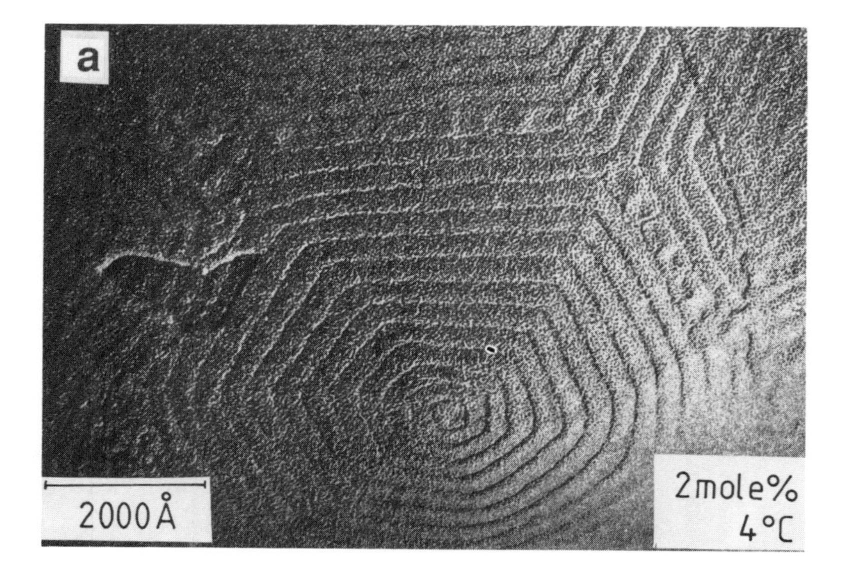

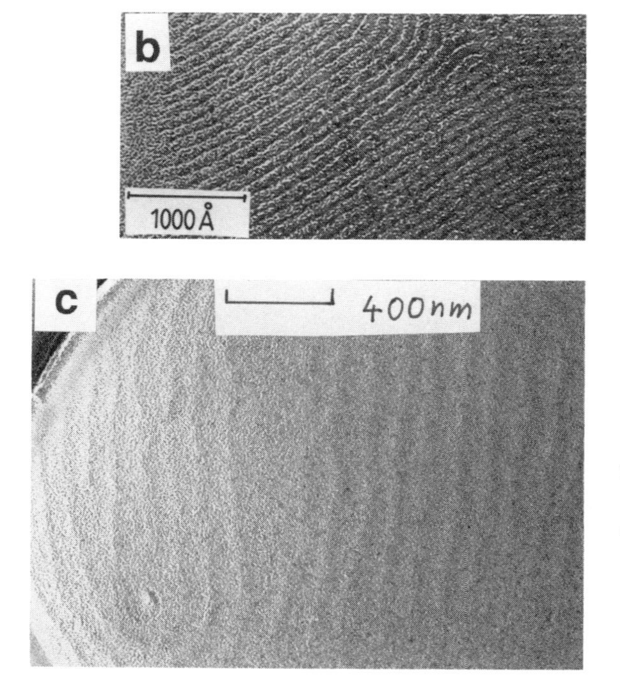

FIGURE 9. The domain structure of a DMPC-cholesterol mixture. (a) Freeze fracture electron micrograpns of DMPC containing 4 mol% cholesterol frozen from the $L_{\beta'}$ phase (4°C, repeat distance 400 Å). (b) Same mixture frozen from $P_{\beta'}$ phase of phospholipid (15°C) showing the well-known Λ/2-ripple superstructure (repeat distance 160 Å). (For the definition of the Λ– and Λ/2 superstructure, see Rüppel and Sackmann, 1983.) (c) Freeze fracture of a mixture containing 28 mol% of cholesterol (frozen from 15°C). (d) Models of a two solid solution of cholesterol in the $P_{\beta'}$-phase of DMPC. (Top: solution of < 4 mol% cholesterol in Λ/2-ripple phase. Bottom: 4:1 stoichiometric DMPC-cholesterol solution.)

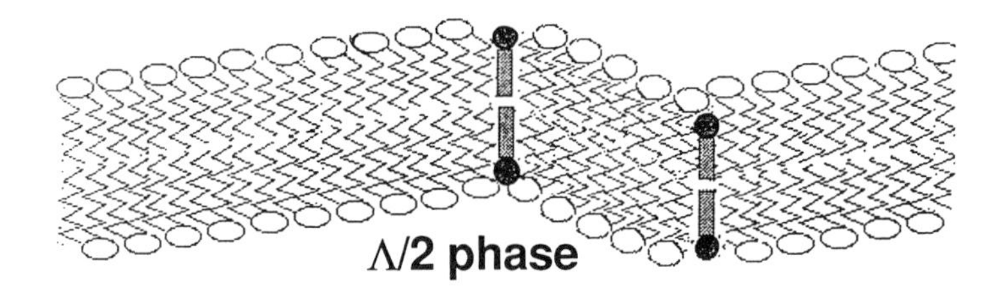

d

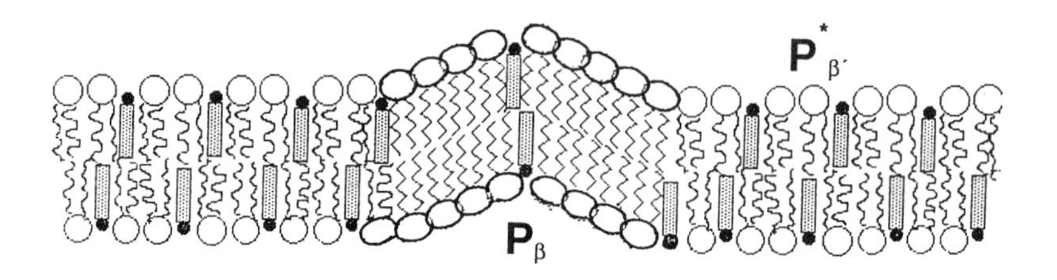

FIGURE 9 (continued).

A. NEUTRON REFLECTION AT INTERFACES (THEORY)

Because of the analogy between reflection of neutrons and of light, one can apply the fundamentals of optics to the phenomenon of specular neutron reflection. For nonadsorbing samples, the neutron refractive index *n* is slightly less than 1 and is given, to within a good approximation, by

$$n = 1 - \frac{\lambda^2 N b}{2\pi} \tag{14}$$

where λ is the wavelength of neutrons, N is the atomic number density, and b is the bound coherent scattering length, which is a property of individual atoms. The value of b can be positive or negative (the latter case corresponds to a phase shift by π due to the scattering). In particular, there is a remarkable difference of b for protons (negative b) and deuterons (positive b). Hence, by isotopic substitution of protons by deuterons, n can be modified and large scattering contrasts can be obtained or adjusted according to the experimental requirements (contrast variation technique).

The simplest case of specular reflection occurs at an interface separating air or vacuum (refractive index $n_1 = 1$) from a solid or liquid (refractive index n_2), where the refraction angle θ_2 is given by the grazing angle of incidence θ_1, according to Snell's law

$$\cos\theta_2 = \left(\frac{1}{n_2}\right)\cos\theta_1 \tag{15}$$

Specular reflection is then observed at incident angles $\theta_1 \geq \theta_c$, the critical angle, where $\cos\theta_c = n_2$. The refractive index profile (or scattering length density profile) normal to an interface can be determined by the measurement of the specular reflection. Hence, reflectivity is measured as a function of momentum transfer normal to the surface,

$$Q_{z,0} = \left(\frac{2\pi}{\lambda}\right)\sin\theta \tag{16}$$

where $\theta = \theta_1$ is the grazing angle of incidence. In a medium i with a neutron scattering length density ρ_i, the normal component of the momentum transfer is modified such that

$$Q_{z,0} = (Q_{z,0}{}^2 - Q_{c,i}{}^2)^{1/2} \tag{17}$$

where $Q_{c,i}$ is the critical value below which total reflection occurs. Thus, specular reflection measurements can be performed either by varying θ_1 at fixed neutron wavelength λ or at various λ (e.g., using polychromatic neutrons and a time of flight analysis), keeping θ_1 fixed.

For a more complicated system consisting of several distinct layers j of different scattering length density ρ_j and thickness d_j, reflection occurs at each interface, and the resulting reflectivity profile is modulated periodically by interference. For real reflectivities, this corresponds to a cosine modulation of the reflectivity, and the distance between two successive minima is $\Delta Q_{z,0}$. From the latter, the average layer thickness $\langle d \rangle$ can be estimated as $\langle d \rangle \approx \pi/\Delta Q_{z,0}$. The value of $\langle d \rangle$ can serve to test the reliabilty of the models applied in the analysis of the reflectivity data. This analysis is often done using the optical matrix method, where the interfacial structure is divided into a series of flat layers with sharp boundaries. A characteristic matrix can be defined for each layer, and for the jth layer, we have

$$M_j = \begin{vmatrix} \cos\beta_j & -(i/p_j)\sin\beta_j \\ -ip_j\sin\beta_j & \cos\beta_j \end{vmatrix} \tag{18}$$

where $p_j = n_j \sin\theta_j$ and $\beta_j = (2\pi/\lambda)n_j d_j \sin\theta_j$ where d_j is the thickness of the layer. For l layers, the characteristic matrices are multiplied: $M_R = [M_1][M_2]\cdots[M_l]$. The reflectivity R is calculated from

$$R = \left|\frac{(m_{11} + m_{12}p_s)p_a - (m_{21} + m_{22})p_s}{(m_{11} + m_{12}p_s)p_a + (m_{21} + m_{22})p_s}\right|^2 \tag{19}$$

where m_{ij} designates the elements of the matrix M_R, and subscripts s and a refer to substrate and air, respectively. In applying the optical matrix method, one begins by specifying a model for the interfacial structure and then proceeds to solve Equation 19 for reflectivity as a function of momentum transfer Q, adjusting the model as necessary to achieve good fits to the data.

In general, there are two important limitations of the method. One is that a real interface is not infinitely sharp, leading to an interfacial roughness or a continous variation of ρ_i near the interface. Second, since the square of the reflectivity amplitude is measured in a reflection experiment, phase information is lost and thus it is not possible to obtain a unique density profile that fits a single set of reflectivity data.

In order to cope with the above limitations, it is important to conduct specular reflection experiments at more than one contrast in scattering length density (contrast variation). For each layer in the model there are two parameters to fit: thickness of the layer, and its scattering length density. Thus, for a two-layer model, such as we use to describe the lipid bilayer, there are four unknown parameters, provided that interfacial roughness and surface roughness are already known. There are essentially two ways of obtaining different scattering contrasts in reflection studies of lipids. One is to change the contrast to the bulk water by using either pure H_2O or D_2O (solvent contrast variation). By mixing both waters at the appropriate ratio, one can obtain a match of the scattering length density of the bulk water to that of air or to that of a solid substrate supporting the lipid bilayer, respectively. In the latter case only the lipids would contribute to the reflectivity. The second method for the variation of the scattering contrast is the application of selectively or totally deuterated phospholipids or of isotopic mixtures of them (inverse contrast variation). This enables the screening of parts of the bilayer such that the number of interfaces contributing to reflectivity is reduced and the model applied can be checked for consistency.

B. APPLICATION OF SRN TO PHOSPHOLIPID BILAYERS WITHOUT AND WITH CHOLESTEROL

The very first application of SRN to phospholipid systems was devoted to the study of the structure of a DMPC/DMPG monolayer at the air/water interface, as a function of the lateral pressure (Bayerl et al., 1990). Recently, the DPPC monolayer structure was studied by SRN, and the data were discussed in the light of previously obtained synchrotron radiation reflection experiments (Vakinin et al., 1991a). A further step toward the application of SRN in membrane biophysics is a study of the interaction of peripheral (extrinsic) proteins and peptides (such as spectrin and polylysine) with electrically charged phospholipid monolayers (Johnson et al., 1991). Recently, recognition processes at a functionalized lipid monolayer were observed using the SRN technique (Vakinin et al., 1991b) However, all these studies were done on phospholipid monolayers at the air/water interface, and their results are not readily applicable to bilayers. Since bilayers cannot be formed at the air/water interface, we recently introduced the use of planar-supported phospholipid bilayers in SRN studies (Johnson et al., 1990). For these experiments, the sample consists of a single lipid bilayer supported by a planar quartz or silicon plate. The sample cell used is depicted in Figure 10. This setup enables the study of a bilayer under conditions of full hydration and well-defined (and adjustable) temperature. Previous studies using mainly NMR methods revealed that the dynamics of such supported bilayers is similar to those of multilamellar bilayer vesicles (Bayerl and Bloom, 1990).

As a first step we performed SRN experiments on a single supported DMPC bilayer at two temperatures, corresponding to the gel phase and to the liquid crystalline phase, respectively. Figure 11 depicts theoretical scattering length density profiles for the three contrasts that we utilized. In case I, the bulk water is contrast matched to the quartz. This allows a sensitive measurement of changes of the bilayer thickness. In case II, heavy water (D_2O) is used as bulk water, providing a strong contrast between the quartz surface and the single bilayer. This arrangement is sensitive to changes of the lipid headgroup hydration and the thickness of the confined water layer. In case III, the bulk water and the lipid chains are matched to the quartz such that the reflectivity is mainly determined by the headgroups. This provides a measure of the thickness of the headgroups. The reflectivity data for two of the three cases of contrast (cases I and II) are shown in Figure 12, and demonstrate the sensitivity of the reflectivity profile to the variation of the contrast.

The analysis of the reflectivity data for case I was done in terms of a one-layer model; for cases II and III a three-layer model and two-layer model, respectively, were applied. In the model fitting, we have constrained the parameters such that the model is self-consistent for all the contrasts. The results are shown in Table 1. They provide unambigous evidence for the existence of an ultrathin water layer between the bilayer and the solid support. The thickness of 30 Å given for this layer in Table 1 does include the 12 to 15 Å surface roughness of the quartz plate, measured in a separate experiment with the plain plate and D_2O as bulk water phase.

A further interesting result is the finding that no $P_{\beta'}$ ("ripple") phase is formed at the transition of the bilayer from the fluid into the gel phase state. We find a thickness change of the bilayer due to the phase transition of 3 Å, a value that is normally observed for the transition from the L_α phase to the $L_{\beta'}$ phase. A likely explanation is that an increased lateral stress that the supported membrane undergoes at the transition from the fluid phase to the gel phase prevents the formation of a $P_{\beta'}$ phase. This is supported by previous results obtained by Needham and Evans (1988) using micromechanical techniques, where it has been shown that a nonrippled phase (denoted as $L^*_{\beta'}$) is formed under conditions of lateral stress. Furthermore, the experiments provide a precise measure for the thickness of the DMPC head group (8 Å) and the acyl chain region. From the measured tail length in the gel phase and the known maximum length of the hydrocarbon tail in an all-*trans* conformation, we can calculate a tilt angle of $\beta = 26 \pm 7°$, in good agreement with values obtained by other methods.

Finally, the analysis of the obtained scattering length densities for case II in Table 1 lets us conclude that there is no significant change of the DMPC

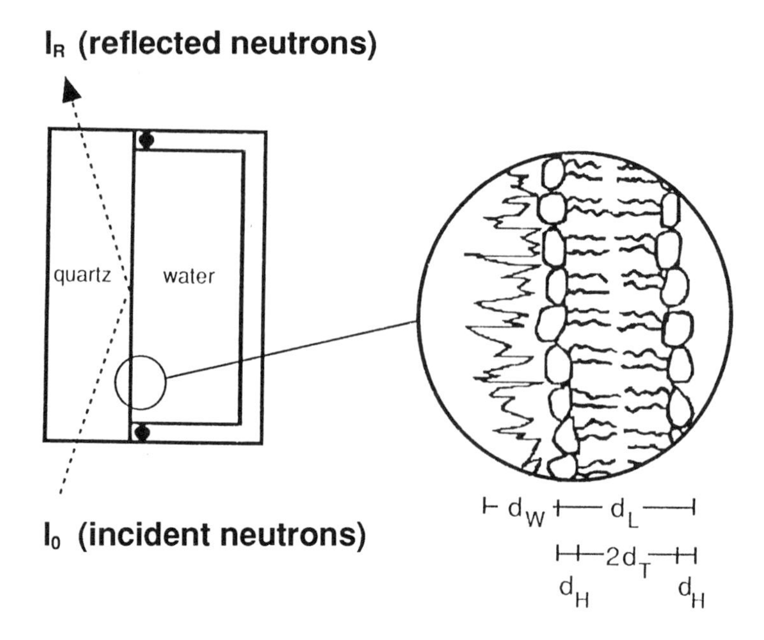

FIGURE 10. Schematic diagram of the sample cell for SRN studies on supported lipid bilayers as seen from above. The incident neutron beam, I_o, passes through the quartz block (dimensions 100 mm × 50 mm × 10 mm) and is reflected at the quartz/water interface as I_R. The enlargement depicts the lipid bilayer supported by the (atomically rough) quartz surface. The thicknesses of the various layers are indicated: d_H is thickness of the polar head group, d_T is thickness of the hydrocarbon tail group, d_W is thickness of the water layer, and d_L is thickness of the entire lipid bilayer.

headgroup hydration between the two phase states studied. The volume fraction α of water in the lipid bilayer for this case can be calculated by

$$\rho_L^{fit} = \alpha\rho_{D_2O} + (1-\alpha)\rho_L^{theory} \quad (20)$$

where ρ_L^{theory} is the theoretical value of ρ for the lipid bilayer without water. This finding indicates a major structural difference between supported bilayers and phospholipid monolayers. For the latter we found a significant change of the hydration between the extended and the condensed phase state (Bayerl et al., 1990).

Having established the structure of a pure bilayer using SRN, we proceeded to study the effect of cholesterol on the structure of a DPPC single supported bilayer at high cholesterol concentrations. This is of particular interest since this mixture is assumed to exhibit a homogenous phase, the so-called liquid ordered (l_o) phase, at cholesterol concentrations ≥ 23 mol% (Vist and Davis, 1990). Moreover, using NMR and FT-IR we recently observed distinct changes of the l_o phase structure in this mixture in the 35 to 45 °C temperature range (Reinl et al., 1991). Therefore, we performed SRN measurements on a mixture containing

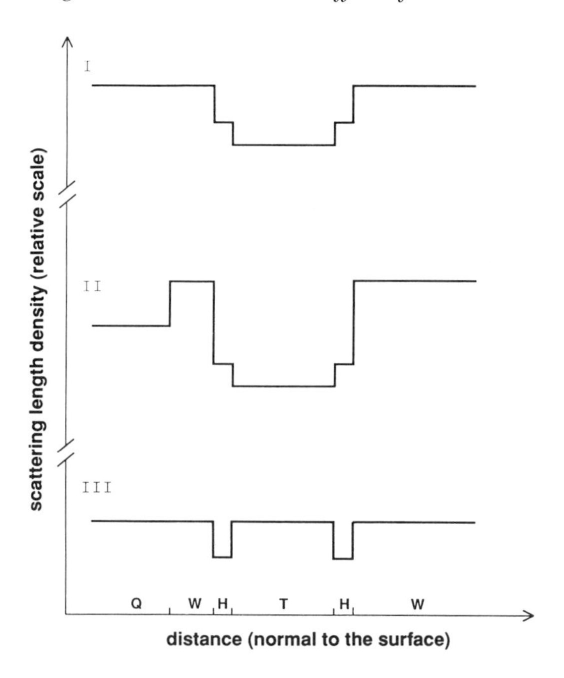

FIGURE 11. Theoretical scattering length density profiles for the three cases of contrast studied. Layers of differing scattering length density are indicated on the abscissa. (Q: quartz plate; W: water; H: polar head group; T: hydrocarbon tail groups.) Case I corresponds to DMPC in an aqueous phase of water contrast-matched to the quartz ("CMQ" water), while the aqueous phase is D_2O in case II. Case III portrays a mixture of DMPC/DMPC-d_{54} having hydrocarbon tails contrast matched to the quartz in an aqueous phase of CMQ water.

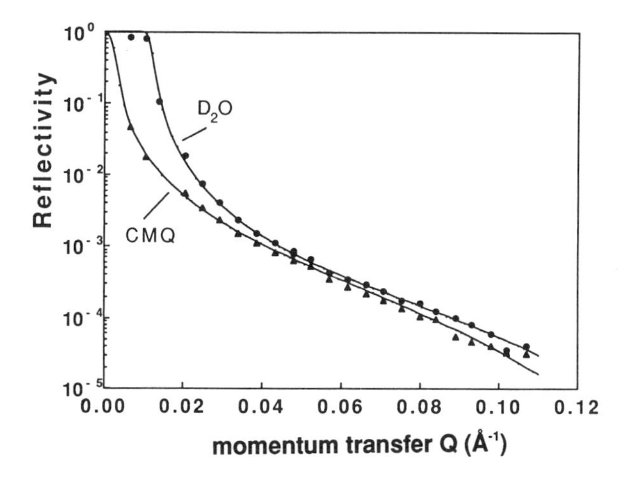

FIGURE 12. Reflectivity data as a function of momentum transfer Q for a bilayer of DMPC at T = 31°C. Data for two different cases are shown: Case I with CMQ water as the aqueous phase (Δ), and case II with D_2O as the aqueous phase (o). The solid lines are best fits to the data, and the parameters used in the fitting routine are reported in Table 1.

TABLE 1
Experimental Results for DMPC

Case	T(°C)	d_L	ρ_L	d_W	ρ_W	d_H	ρ_H	d_T	ρ_T
I	31	43	0.7	—	—	—	—	—	—
II	31	43	2.0	30	3.7	—	—	—	—
III	31	43	—	—	—	8	1.6	13.5	3.5
I	20	46	0.7	—	—	—	—	—	—
II	20	46	2.1	30	3.7	—	—	—	—

[a] Lipid dimensions d_L, d_H, and d_T (cf. Figure 11 for depiction of these dimensions) are reported in Å with an error of ±1.5 Å, while the error in the water layer thickness d_W is ±10 Å. The scattering length density r is reported as 10^{-6} Å^{-2} with an error of ±0.2 ¥ 10^{-6} Å^{-2}. The three cases I, II, and III representing different constrasts are defined in the text. The dashes indicate that the corresponding quantity could not be determined under the contrast of that particular case.

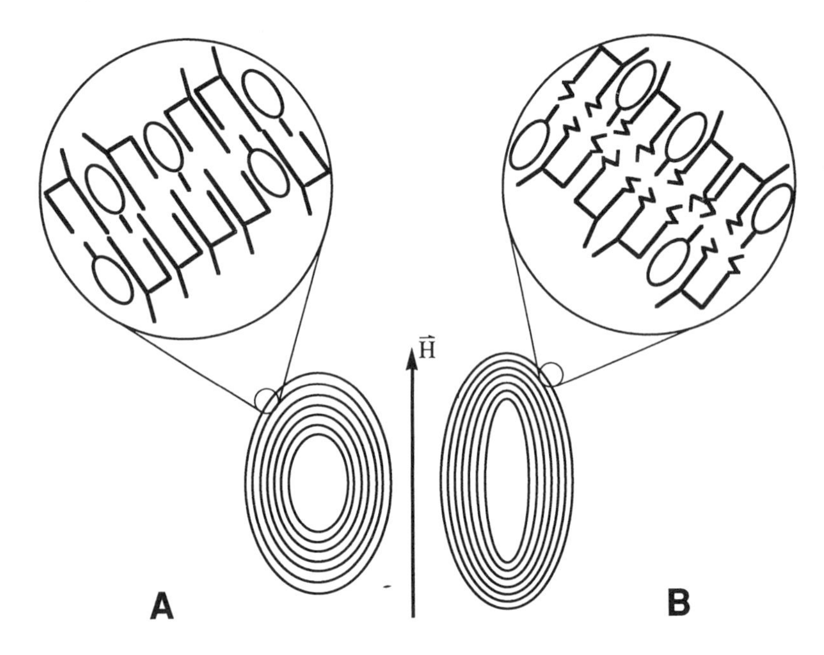

FIGURE 13. Schematic drawing of the effect of temperature on the bilayer structure and shape of DPPC multilamellar vesicles containing 25 mol% cholesterol. The shape change is a result of the macroscopic orientation of the vesicles in the field of NMR magnet [9.6 Tesla, the arrow indicates the direction of the magnetic field (Brumm et. al., 1992) while the structural change is independent on the magnetic field]. A: T = 30°C; B: T = 45°C.

25 mol% cholesterol at two temperatures (below and above the anticipated structural change) and at two bulk water contrasts (corresponding to cases I and II in Figure 11). The analysis of the SRN data yielded bilayer thicknesses of d_L = (50.5 ± 1) Å at 30°C and d_L = (46 ± 1) Å at 45°C. The scattering length density ρ_L of the single bilayer increases by 8 (±3)% with increasing temperature in the

range 30 to 45°C. Since we found that the thickness of the water layer between the bilayer and the quartz surface does not change with temperature, the observed changes indicate a significant alteration of the internal structure of the DPPC/cholesterol bilayer in this temperature range.

Using the value of $d_H = 8\ (\pm 1.5)$ Å from Table 1, we can calculate the thickness of the tail region per monolayer d_T as $d_T = (d_b - 2d_h)/2 = 17.3\ (\pm 1.5)$ Å at 30 °C, and $d_T = 15\ (\pm 1.5)$ Å at 45°C. This provides evidence that the acyl chains are not, even at 30°C, in a complete all-*trans* conformational state, since the theoretical length of an all-*trans* palmitoyl chain is 19.15 Å. The 2 Å difference can account for an average of 1.5 gauche conformers per acyl chain at 30°C.

Combining the SRN data with our NMR and FT-IR results obtained in cholesterol/DPPC mixtures in the l_o phase, we suggest that cholesterol undergoes a reversible change of its site in the bilayer in the 30 to 45°C temperature range. In this model a temperature increase drives the cholesterol out of the bilayer center towards the interface by about 4 Å (Reinl et al., 1991). This has consequences for the elastic properties of the membrane, which are manifested by a concomitant change of the vesicle shape. This situation is depicted in Figure 13.

REFERENCES

Albrecht, O., H. Gruler, and E. Sackmann. 1981. Polymorphism of phospholipid monolayers. *J. Coll. Interface Sci.*, 79:319–338.

Bayerl, T. M. and M. Bloom. 1990. Physical properties of single phospholipid bilayers adsorbed to micro glass beads. *Biophys. J.*, 58:357–362.

Bayerl, T. M., R. K. Thomas, J. Penfold, A. Rennie, and E. Sackmann. 1990. Specular reflection of neutrons at phospholipid monolayers. *Biophys. J.*, 57:1095–1098.

Brown, M. F. and J. Seelig. 1978. Influence of cholesterol on the polar region of phosphatidylcholine and phosphatidylethanolamine bilayers. *J. Am. Chem. Soc.*, 17(2):381-384.

Brumm, T., A. Möps, C. Dolainsky, S. Brückner, and T. M. Bayerl. 1992. Macroscopic orientation effects in broadline NMR-spectra of model membranes at high magnetic field strength. A method preventing such effects. *Biophys. J.*, in press.

Cadenhead, D. A. and F. Müller-Landau. 1984. Molecular packing in steroid-lecitin monolayers. *Can. J. Biochem. Cell Biol.*, 62:732.

de Kruijff, B., P. R. Cullis, and A. J. Verkleij. 1980. Non-bilayer lipid structures in model and biological membranes. *TIBS*, March 1980:79–81.

Evans, E. and D. Needham. 1986. Giant vesicle bilayers composed of mixtures of lipids, cholesterol and polypeptides. *Faraday Discuss. Chem. Soc.*, 81:267–280.

Finegold, L. and M. A. Singer. 1992. Cholesterol/phospholipid interactions studied by DSC: effect of acyl chain length and role of the C(17) sterol side group. *Cholesterol in Membrane Models*, CRC Press, Boca Raton, FL.

Galla, H. J., W. Hartmann, U. Theilen, and E. Sackmann. 1979. On two-dimensional passive random walk in lipid bilayers and fluid pathways in biomembranes. *J. Membr. Biol.*,48:215–236.

Hatta, I. and S. Imaizumi. 1985. AC calorimetric study of phospholipid-cholesterol systems and their structures. *Mol. Cryst. Liq. Cryst.*,124:219–224.

Helfrich, W. 1978. Steric interaction of fluid membranes in multilayer systems. *Z. Naturforschung.*, 32a:305.

Hicks, A., M. Dinda, and M. A. Singer. 1987. The ripple phase of phosphatidylcholines: effect of chain length and cholesterol. *Biochim. Biophys. Acta,* 903:177–185.

Ipsen, J. H., G. Karlström, O. G. Moritsen, H. Wennerström, and M. J. Zuckermann. 1987. Phase equilibria in the phosphatidylcholine-cholesterol system. *Biochim. Biophys. Acta,* 905:162–172.

Ipsen, J. H., O. G. Mouritsen, and M. J. Zuckermann. 1989. Theory of thermal anomalies in the specific heat of lipid bilayers containing cholesterol. *Biophys. J.*, 56:661–667.

Jacrot, B. 1976. Neutron scattering from biological structures. *Rep. Progr. Phys.*, 39:911.

Johnson, S. J., T. M. Bayerl, D. C. McDermott, G. W. Adam, A. R. Rennie, R. K. Thomas, and E. Sackmann. 1990. Structure of an adsorbed DMPC bilayer measured with specular reflection of neutrons. *Biophys. J.*, 59:289–294.

Johnson, S. J., T. M. Bayerl, W. Weihan, H. Noack, J. Penfold, R. K. Thomas, D. Kanellas, A. R. Rennie, and E. Sackmann. 1991. Coupling of spectrin and polylysine to phospholipid monolayers studied by specular reflection of neutrons. *Biophys. J.*, 60:1017–1025.

Knoll, W., K. Ibel, and E. Sackmann. 1981a. SANS study of lipid phase diagrams by the contrast variation method. *Biochemistry*, 20:6379–6383.

Knoll, W., J. Haas, H. B. Stuhrmann, H. Vogel, H. Füldner, and E. Sackmann. 1981b. SANS of aqueous dispersions of lipids and lipid mixtures. A contrast variation study. *J. Appl. Cryst.*, 14:191–202.

Knoll, W., G. Schmidt, and K. Ibel. 1985a. The inverse contrast variation in SANS: a sensitive technique for the evaluation of lipid phase diagrams. *J. Appl. Cryst.*, 18:65–70.

Knoll, W., G. Schmidt, K. Ibel, and E. Sackmann. 1985b. SANS study of lateral phase separation in DMPC — cholesterol mixed membranes. *Biochemistry*, 24:5240–5246.

Knoll, W., G. Schmidt, H. Rötzer, T. Henkel, W. Pfeiffer, E. Sackmann, S. Mittler-Neher, and J. Spinke. 1991. Lateral order in binary lipid alloys and its coupling to membrane function. *Chem. Phys. Lipids.*, 57:363–374.

König, S., W. Pfeiffer, T. M. Bayerl, D. Richter, and E. Sackmann. 1992. Molecular dynamics of lipid bilayers studied by incoherent quasi-elastic neutron scattering. *J. Phys. (France),* submitted.

Lee, A. G., 1977. Lipid phase transitions and phase diagrams. *Biochim. Biophys. Acta,* 472:285–300.

Leibler, S. and D. Andelman. 1987. Ordered and curved mesocopic structures in membranes and amphiphilic films. *J. Phys. (France),* 48:2013.

Mabrey, S., P. L. Mateo, and J. M. S. Sturtevant. 1978. High sensitivity DSC study of mixtures of cholesterol with DMPC and DPPC. *Biochemistry*, 17:2464–2468.

Maksymiw, R., S. Senfang Sui, H. Gaub, and E. Sackmann. 1987. Electrostatic coupling of spectrin dimers to phosphatidylserine containing lipid lamellae. *Biochemistry*, 26:2983.

Marsh, D. and G. Cevc. 1987. *Phospholipid Bilayers,* Wiley Interscience, New York.

Mortensen, K., W. Pfeiffer, E. Sackmann, and W. Knoll. 1988. Structural properties of a phosphatidylcholine-cholesterol system as studied by SANS: ripple structure and phase diagram. *Biochim. Biophys. Acta,* 945, 221–245.

Needham, D. and E. Evans. 1988. Structure and mechanical properties of giant DMPC vesicle bilayers from 20°C below to 10°C above the liquid crystal-crystalline phase transition at 24°C. *Biochemistry*, 27:8261–8269.

Pfeiffer, W., G. Schlossbauer, W. Knoll, B. Farago, A. Steyerl, and E. Sackmann. 1988. Ultracold neutron scattering study of local mobility in bilayer membranes. *J. Phys. (France)*, 49:1077.

Pfeiffer, W., T. Henkel, E. Sackmann, W. Knoll, and D. Richter. 1989. Local dynamics of lipid bilayers studied by incoherent quasi-elastic neutron scattering. *Europhys. Lett.*, 8:201.

Recktenwald, D. and H. M. McConnell. 1981. Phase equilibria in binary mixtures of phosphatidylcholine and cholesterol. *Biochemistry,* 20:4505–4510.

Reinl, H., T. Brumm, and T. M. Bayerl. 1991. Changes of the physical properties in the liquid-ordered phase with temperature in binary mixtures of DPPC with cholesterol: a 2H-NMR, FT-IR, DSC and neutron scattering study. *Biophys. J.*, in press.

Rubenstein, J. L. R., B. A. Smith, and H. M. McConnell. 1979. Lateral diffusion in binary mixtures of cholesterol and phosphatidylcholines. *Proc. Natl. Acad. Sci. U.S.A.*, 76:15–18.

Rüppel, D. and E. Sackmann. 1983. On defects in different phases of two dimensional lipid bilayers. *J. Phys. (France)*, 44:1025.

Sackmann, E., D. Rüppel, and C. Gebhardt. 1980. Domain structure in model membranes. Springer Series in *Chem. Phys.*, 11:309.

Sackmann, E. 1984. Physical basis of trigger processes and membrane structures. In *Biomembranes*, vol. 5, Chapmann, D., Ed., Academic Press, London, pp. 105–143.

Sackmann, E. 1990. Molecular and global structure and dynamics of membranes and lipid bilayers. *Can. J. Phys.*, 68:999.

Sackmann, E. 1983. Physical foundations of the molecular organization and dynamics of membranes. In *Biophysics*, Hoppe, W., Ed., Springer-Verlag, Berlin, pp. 425–457.

Sankaram, M. B. and T. E. Thompson. 1990a. Modulation of Phospholipid acyl chain order by cholesterol. A solid state 2H-NMR study. *Biochemistry*, 29:10676–10684.

Sankaram, M. B. and T. E. Thompson. 1990b. Interaction of cholesterol with various glycerophospholipids and sphingomyelin. *Biochemistry*, 29, 10670–10675.

Stuhrmann, H. B. and E. D. Duee. 1975. Neutron scattering on binary mixtures of phosphatidylcholine and cholesterol. *J. Appl. Cryst.*, 8:538–546.

Vakinin, D., K. Kjaer, J. Als-Nielsen, and M. Lösche. 1991a. Structural properties of phosphatidylcholine in a monolayer at the air/water interface. *Biophys. J.*, 59:1325–1332.

Vakinin, D., J. Als-Nielsen, M. Piepenstock, and M. Lösche. 1991b. Recognition processes at a functionalized lipid surface observed with molecular resolution. *Biophys. J.*, 60:1545–1552.

Vist, M. R. and J. H. Davis. 1990. Phase equilibria of cholesterol/DPPC mixtures: 2H-NMR and DSC. *Biochemistry*, 29:451–464.

Chapter 3

A REVIEW OF THE KINETICS OF CHOLESTEROL MOVEMENT BETWEEN DONOR AND ACCEPTOR BILAYER MEMBRANES

Robert Bittman

TABLE OF CONTENTS

0-8493-4207-4/93/$0.00+$.50

ABSTRACT

Spontaneous sterol exchange between unilamellar vesicles is reviewed. The effects of variation in lipid composition, vesicle geometry, sterol structure, and addition of water-soluble proteins on the rate of sterol movement are discussed. It appears that the molecular interactions of cholesterol with neighboring phospholipids in the *outer* leaflet of the bilayer are particularly important in determining the cholesterol exchange rate. At temperatures above the transition temperature of the phospholipid components, the exchange rate follows the order unsaturated phosphatidylcholine (PC) > saturated PC > sphingomyelin (SPM). The exchange rate is not affected by conformational constraints in the glycerol backbone of phospholipids. The ability of SPM to lower the rate of cholesterol exchange from donor to acceptor species is explained by the greater lateral packing density in the lipid-water interface when SPM is present; van der Waals attractive forces are apparently stronger between SPM and cholesterol than between PCs and cholesterol. The hydroxy group of SPM does not play an important role in decreasing the cholesterol exchange rate. Sterols that are more hydrophilic than cholesterol, as estimated by reversed-phase high-performance liquid chromatography, undergo faster exchange between vesicles than cholesterol, whereas sterols that are more hydrophobic undergo slower exchange. However, the rates of [^{14}C]sterol desorption from the lipid-water interface are not correlated with the relative sterol hydrophobicity, and the interaction of sterols with phospholipids appears to provide the principal physical-chemical basis for determining the rates of spontaneous exchange of sterols between bilayers. Cholesteryl amines, which bear a positive charge at the 3 position of the sterol nucleus, undergo full exchange, but with biphasic kinetics; the slow phase is considered to reflect the high activation energy for inner-to-outer leaflet movement of the charged lipid. The cholesterol exchange rate is enhanced by incorporation of at least five molecules of various apolipoproteins (apoA-I, apoA-II, and apo C) into small unilamellar vesicles, suggesting that nonspecific lipid-protein interactions are responsible for the rate enhancement. Sterol carrier protein-2 from rat liver enhanced the intervesicle exchange rate of cholesterol when the vesicles contained an acidic phospholipid. The finding that the cholesterol exchange rate is slower with large unilamellar vesicles as donors than with small unilamellar vesicles suggests that cholesterol transfer from a planar-like plasma membrane of biological cells to internal cell membranes may be a slow process.

I. INTRODUCTION

Cholesterol is the principal neutral lipid component of many mammalian cells. Cholesterol affects the physical state and the packing density of membrane phospholipids, modulates the activities of several membrane proteins, and regulates membrane permeability behavior, as discussed elsewhere in this volume (Yeagle, 1992; Hui, 1992). Many studies have focused on the effects of variation of lipid bilayer composition on the structural and functional properties of membranes. Spontaneous interbilayer transfer of lipids such as cholesterol is one process by which the lipid composition of membranes is varied. Free (unesterified) cholesterol has the highest mobility of the principal membrane components of cell membranes, both with respect to its rate of transbilayer migration and its rate of movement between membranes. Efflux of cholesterol takes place from the outer half of the donor bilayer. The cholesterol molecules taken up by an acceptor species must first be transferred into the external leaflet of the acceptor membrane before they may move across the bilayer of the acceptor. In eukaryotic cells, uptake of cholesterol by membranes of organelles is a complicated process because the membranes have different compositions and different geometrical curvatures and because proteins may influence the rate of cholesterol movement between membranes by binding with or catalyzing reactions involving cholesterol. Furthermore, surface transfer or exchange of cholesterol associated with lipoproteins is complicated by receptor-mediated uptake.

It is well known that redistribution of cholesterol takes place between cell membranes and lipoproteins, leading to changes in the molar ratio of cholesterol/phospholipid in the lipid compartments (Rottem et al., 1978; Lund-Katz et al., 1982). In fact, the phenomenon of spontaneous exchange between cholesterol in cell membranes and cholesterol present in an exogenous species, such as lipoproteins and red blood cells, was established more than 40 years ago (Hagerman and Gould, 1951). The time course of transfer or exchange of cholesterol between membrane structures such as vesicles, erythrocytes, lipoproteins, mycoplasmas, and mammalian cells in tissue culture in the absence of proteins is relatively rapid (with a half-time of about 1 to several hours) (reviewed by Bittman et al., 1983; Phillips et al., 1987; Dawidowicz, 1987; Bittman, 1988). Nevertheless, the distribution of cholesterol among the membranes of mammalian cells is highly asymmetric; for example, high cholesterol contents are found in plasma membranes and secretory vesicles, but very low concentrations are found in the endoplasmic reticulum and mitochondria.

The implications of the net transfer of cholesterol between the outer leaflets of different membrane surfaces in cultured cells are strikingly important, with changes in intracellular lipid metabolism occurring (Slotte et al., 1989, 1991; Pörn and Slotte, 1990), along with alterations in membrane transport and low-density lipoprotein (LDL) binding (Robertson and Poznansky, 1985). Enzyme-catalyzed degradation of SPM in the fibroblast cell surface induced a flow of cholesterol from the plasma membrane into the cell, even when exogenous high-density lipoprotein (HDL) was provided as a possible acceptor species (Slotte et al., 1990). Internalization of cholesterol in response to SPM depletion results in an increased formation of cholesteryl esters and a downregulation of cholesterol biosynthesis. Thus, some cells, when exposed to lipid surfaces rich in cholesterol, possess a mechanism that enables them to accumulate esterified cholesterol within the cell rather than accumulate free cholesterol into the cell surface. The process of transfer of cholesterol in the opposite direction, i.e., from intracellular membranes to the plasma membrane, was observed after binding of HDL to fibroblasts and after restoration of the SPM content in sphingomyelinase-treated baby kidney cells (Slotte et al., 1990).

There has been considerable interest in recent years in the spontaneous movement of unesterified cholesterol between unilamellar vesicles. This chapter presents an update of the information about the kinetics of cholesterol interbilayer movement between vesicles since the topic was last reviewed in 1987 (Phillips et al., 1987; Dawidowicz, 1987). Other reviews cover studies of cholesterol exchange between lipoproteins, mycoplasmas, and mammalian cells (Phillips et al., 1987; Dawidowicz, 1987; Bittman, 1988; Johnson et al., 1991). Unilamellar vesicles have been studied extensively to elucidate the factors that affect the rate of nonreceptor-mediated cholesterol intermembrane movement, since the composition and physical state of their lipids and their size can be manipulated experimentally. Multilamellar vesicles are less attractive than unilamellar vesicles because cholesterol can move between lamellae. Therefore, efflux data obtained with unilamellar vesicles are useful in understanding the more complicated process of cell cholesterol efflux. Exchange, rather than net transfer, of cholesterol was examined in most studies, since the experimental design was to use the same ratio of cholesterol/total phospholipid in the donor and acceptor vesicles; transfer implies a change in cholesterol/phospholipid molar ratio in the donor and acceptor bilayers during the course of the experiment.

II. ASSAY SYSTEMS

Unilamellar vesicles for exchange studies are generally prepared by sonication of aqueous multilamellar dispersions of lipids in buffered 0.5 to 1 m*M* EDTA and 0.02 to 0.05% w/v sodium azide. The lipid film is dispersed completely in the desired volume of buffer by vortexing at a temperature above the main transition temperature of the highest-melting phospholipid. Sonication is carried out under nitrogen or argon in a water bath; during the sonication, resting periods of several minutes duration are included to avoid local heating effects. A sonicator equipped with a broad-tip probe, a tapered microtip, or a cup horn is frequently used. After sonication the vesicle suspension is centrifuged (e.g., at 15,000 *g* for 10 to 20 min) to remove large lipid particles (and titanium particles released from the probe, if a tip probe is used), and the supernatant is used for the exchange experiments. In many studies the sizes and homogeneities of the vesicles have been examined by photon correlation spectroscopy, and negative-staining electron microscopy of at least 50 to 100 particles in each preparation has been used to determine whether the vesicles are unilamellar and to estimate their sizes. Trapped volume measurements and NMR estimations of PC distribution have also been used to characterize vesicle size. Large unilamellar vesicles are generally prepared by the extrusion method in which multilamellar dispersions are filtered through polycarbonate filters of different pore diameters in an extrusion apparatus. Large unilamellar vesicles with diameters >200 nm have been made by freeze-thawing of the dispersion during the extrusion procedure, but multilamellar vesicles may be present (Thomas and Poznansky, 1988b). Large unilamellar vesicles made by the slow detergent dialysis procedure have also been used in exchange experiments (Fugler et al., 1985; Thomas and Poznansky, 1988a).

The same phospholipid is generally used in both donor and acceptor vesicles in order to avoid a change in the phospholipid composition during the cholesterol exchange process. Although the half-time for the spontaneous exchange of long-chain phosphatidylcholine (PC) between vesicles is more than a factor of ten higher than that of cholesterol (Bloj and Zilversmit, 1977; McLean and Phillips, 1981), long periods of incubation of the donor and acceptor species may be required to follow the kinetics of sterol exchange when large vesicles or SPM-rich donors are used. Many different types of acceptors have been used, including lipid vesicles, discoidal apoprotein-PC complexes, lipoproteins, erythrocytes and erythrocyte ghosts, and cultured mammalian cells. The donor and acceptor vesicles are allowed to equilibrate separately under nitrogen at the temperature of the exchange experiments before they are mixed to initiate the kinetic experiments.

Several methods are available for separation of two vesicle populations (i.e., donor and acceptor):

1. Anion-exchange chromatography using diethylaminoethyl (DEAE)-Sepharose CL-6B or DEAE-Sephacel columns is used when the vesicle-vesicle exchange system consists of either negatively charged donors and neutral acceptors or negatively charged acceptors and neutral donors. In order to impart a substantial negative charge to one population of the vesicles, 15 mol% of negatively charged lipid (e.g., dicetyl phosphate or phosphatidylglycerol) is usually used, assuring efficient binding of the vesicles to the column. The donor vesicles contain a trace (~1 μCi/mg of donor phospholipid) of [^{14}C]cholesterol; generally, [4-^{14}C]cholesterol is used. In some studies, [1,2-^{3}H]cholesterol has been used. It is worth noting that one commercial preparation of [7-^{3}H]cholesterol obtained from an unidentified vendor was found not to be cholesterol (Mahlberg et al., 1990). In addition, the 7 position of cholesterol is prone to oxidation because of its allylic nature; thus, [1,2-^{3}H]cholesterol is the preferred tritiated cholesterol.

   The charged vesicle population is separated rapidly on short columns (e.g., 0.6 × 3 cm) from the neutral vesicles, which are readily eluted with buffer. In order to monitor recovery of the neutral vesicles, a trace (e.g., ~0.1 μCi/mg of acceptor phospholipid) of nonexchangeable marker, such as [^{3}H]triolein (when [^{14}C]cholesterol exchange is monitored) or [^{14}C]cholesteryl oleate (when [^{3}H]cholesterol exchange is monitored), is added to the lipid film when the acceptor dispersions are prepared. To minimize back-exchange of radiolabeled cholesterol during the exchange process, a tenfold excess of acceptor vesicles is generally used; a donor lipid concentration of about 1 m*M* (total lipid) and an acceptor lipid concentration of about 10 m*M* are typical. As stated above, the donor and acceptor vesicles are prepared with the same molar ratio of cholesterol/phospholipid for exchange studies. In most studies the temperature of the incubation is above the phase transition temperature of the pure phospholipids used in order to mimic the liquid-crystalline phase of most biomembranes. Obviously, it is essential that no appreciable aggregation or fusion of vesicles takes place during the exchange. It is important to use at least two separate preparations of donor and acceptor vesicles of each composition. It is advisable to analyze whether oxidation of cholesterol or phospholipids has occurred if prolonged incubation periods are necessary; gas chromatography and thin-layer chromatography have been used (Kan and Bittman, 1991).

2. A lectin-induced agglutination separation of glycolipid-containing donors or acceptors, developed by Curatolo et al. (1978), Backer and Dawidowicz (1979), and Kasper and Helmkamp (1981), has also been used for the measurement of radiolabeled sterol exchange (Kan et al., 1992). The galactosyl-binding lectin *Ricinus communis* agglutin I (7.5 μg) was added to 200-μL aliquots of the mixture of vesicles undergoing exchange; the donor vesicles contained 20% w/v *N*-palmitoyldihydrolactocerebroside. The donor vesicles are removed by centrifugation for 2 min in a table-top centrifuge, and the acceptor vesicles in the supernatant are counted. The recovery, estimated from the ^{3}H disintegrations per minute in the acceptor fraction, is similar (>80%) to that found in the anion-exchange assay method.
3. An ultracentrifugation method has been used to separate small unilamellar vesicles from large unilamellar vesicles (Yeagle and Young, 1986; Thomas and Poznansky, 1988a). Conditions have been reported in which the latter are pelleted during ultracentrifugation, whereas the small unilamellar vesicles remain in the supernatant. The pellet and the supernatant can both be counted to monitor the exchange kinetics.
4. Fluorescence polarization has been used to monitor lipid transfer between vesicles without separation of the vesicle populations. This type of assay offers the advantage that kinetic points are easily obtained at short times of incubation, allowing detection of fast kinetic phases. This assay method was used to follow protein-mediated PC transfer between vesicles by use of an exogenous probe (Xü et al., 1983) and to follow the spontaneous transfer of the cholesterol analog dehydroergosterol (e.g., Rogers et al., 1979) from donor vesicles to cholesterol-containing acceptor vesicles

(Nemecz and Schroeder, 1988; Nemecz et al., 1988). The fluorescent sterol exchange assay is based on the self-quenching of dehydroergosterol in donor vesicles. Self-quenching causes decreased fluorescence polarization of the fluorescent sterol in the donor vesicles; transfer of dehydroergosterol to the acceptor vesicles that contain cholesterol is accompanied by an increase in the fluorescence polarization of dehydroergosterol, since the nonfluorescent cholesterol moves into the donor bilayer. The increase in fluorescence polarization is converted to dehydroergosterol mass by using a standard curve.

In addition to the assays of bilayer-bilayer exchange, it should be noted that the kinetics of exchange of sterols between donor micelles and acceptor micelles have also been measured (Kan and Bittman, 1991). Lipid exchange between bilayers and monolayers has also been measured experimentally (Jähnig, 1984). Cholesterol transfer from cell monolayers has been reviewed (Phillips et al., 1987).

III. ESTIMATION OF KINETIC PARAMETERS

The first-order rate constants and half-times for the exchange of [^{14}C]cholesterol between donor and acceptor vesicles are calculated from the dual-labeling data, as described previously (e.g., Kan et al., 1991a). Aliquots are taken in duplicates or triplicates from the incubation mixture at various times, including those approaching equilibrium, and the fraction of exchange at each time is calculated from the dual-labeling data as described (e.g., Kan and Bittman, 1991). Examples of first-order plots for cholesterol exchange between vesicles are given in many of the papers cited in this review. Briefly, the fraction of radiolabeled cholesterol that undergoes exchange at time t is defined as α_t and is calculated as follows:

$$\alpha_t = [^{14}C/^{3}H]_t / [^{14}C/^{3}H]_{mix}$$

where $[^{14}C/^{3}H]_t$ and $[^{14}C/^{3}H]_{mix}$ represent the ratio of [^{14}C]cholesterol to [^{3}H]triolein in the eluate at time t and in the donor-acceptor vesicles without separation, respectively. The initial values of exchange of labeled cholesterol at times zero and infinity are α_0 and α_∞, respectively. These values are estimated by using a downhill simplex algorithm (Nelder and Mead, 1965; Noggle, 1985; Press et al., 1986). All of the kinetic data (α_t) including experimental points obtained at times approaching the equilibrium value are fit to a function of time, F(t), according to Equation 1:

$$F(t) = \alpha_\infty + [(\alpha_0 - \alpha_\infty)\exp(\text{slope} \times t)] \qquad (1)$$

The pseudo first-order rate constants (k) and half-times of exchange ($t_{1/2}$) are calculated using the following relationships (for a tenfold excess of acceptor vesicles)

$$k = -\text{slope}/1.1 \tag{2}$$

$$t_{1/2} = \ln 2/k \tag{3}$$

The size of the exchangeable pool of labeled cholesterol, X_{xch}, can be calculated from the relationship

$$X_{xch} = [(\alpha_{\infty} - \alpha_0) / (1 - \alpha_0)] \times 1.1 \tag{4}$$

The factor of 1.1 is used to correct the rate constant of labeled sterol movement from donor to acceptor particles for back exchange of label (Equation 2) and to normalize the equilibrium value of 90.9% expected for random distribution of the label between donor and acceptor particles (acceptor/donor ratio, 10:1) to 100% exchange of ^{14}C-sterol at equilibrium (Equation 4). When two kinetic phases are observed, Equation 5 is used to obtain the rates of the slow (s) and fast (f) phases:

$$\begin{aligned} F(t) = \ & [\alpha_{\infty s} + [(\alpha_{0s} - \alpha_{\infty s})\exp(\text{slope}_s \times t)] \\ & + [\alpha_{\infty f} + (\alpha_{0f} - \alpha_{\infty f})\exp(\text{slope}_f \times t)] \end{aligned} \tag{5}$$

IV. EFFECTS OF MEMBRANE LIPIDS ON THE RATE OF CHOLESTEROL EFFLUX FROM THE SURFACE OF DONOR VESICLES

A. VARYING PC STRUCTURE

The rate of spontaneous transfer of cholesterol between membranes is highly sensitive to the degree of saturation of the phospholipid fatty acyl chains and to the extent of the interactions between cholesterol and nearest-neighbor molecules in the bilayer. Disaturated PCs impede the exchange rate compared with branched-chain or unsaturated PCs (Fugler et al., 1985). The rate of cholesterol exchange between vesicles prepared from cyclopentano analogs of phospholipids, which are conformationally restricted in the glycerol backbone region, was similar to that found in vesicles prepared with glycerophospholipids (Bittman et al., 1992). Therefore, the geometrical orientation of the acyl chains or head group does not exert a significant influence on the rate of cholesterol desorption from the lipid-water interface.

The lipid packing density is greater in bilayers of cholesterol and disaturated PCs than in bilayers containing cholesterol and the sterically bulky branched-chain or unsaturated PCs. Since desorption of cholesterol takes place from the outer leaflet of the membrane, it appears that the molecular packing of cholesterol with neighboring molecules in the *outer* half of the bilayer is a particularly important determinant of the cholesterol exchange rate.

B. VARIATION IN SPM CONTENT OF THE BILAYER

The SPM content of the membrane dramatically affects the exchange rate, especially in lipid vesicles, but also in biological membranes. As the SPM content of the bilayer of the donor vesicles increases, the rate of cholesterol exchange between vesicles decreases (Fugler et al., 1985; Bhuvaneswaran and Mitropoulos, 1986; Yeagle and Young, 1986; Lund-Katz et al., 1988; Thomas and Poznansky, 1988a; Kan et al., 1991a,b). However, there is one report that shows a rapid exchange of dehydroergosterol between vesicles in the presence of 10 mol% SPM; only at 33 mol% SPM was the exchange rate decreased (Schroeder et al., 1991).

With biological membranes and lipoproteins used as the donor species, a correlation has been noted between a high SPM content and a slow cholesterol exchange rate (Lund-Katz et al., 1982; Clejan and Bittman, 1984c; Gold and Phillips, 1990).

The ability of SPM to lower the rate of cholesterol exchange from donor to acceptor species is explained by the greater lateral packing density in the lipid-water interface when SPM is present (Lund-Katz et al., 1988; Kan et al., 1991a). The activation energy for cholesterol desorption from cholesterol/SPM vesicles is higher by 45% than that from cholesterol/PC vesicles, reflecting the higher kinetic barrier for release of cholesterol from SPM-rich surfaces (Bhuvaneswaran and Mitropoulos, 1986). Van der Waals attractive forces appear to be significantly greater between SPM and cholesterol than between PCs and cholesterol (Lund-Katz et al., 1988), and the increased interactions of lipid molecules results in a tightly packed outer leaflet of the donor species. Thus, the cholesterol exchange rate decreases when the SPM/PC ratio is increased. The expected relationship between the relative molecular area and the exchange rate was not found in vesicles prepared with 3-deoxy- and 3-methoxy-SPM (Kan et al., 1991a), suggesting that factors other than van der Waals interactions may be significant. Hydrogen bonding involving the 3-hydroxy group of SPM is not critical to the cholesterol-SPM interaction, as judged by cholesterol exchange kinetics between vesicles containing chemically modified SPM analogs (Kan et al., 1991a) and by monolayer experiments (Grönberg et al., 1991).

C. OTHER LIPIDS

Incorporation of bovine-brain gangliosides into egg PC/cholesterol vesicles resulted in only a slight increase in the rate of cholesterol efflux from vesicles, whereas incorporation of 50 mol% of egg phosphatidylethanolamine (PE) decreased the efflux rate significantly (Thomas and Poznansky, 1988a). Incorporation of 20 mol% of PE or bovine-brain phosphatidylserine into PC/cholesterol vesicles had no effect on the efflux rate (Thomas and Poznansky, 1988a).

Incorporation of chemically crosslinked PE into vesicles decreased the rate of cholesterol exchange between vesicles; the coupling agents used to crosslink the amino groups of PE did not decrease the exchange rate, as judged from efflux data obtained with vesicles containing monomeric PE bearing the coupling moiety (Bittman et al., 1985). The decrease in cholesterol efflux rate in the presence of crosslinked, dimeric PE may result from an increased lipid order and increased interactions between cholesterol and its nearest neighbors in the surface of the donor vesicles. Since crosslinking of the amino groups of phospholipids apparently takes place in the process of cell aging (Jain and Shohet, 1984), these results with vesicles indicate that degenerative transformations of phospholipids associated with cellular aging contribute to inhibition of cholesterol exchangeability in biological membranes.

D. CHOLESTEROL CONTENT OF THE BILAYER

The cholesterol exchange rate between donor and acceptor vesicles and the activation energy were not markedly affected by the cholesterol/phospholipid molar ratio (McLean and Phillips, 1982). This observation was confirmed in studies using palmitoyloleoyl-PC (POPC) vesicles (Bar et al., 1986; Nemecz et al., 1988). A conflicting result was obtained, however, in a study of cholesterol transfer between POPC/cholesterol vesicles and erythrocyte ghosts as acceptors; the activation energy decreased when the mol% of cholesterol decreased from 33–48 mol% to 20 mol%, and the initial rate of cholesterol transfer increased when the cholesterol content was increased from 19 to 48 mol% (Poznansky and Czekanski, 1982). The latter observation suggests that altered interactions in the lipid matrix of the donor at high cholesterol (33 to 48 mol%) facilitated efflux.

Evidence in favor of the opposite dependence of exchange rate on cholesterol content was reported with mycoplasma membranes. The rate of cholesterol exchange from *M. gallispecticum* cell membranes to lipid vesicles decreased dramatically with an increase in cholesterol/phospholipid molar ratio in the range of 0.25 to 0.36 (Clejan and Bittman, 1984a). The ability of cholesterol to modulate its own exchange rate as well as that of phospholipids in these cells is related to the unique ability of this sterol to modify the degree of lipid order; as the bilayer becomes more tightly packed, the lipid exchange rate decreases.

A very slow rate of sterol exchange was reported in POPC vesicles at high sterol content; this process was proposed to reflect a slow desorption from sterol-rich clusters (Nemecz et al., 1988). In contrast, dipalmitoyl-PC vesicles containing very high levels of cholesterol were found to give fast cholesterol exchange rates (McLean and Phillips, 1982; Kan and Bittman, 1991). No such rate enhancement was found in egg PC vesicles at comparably high cholesterol content (Kan and Bittman, 1991). Phase boundaries may be present in dipalmitoyl-PC bilayers rich in cholesterol, and desorption may be facilitated at the phase boundaries.

V. EFFECTS OF MEMBRANE PROTEINS ON THE RATES OF CHOLESTEROL EXCHANGE

Local domain structure, as influenced by membrane proteins, also affects the rate of cholesterol transfer from cell membranes to acceptor species (Clejan and Bittman, 1984a; Bellini et al., 1984). Until now there have been no reports of cholesterol exchange between unilamellar vesicles containing integral membrane proteins.

VI. EFFECTS OF WATER-SOLUBLE PROTEINS ON THE RATES OF CHOLESTEROL EXCHANGE

Fatty acid free bovine serum albumin at high concentrations (2% w/v) enhanced the rate of cholesterol movement between mycoplasma membranes and lipid vesicles (Rottem et al., 1981; Clejan and Bittman, 1984b). Albumin (1% w/v) increased the rate of cholesterol transfer between lipoproteins and fibroblasts (Lundberg and Suominen, 1985). The addition of 2% w/v albumin caused the exchange rates of cholesterol and sitosterol to decrease by factors of 2 and 4, respectively, between phospholipid/sterol vesicles (Kan and Bittman, 1991).

Sterol carrier protein-2 (SCP_2) from rat liver stimulated the rate of transfer of the cholesterol analog dehydroergosterol from small unilamellar vesicles containing phosphatidylserine/PC/dehydroergosterol to small unilamellar vesicles containing phosphatidylserine/PC/cholesterol (Schroeder et al., 1990a,b). The initial rate of sterol transfer was increased by SCP_2 when the vesicles contained phosphatidylserine, phosphatidylinositol, or cardiolipin (Butko et al., 1990). Since the rate of SCP_2-mediated dehydroergosterol exchange was strongly inhibited by the addition of 0.6 *M* KCl and by low concentrations (1 to 10 μ*M*) of the polycation neomycin, it was concluded that the interaction of SCP_2 with acidic membrane phospholipids is electrostatic in nature (Schroeder et al., 1991). SCP_2 forms a complex with dehydroergosterol (Schroeder et al., 1990b), so the enhanced rate of dehydroergosterol intervesicle transfer may arise because of an increase in the critical micelle concentration of the sterol upon binding to the protein. An alternative explanation to the carrier-type mechanism is a collisional-type mechanism in which SCP_2 may enhance the sterol transfer rate by binding to the donor membrane (van Amerongen et al., 1989).

The effects of incorporating human apolipoproteins on the rate of cholesterol exchange were examined recently. A marked increase in cholesterol exchange rate between vesicles was observed when at least five molecules of apoA-I were incorporated per egg PC/cholesterol donor vesicle (Letizia and Phillips, 1991). Similar increases in exchange rates were found in the presence of similar numbers of apoA-II and apo-C, suggesting that specific lipid-protein interactions are not responsible for the rate enhancement. Instead, nonspecific disruption of the packing involving cholesterol and phospholipid molecules in the lipid-water interface is considered to occur on incorporation of apoprotein molecules into the surface of the donor vesicles.

VII. EFFECTS OF BILAYER CURVATURE ON CHOLESTEROL EXCHANGE RATES

The geometrical curvature of the vesicle surface from which desorption takes place also affects the rate of cholesterol exchange, with higher rates observed in small donor species than in large donor species (McLean and Phillips, 1984a; Fugler et al., 1985; Thomas and Poznansky, 1988b). Looser packing of cholesterol molecules in the surface of highly curved small unilamellar vesicles is consistent with an enhanced rate of efflux. There is also evidence that packing defects in the mycoplasma cell membrane facilitate cholesterol efflux to acceptor vesicles (Bittman et al., 1990). The observation of slow cholesterol desorption from large, planar-like vesicles is of interest in seeking an explanation of why plasma membranes and intracellular membranes have widely disparate cholesterol/phospholipid molar ratios. The effect of membrane curvature of the donor species, along with preferential interactions between cholesterol and phospholipid classes preferentially localized in the outer leaflet of the plasma membrane, may contribute to the slow spontaneous movement of cholesterol from plasma to intracellular membranes.

VIII. MECHANISM OF CHOLESTEROL EXCHANGE BETWEEN VESICLES

In view of the importance of cholesterol in modulating the physical state of membrane lipids, the mechanisms underlying the intermembrane redistribution of cholesterol have attracted extensive interest. Nevertheless, the principal mechanisms by which cholesterol undergoes spontaneous intermembrane movement are not fully established. The two limiting mechanisms that have been proposed are (1) the aqueous diffusion pathway in which cholesterol diffuses through the aqueous medium separating donor and acceptor species and then is resorbed into the acceptor membrane, and (2) the collisional complex pathway in which cholesterol does not enter into the aqueous medium, but instead transfers during transient collision of the membranes. Strong evidence has been presented in favor of the aqueous diffusion pathway for cholesterol movement from small unilamellar vesicles to a large excess of acceptor vesicles (McLean and Phillips, 1981, 1982, 1984b; Backer and Dawidowicz, 1981); the data that support the aqueous diffusion model have been reviewed (Phillips et al., 1987; Dawidowicz, 1987). In this mechanism the rate-determining step is the release of cholesterol from the surface of the donor particle (see Pownall et al., 1983; Phillips et al., 1987; Dawidowicz, 1987; Bittman, 1988). The following features are consistent with this model: (1) first-order kinetics with respect to cholesterol efflux from the donor species, (2) zero-order dependence with respect to acceptor lipid concentration when an excess of acceptor species is present and donor lipid concentration is about 1 m*M*, (3) a marked influence of phospholipid structure in the donor vesicles on exchange rate (e.g., increased rates on introduction of branching and unsaturation into PC acyl chains and on decreasing SPM content), (4) a large temperature dependence of the rate constant because of the high activation energy associated with desorption, and (5) examples of exchange continuing (at a decreased rate) when a physical barrier such as a dialysis membrane is placed between donor and acceptor vesicles.

However, a collisional complex between donor and acceptor particles has been postulated for cholesterol exchange in some systems (Mütsch et al., 1986; Steck et al., 1988). In systems having a cell glycocalyx, close contact between donor and acceptor species is prevented. Kinetic arguments have been presented in favor of a collision-induced transfer of cholesterol from aqueous dispersions of lyso-PC and egg PC to brush border membranes (Thurnhofer and Hauser, 1990). In this system, the rate-limiting step in the transfer of cholesterol from egg PC small unilamellar vesicles was the adsorption of lipid monomers by the acceptor membrane vesicles (Mütsch et al., 1986). The brush border membranes were found to contain an endogenous membrane-bound protein that binds cholesterol; when the protein(s) was (were) subjected to proteolysis, the kinetic data became consistent with the aqueous diffusion pathway (Thurnhofer and Hauser, 1990).

In addition to the interactions experienced by a cholesterol molecule with its neighboring lipids in the lipid matrix of the donor species, the rate of cholesterol transfer between membrane structures is sensitive to factors that modify the aqueous-phase solubility of cholesterol, such as the presence of chaotropic salts (Clejan and Bittman, 1984a), bile salts (Vlahcevic et al., 1990), and polar water-miscible organic solvents (Bruckdorfer and Green, 1967; Quarfordt and Hilderman, 1970; Bruckdorfer and Sherry, 1984). A comparison of the rates of cholesterol and sitosterol exchange suggested that the rate of sterol exchange between vesicles decreases with increasing sterol hydrophobicity, since the presence of the 24α-ethyl group in sitosterol represents a constraint to the rate of intermembrane movement (Kan and Bittman, 1990, 1991). Because of the insertion of an ethyl group at C-24 in place of a hydrogen atom, sitosterol has a slightly more hydrophobic character than cholesterol, as estimated by the logarithm of the HPLC capacity factor (k′) (Kan et al., 1992). However, the ratio of the sitosterol/cholesterol exchange half-times (Kan and Bittman, 1991) is much greater than the difference in their log k′ values using mixtures of acetonitrile/2-propanol/water as eluent. Additional van der Waals interactions involving the hydrophobic sterol side chain and phospholipid acyl chains may be important in determining the off-rate constants from lipid-water interfaces. Indeed, a model of the transition state for cholesterol desorption proposes that the sterol is held to the surface of the donor species by its side chain (Phillips et al., 1987). It was concluded that decreasing the affinity of cholesterol for phospholipids in the donor membranes appears to stimulate the rate of movement of cholesterol between membranes to a greater extent than increasing the sterol's aqueous solubility (Kan et al., 1992).

IX. INFLUENCE OF STRUCTURAL MODIFICATIONS OF STEROLS ON THE RATE OF STEROL EXCHANGE BETWEEN VESICLES

The rates of exchange of many different sterols between membranes have been studied. Figure 1 shows the structures of the sterols that have been used in these studies. Sterols having oxygen-containing functions at the 7 position underwent exchange more rapidly than did cholesterol (Kan et al., 1992). The half-time of exchange of [^{14}C]7-ketocholesterol between DPPC/sterol vesicles

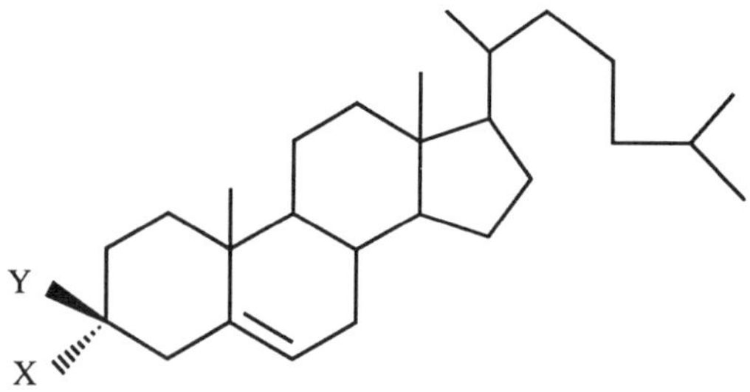

Y

X

X	Y	
OH	H	epicholesterol
NH_3^+	H	3α-aminocholesterol
H	NH_3^+	3β-aminocholesterol
H	$\mathrm{OC(=O)CH{=}\overset{+}{N}{=}\overset{-}{N}}$	cholesteryl diazoacetate
$O(CH_2)_2O(CH_2)_2O(CH_2)_2OH$	H	triethoxycholesterol

FIGURE 1. Structures of the sterols used in exchange experiments.

at 50°C was about 25-fold lower than that of [^{14}C]cholesterol; [^{14}C]7α-hydroxycholesterol underwent exchange between dipalmitoyl-PC/sterol vesicles at 37°C about 115 times faster than did [^{14}C]cholesterol, and the rate of exchange of [^{14}C]7β-hydroxycholesterol was even higher. The initial rates of oxysterol transfer from monolayers to a subphase containing vesicles or lipoproteins followed the order of 7α-hydroxycholesterol > 7-ketocholesterol > cholesterol (Theunissen et al., 1986; van Amerongen et al., 1989). The rapid rates of transfer of 7-oxygen-substituted sterols reflect weakened interactions between these sterols and PC compared with the interactions between PC and cholesterol, as estimated by a variety of physical methods (Bruckdorfer et al., 1968; Nakagawa et al., 1980; Demel et al., 1972a,b; Theunissen et al., 1986; Rooney et al., 1986).

Inversion of the configuration at the 3 position and alkylation with a hydrophilic chain also led to a dramatic enhancement in exchange rate. The half-times of exchange of [^{14}C]epicholesterol and [^{14}C]3α-triethoxycholesterol between dipalmitoyl-PC/sterol vesicles at 50°C were about 15-fold lower than that of [^{14}C]cholesterol (Kan et al., 1992). Epicholesterol (the 3α-hydroxy stereoisomer of cholesterol) does not interact with PCs as tightly as cholesterol (Bittman and Blau, 1972; Demel et al., 1972a,b; Demel and DeKruyff, 1976; Clejan et al.,

1979; Murari et al.,1986). 3-Oxyethylene derivatives of cholesterol also experience weaker lipid packing in PC bilayers (Goodrich et al., 1988).

Introduction of an ionic group (NH_3^+) into the 3 position of the sterol molecule in place of the uncharged hydroxy group resulted in a change in the kinetics of exchange. The data for the exchange of [^{14}C]3α- and 3β-aminocholesterol were fit to a two-exponential function, i.e., a fast and slow phase (Kan et al., 1992), whereas only one kinetic phase was found for exchange of all of the other ^{14}C-labeled sterols having a 3-hydroxy group. The two phases do not arise because a pool of positively charged 3-aminocholesterol interacts tightly with dicetyl phosphate used in the negative donor/neutral acceptor assay system, which would impede its movement across the bilayer and desorption from the external surface. When the lectin-based separation system was used (incorporation of glycosphingolipid, avoiding the use of dicetyl phosphate), two kinetic pools were still observed. The observation that the exchange or transfer data for [^{14}C]3α- and 3β-aminocholesterol between vesicles were best fit by a two-exponential function in two different assay systems suggests that transbilayer movement of the positively charged amino-sterol becomes rate limiting relative to desorption. The slow phase of [^{14}C]3α- and 3β-aminocholesterol exchange is thought to result from the high free energy required for the inner-to-outer leaflet movement ("flip-flop") of the positively charged sterol, since at least transient disruption of lipid packing takes place between the sterol polar group and phospholipid acyl chains during flip-flop. In agreement with the hypothesis that sterols bearing a charged polar head group disrupt membrane packing during transverse migration to a greater extent than does cholesterol, it was found that cholesteryl diazoacetate (which is zwitterionic) also underwent biphasic exchange kinetics (Kan et al., 1992). In contrast, [^{14}C]cholesterol underwent exchange in a single kinetic pool in both assay systems, indicating that transbilayer movement of cholesterol is not rate limiting for desorption from the surface of the donor particle.

The studies summarized above indicate that sterols that are more hydrophilic than cholesterol, as estimated by reversed-phase HPLC, undergo faster exchange than cholesterol, whereas sterols that are more hydrophobic undergo exchange more slowly. However, the [^{14}C]sterol exchange rates between lipid vesicles were not correlated with the relative hydrophobicity of the sterol (Kan et al., 1992). The interaction of sterols with phospholipids in the donor bilayer provides the principal physical-chemical basis for determining the rates of spontaneous desorption of sterols from a host phospholipid bilayer.

X. EXTENT OF CHOLESTEROL EXCHANGE AND PRESENCE OF MULTIPLE KINETIC POOLS

Previous reports of a lack of full cholesterol exchangeabilty from lipid vesicles as donors (e.g., Poznansky and Lange, 1976) were not reproduced (Bloj and Zilversmit, 1977). Full sterol exchangeability was found in many earlier reports of cholesterol exchange between PC vesicles (cited in Phillips et al.,

1987). On reinvestigation, however, some investigators detected a sizeable nonexchangeable pool of cholesterol (~40 to 50% of the total cholesterol) in donor vesicles prepared with POPC (Bar et al., 1986; Nemecz et al., 1988) and dimyristoyl-PC (Bar et al., 1987). A nonexchangeable cholesterol pool of 60% and 27% was found at 50°C in vesicles prepared with SPM and 10 mol% cholesterol by Bar et al. (1987) and Kan and Bittman (1991), respectively. Dehydroergosterol, a fluorescent analog of cholesterol, was also reported to show a large nonexchangeable pool in POPC vesicles (Nemecz et al., 1988; Nemecz and Schroeder, 1988; Schroeder et al., 1991). The nonexchangeable pool was considered to arise from a laterally segregated sterol domain. The nonexchangeable sterol pool was not detected in vesicles prepared with 33 mol% POPC, 33 mol% bovine-brain SPM, and 34 mol% dehydroergosterol (Schroeder et al., 1991). However, Bar et al. (1989) reported that 47% of the total dehydroergosterol was nonexchangeable in bovine-brain SPM vesicles at 50°C.

The possible reasons for the discrepancy in exchangeability results obtained in PC vesicles have been discussed recently (Kan and Bittman, 1991). One source of the differences in the extent of exchange may be the failure to accumulate sufficient data points at times approaching the equilibrium position.

Cholesterol, cholestanol, and epicholesterol underwent full exchange in a single kinetic pool, and 90% of the 3α-triethoxycholesterol was exchangeable in one pool (Kan et al., 1992). Biphasic kinetics with full exchangeability were observed for cholesteryl amines, which bear a positive charge at the 3 position; the slow phase reflects the high activation energy for inner-to-outer leaflet movement of the charged lipid. Biphasic kinetics were also found for cholesteryl diazoacetate, indicating that this analog and cholesterol have different mechanisms of transfer.

XI. FUTURE OUTLOOK

It is likely that lipid vesicles will continue to be used to examine what constraints exist to spontaneous cholesterol transfer between membranes. The current view is that the asymmetric distribution of phospholipids in membranes, together with the small radius of curvature of many natural membranes known to be cholesterol rich, are primarily responsible for the nonrandom distribution of cholesterol between membranes of cells. The effects of transmembrane potential, polymorphic lipid phase transitions, and distribution of sterol between different domains on the rate of cholesterol movement have not been studied. Information is also lacking about the influence of integral membrane proteins on the kinetics of cholesterol exchange. Little is known about the roles of several other membrane components, such as sphingoglycolipids, in the exchange of cholesterol. Continued progress should be forthcoming in more clearly defining the effects of apolipoproteins on cholesterol desorption from membrane domains (Rothblat et al., 1992). Further studies of the effects of enzyme-induced changes in the lipid composition of the vesicle surface on cholesterol movement between membranes will help unravel the interrelationship between lipid turnover and cholesterol transfer. Thus, lipid vesicles should continue to be useful in understanding how cholesterol movement between cell membranes is affected by changes in membrane composition.

ACKNOWLEDGMENT

The work cited in this chapter from the author's laboratory was supported in part by National Institutes of Health Grant HL 16660.

REFERENCES

Backer, J. M. and E. A. Dawidowicz. 1979. The rapid transbilayer movement of cholesterol in small unilamellar vesicles. *Biochim. Biophys. Acta,* 551:260-270.

Backer, J. M. and E. A. Dawidowicz. 1981. Mechanism of cholesterol exchange between phospholipid vesicles. *Biochemistry,* 20:3805–3810.

Bar, K. L., Y. Barenholz, and T. E. Thompson. 1986. Fraction of cholesterol undergoing spontaneous exchange between small unilamellar phosphatidylcholine vesicles. *Biochemistry,* 25:6701–6705.

Bar, L. K., Y. Barenholz, and T. E. Thompson. 1987. Dependence on phospholipid composition of the fraction of cholesterol undergoing spontaneous exchange between small unilamellar vesicles. *Biochemistry,* 26:5460-5465.

Bar, K. L., P. L.-G. Chong, Y. Barenholz, and T. E. Thompson. 1989. Spontaneous transfer between phospholipid bilayers of dehydroergosterol, a fluorescent cholesterol analog. *Biochim. Biophys. Acta,* 983:109–112.

Bellini, F., M. C. Phillips, C. Pickell, and G. H. Rothblat. 1984. Role of the plasma membrane in the mechanism of cholesterol efflux from cells. *Biochim. Biophys. Acta,* 777:209–215.

Bhuvaneswaran, C. and K. A. Mitropoulos. 1986. Effect of liposomal phospholipid composition on cholesterol transfer between microsomal and liposomal vesicles. *Biochem. J.,* 238:647–652.

Bittman, R. 1988. Sterol exchange between mycoplasma membranes and vesicles. In *Biology of Cholesterol,* Yeagle, P. L., Ed., CRC Press, Boca Raton, FL, pp. 173–195.

Bittman, R. and L. Blau. 1972. The phospholipid-cholesterol interaction. Kinetics of water permeability in liposomes. *Biochemistry,* 11:4831–4839.

Bittman, R., S. Clejan, and S. W. Hui. 1990. Increased rates of lipid exchange between *Mycoplasma capricolum* membranes and vesicles in relation to the propensity of forming nonbilayer lipid structures. *J. Biol. Chem.,* 265:15110-15117.

Bittman, R., S. Clejan, B. P. Robinson, and N. M. Witzke. 1985. Kinetics of cholesterol and phospholipid exchange from membranes containing cross-linked proteins or cross-linked phosphatidylethanolamines. *Biochemistry,* 24:1403–1409.

Bittman, R., S. Clejan, and S. Rottem. 1983. Transbilayer distribution of sterols in mycoplasma membranes: a review. *Yale J. Biol. Med.,* 56:397–403.

Bittman, R., L. Fugler, S. Clejan, M. D. Lister, and A. J. Hancock. 1992. Interaction of cholesterol with conformationally restricted phospholipids in vesicles. *Biochim. Biophys. Acta,* 1106:40–44.

Bloj, B. and D. B. Zilversmit. 1977. Heterogeneity of rabbit intestine brush border plasma membrane cholesterol. *J. Biol. Chem.,* 257:7608–7614.

Bruckdorfer, K. R. and C. Green. 1967. The exchange of unesterified cholesterol between human low-density lipoproteins and rat erythrocyte ghosts. *Biochem. J.*, 104:270-277.

Bruckdorfer, K. R., J. M. Graham, and C. Green. 1968. The incorporation of steroid molecules into lecithin sols, β-lipoproteins and cellular membranes. *Eur. J. Biochem.*, 4:512–518.

Bruckdorfer, K. R. and M. K. Sherry. 1984. The solubility of cholesterol and its exchange between membranes. *Biochim. Biophys. Acta*, 769:187–196.

Butko, P., I. Hapala, T. J. Scallen, and F. Schroeder. 1990. Acidic phospholipids strikingly potentiate sterol carrier protein 2 mediated intermembrane sterol transfer. *Biochemistry*, 29:4070-4077.

Clejan, S. and R. Bittman. 1984a. Kinetics of cholesterol and phospholipid exchange between *Mycoplasma gallisepticum* cells and lipid vesicles. Alterations in membrane cholesterol and protein content. *J. Biol. Chem.*, 259:441–448.

Clejan, S. and R. Bittman. 1984b. Distribution and movement of sterols with different side chain structures between the two leaflets of the membrane bilayer of mycoplasma cells. *J. Biol. Chem.*, 259:449–455.

Clejan, S. and R. Bittman. 1984c. Decreases in rates of lipid exchange between *Mycoplasma gallisepticum* cells and unilamellar vesicles by incorporation of sphingomyelin. *J. Biol. Chem.*, 259:10823–10826.

Clejan, S., R. Bittman, P. W. Deroo, Y. A. Isaacson, and A. F. Rosenthal. 1979. Permeability properties of sterol-containing liposomes from analogs of phosphatidylcholine lacking acyl groups. *Biochemistry*, 18:2118–2125.

Curatolo, W., A. O. Yau, D. M. Small, and B. Sears. 1978. Lectin-induced agglutination of phospholipid/glycolipid vesicles. *Biochemistry*, 17:5740-5744.

Dawidowicz, E. A. 1987. Lipid exchange: transmembrane movement, spontaneous movement, and protein-mediated transfer of lipids and cholesterol. *Curr. Top. Membr. Transp.*, 29:175–202.

Demel, R. A., K. R. Bruckdorfer, and L. L. M. van Deenen. 1972a. Structural requirements of sterols for the interaction with lecithin at the air-water interface. *Biochim. Biophys. Acta*, 255:311–320.

Demel, R. A., K. R. Bruckdorfer, and L. L. M. van Deenen. 1972b. The effect of sterol structure on the permeability of liposomes to glucose, glycerol and Rb^+. *Biochim. Biophys. Acta*, 255:321–330.

Demel, R. A. and B. DeKruyff. 1976. The function of sterols in membranes. *Biochim. Biophys. Acta*, 457:109–132.

Fugler, L., S. Clejan, and R. Bittman. 1985. Movement of cholesterol between vesicles prepared with different phospholipids or sizes. *J. Biol. Chem.*, 260:4098–4102.

Gold, J. C. and M. C. Phillips. 1990. Effects of membrane lipid composition on the kinetics of cholesterol exchange between lipoproteins and different species of red blood cells. *Biochim. Biophys. Acta*, 1027:85–92.

Goodrich, R. P., T. M. Handel, and J. D. Baldeschwieler. 1988. Modification of lipid phase behavior with membrane-bound cryoprotectants. *Biochim. Biophys. Acta*, 938:143–154.

Grönberg, L., Z.-S. Ruan, R. Bittman, and J. P. Slotte. 1991. Interaction of cholesterol with synthetic sphingomyelin derivatives in mixed monolayers. *Biochemistry*, 30:10746–10754.

Hagerman, J. S. and R. G. Gould. 1951. The in vitro interchange of cholesterol between plasma and red cells. *Proc. Soc. Exp. Biol. Med.*, 78:329–332.

Hui, S. W. 1992. Visualization of cholesterol domains in model membranes. In *Cholesterol in Membrane Models*, Finegold, L., Ed., CRC Press, Boca Raton, FL.

Jähnig, F. 1984. Lipid exchange between membranes. *Biophys. J.*, 46:687–694.

Jain, S. K. and S. B. Shohet. 1984. A novel phospholipid in irreversibly sickled cells: evidence for *in vivo* peroxidative membrane damage in sickle cell disease. *Blood*, 63:362–367.

Johnson, W. J., F. H. Mahlberg, G. H. Rothblat, and M. C. Phillips. 1991. Cholesterol transport between cells and high-density lipoproteins. *Biochim. Biophys. Acta*, 1085:273–298.

Kan, C.-C. and R. Bittman. 1990. Constraint of the spontaneous intermembrane movement of sitosterol by its 24α-ethyl group. *J. Am. Chem. Soc.*, 112:884–886.

Kan, C.-C. and R. Bittman. 1991. Spontaneous rates of sitosterol and cholesterol exchange between phospholipid vesicles and between lysophospholipid dispersions: evidence that desorption rate is impeded by the 24α-ethyl group of sitosterol. *J. Am. Chem. Soc.*, 113:6650-6656.

Kan, C.-C., R. Bittman, and J. Hajdu. 1991b. Phospholipids containing nitrogen- and sulfur-linked chains: kinetics of cholesterol exchange between vesicles. *Biochim. Biophys. Acta*, 1066:95–101.

Kan, C.-C., Z.-S. Ruan, and R. Bittman. 1991a. Interaction of cholesterol with sphingomyelin in bilayer membranes: evidence that the hydroxy group of sphingomyelin does not modulate the rate of cholesterol exchange between vesicles. *Biochemistry*, 30:7759–7766.

Kan, C.-C., J. Yan, and R. Bittman. 1992. Rates of spontaneous exchange of synthetic radiolabeled sterols between lipid vesicles. *Biochemistry*, 31:1866–1874.

Kasper, A. M. and G. M. Helmkamp, Jr. 1981. Protein-catalyzed phospholipid exchange between gel and liquid-crystalline phospholipid vesicles. *Biochemistry*, 20:146–151.

Letizia, J. Y. and M. C. Phillips. 1991. Effects of apolipoproteins on the kinetics of cholesterol exchange. *Biochemistry*, 20:866–873.

Lund-Katz, S., B. Hammerschlag, and M. C. Phillips. 1982. Kinetics and mechanism of free cholesterol exchange between human serum high- and low-density lipoproteins. *Biochemistry*, 21:2964–2969.

Lund-Katz, S., H. M. Laboda, L. R. McLean, and M. C. Phillips. 1988. Influence of molecular packing and phospholipid type on rates of cholesterol exchange. *Biochemistry*, 27:3416–3423.

Lundberg, B. B. and L. A. Suominen. 1985. Physicochemical transfer of [^{3}H]cholesterol from plasma lipoproteins to cultured human fibroblasts. *Biochem. J.*, 228:219–225.

Mahlberg, F. H., A. Rodriguez-Oquendo, D. W. Bernard, J. M. Glick, and G. H. Rothblat. 1990. Potential problems in the use of commercial preparations of radiolabeled cholesterol. *Atherosclerosis*, 84:95–100.

McLean, L. R. and M. C. Phillips. 1981. Mechanism of cholesterol and phosphatidylcholine exchange or transfer between unilamellar vesicles. *Biochemistry*, 20:2893–2900.

McLean, L. R. and M. C. Phillips. 1982. Cholesterol desorption from clusters of phosphatidylcholine and cholesterol in unilamellar vesicle bilayers during lipid transfer or exchange. *Biochemistry*, 21:4053–4059.

McLean, L. R. and M. C. Phillips. 1984a. Cholesterol transfer from small and large unilamellar vesicles. *Biochim. Biophys. Acta*, 776:21–26.

McLean, L. R. and M. C. Phillips. 1984b. Kinetics of phosphatidylcholine and lysophosphatidylcholine exchange between unilamellar vesicles. *Biochemistry*, 23:4624–4630.

Murari, R., P. Murari, and W. J. Baumann. 1986. Sterol orientations in phosphatidylcholine liposomes as determined by deuterium NMR. *Biochemistry*, 25:1062–1067.

Mütsch, B., N. Gains, and H. Hauser. 1986. Interaction of intestinal brush border membrane vesicles with small unilamellar phospholipid vesicles. Exchange of lipids between membranes is mediated by collisional contact. *Biochemistry*, 25:2134–2140.

Nakagawa, Y., K. Inoue, and S. Nojima. 1980. Transfer of cholesterol between liposomal membranes. *Biochim. Biophys. Acta,* 553:307–319.

Nelder, J. A. and R. Mead. 1965. A simplex method for function minimization. *Comput. J.,* 7:308–313.

Nemecz, G., R. N. Fontaine, and F. Schroeder. 1988. A fluorescence and radiolabel study of sterol exchange between membranes. *Biochim. Biophys. Acta,* 943:511–521.

Nemecz, G. and F. Schroeder. 1988. Time-resolved fluorescence investigation of membrane cholesterol heterogeneity and exchange. *Biochemistry,* 27:7740-7749.

Noggle, J. H. 1985. *Physical Chemistry on a Microcomputer,* Little, Brown, Boston, chap. 8.

Phillips, M. C., W. J. Johnson, and G. H. Rothblat. 1987. Mechanisms and consequences of cellular exchange and transfer. *Biochim. Biophys. Acta,* 906:223–276.

Pörn, M. I. and J. P. Slotte. 1990. Reversible effects of sphingomyelin degradation on cholesterol distribution and metabolism in fibroblasts and transformed neuroblastoma cells. *Biochem. J.,* 271:121–126.

Pownall, H. J., D. L. Hickson, and L. C. Smith. Transport of biological lipophiles: effect of lipophile structure. *J. Am. Chem. Soc.,* 105:2440-2445.

Poznansky, M. J. and S. Czekanski. 1982. Cholesterol movement between human skin fibroblasts and phosphatidylcholine vesicles. *Biochim. Biophys. Acta,* 685:182–190.

Poznansky, M. J. and Y. Lange. 1976. Transbilayer movement of cholesterol in dipalmitoyllecithin-cholesterol vesicles. *Nature,* 259:420-421.

Press, W. H., B. P. Flannery, S. A. Tevkolsky, and W. T. Vetterling. 1986. Downhill simplex method in multidimensions. In *Numerical Recipes: The Art of Scientific Computing,* Cambridge University Press, Cambridge, pp. 289–293.

Quarfordt, S. H. and H. L. Hilderman. 1970. Quantitation of the *in vitro* free cholesterol exchange of human red cells and lipoproteins. *J. Lipid Res.,* 11:528–535.

Robertson, D. L. and M. J. Poznansky. 1985. The effect of non-receptor-mediated uptake of cholesterol on intracellular cholesterol metabolism in human skin fibroblasts. *Biochem. J.,* 232:553–557.

Rogers, J., A. G. Lee, and D. C. Wilton. 1979. The organization of cholesterol and ergosterol in lipid bilayers based on studies using non-perturbing fluorescent sterol probes. *Biochim. Biophys. Acta,* 552:23–37.

Rooney, M., W. Tamura-Lis, L. J. Lis, S. Yachnin, O. Kucuk, and J. W. Kauffman. 1986. The influence of oxygenated sterol compounds on dipalmitoylphosphatidylcholine bilayer structure and packing. *Chem. Phys. Lipids,* 41:81–92.

Rothblat, G. H., F. H. Mahlberg, W. J. Johnson, and M. C. Phillips. 1992. Apolipoproteins, membrane cholesterol domains, and the regulation of cholesterol efflux. *J. Lipid Res.,* 33:1091–1097.

Rottem, S., D. Shinar, and R. Bittman. 1981. Symmetrical distribution and rapid transbilayer movement of cholesterol in *Mycoplasma gallisepticum* membranes. *Biochim. Biophys. Acta,* 649:572–580.

Rottem, S., G. M. Slutzky, and R. Bittman. 1978. Cholesterol distribution and movement in the *Mycoplasma gallisepticum* cell membrane. *Biochemistry,* 17:2723–2726.

Schroeder, F., P. Butko, I. Hapala, and T. J. Scallen. 1990a. Intermembrane cholesterol transfer: role of sterol carrier proteins and phosphatidylserine. *Lipids,* 25:669–674.

Schroeder, F., P. Butko, G. Nemecz, and T. J. Scallen. 1990b. Interaction of fluorescent $\Delta^{5,7,9(11),22}$-ergostatetraen-3β-ol with sterol carrier protein-2. *J. Biol. Chem.*, 265:151–157.

Schroeder, F., J. R. Jefferson, A. B. Kier, J. Knittel, T. J. Scallen, W. G. Wood, and I. Hapala. 1991. Membrane cholesterol dynamics: cholesterol domains and kinetic pools. *Proc. Soc. Exp. Biol. Med.*, 196:235–252.

Slotte, J. P., G. Härmälä, C. Jansson, and M. I. Pörn. 1991. Rapid turnover of plasma membrane sphingomyelin and cholesterol in baby kidney cells after exposure to sphingomyelinase. *Biochim. Biophys. Acta*, 1030:251–257.

Slotte, J. P., G. Hedström, S. Rannström, and S. Ekman. 1989. Effects of sphingomyelin degradation on cell cholesterol oxidizability and steady-state distribution between the cell surface and the cell interior. *Biochim. Biophys. Acta*, 985:90-96.

Slotte, J. P., J. Tenhunen, and I. Pörn. 1990. Effects of sphingomyelin degradation on cholesterol mobilization and efflux to high-density lipoproteins in cultured fibroblasts. *Biochim. Biophys. Acta*, 1025:152–156.

Steck, T. L., F. J. Kezdy, and Y. Lange. 1988. An activation-collision mechanism for cholesterol transfer between membranes. *J. Biol. Chem.*, 263:13023–13031.

Theunissen, J. J. H., R. L. Jackson, H. J. M. Kempen, and R. A. Demel. 1986. Membrane properties of oxysterols. Interfacial orientation, influence on membrane permeabilty and redistribution between membranes. *Biochim. Biophys. Acta*, 860:66–74.

Thomas, P. D. and M. J. Poznansky. 1988a. Cholesterol exchange between lipid vesicles. Effect of phospholipids and gangliosides. *Biochem. J.*, 251:55–61.

Thomas, P. D. and M. J. Poznansky. 1988b. Effect of surface curvature on the rate of cholesterol transfer between lipid vesicles. *Biochem. J.*, 254:155–160.

Thurnhofer, H. and H. Hauser. 1990. Uptake of cholesterol by small intestinal brush border membrane is protein-mediated. *Biochemistry*, 29:2142–2148.

van Amerongen, A., R. A. Demel, J. Westerman, and K. W. A. Wirtz. 1989. Transfer of cholesterol and oxysterol derivatives by the nonspecific lipid transfer protein (sterol carrier protein 2): a study on its mode of action. *Biochim. Biophys. Acta*, 1004:36–43.

Vlahcevic, Z., E. C. Gurley, D. M. Heuman, and P. B. Hylemon. 1990. Bile salts in submicellar concentrations promote bidirectional cholesterol transfer (exchange) as a function of their hydrophobicity. *J. Lipid Res.*, 31:1063–1071.

Xü, Y.-H., K. Gietzen, H.-J. Galla, and E. Sackmann. 1983. A simple assay to study protein-mediated lipid exchange by fluorescence polarization. *Biochem. J.*, 209:257–260.

Yeagle, P. L. 1992. The biophysics and cell biology of cholesterol: a hypothesis for the essential role of cholesterol in mammalian cells. In *Cholesterol in Membrane Models*, Finegold, L., Ed., CRC Press, Boca Raton, FL.

Yeagle, P. L. and J. E. Young. 1986. Factors contributing to the distribution of cholesterol among phospholipid vesicles. *J. Biol. Chem.*, 261:8175–8181.

Chapter 4

THE MOLECULAR DYNAMICS, ORIENTATIONAL ORDER, AND THERMODYNAMIC PHASE EQUILIBRIA OF CHOLESTEROL/PHOSPHATIDYLCHOLINE MIXTURES: ^{2}H NUCLEAR MAGNETIC RESONANCE

James ^{2}H. Davis

TABLE OF CONTENTS

0-8493-4207-4/93/$0.00+$.50

ABSTRACT

The biological role of cholesterol in membranes may be closely linked to the dramatic changes in the physical properties of the membrane induced by the presence of physiological concentrations of cholesterol. A wide variety of physical techniques has been used to determine the nature of the physical properties of membranes and to quantitate the changes induced by cholesterol. In order to understand how cholesterol affects the physical properties of membranes, it is necessary to study these effects systematically.

^2H nuclear magnetic resonance has been a particularly useful tool for studying the molecular orientational order and dynamics of both cholesterol and phospholipids in a variety of model and biological membranes. Careful measurement of the temperature and orientation dependence of nuclear spin lattice relaxation using oriented samples can distinguish among the different models proposed for molecular reorientation of cholesterol itself in model membranes. Due to the large degree of internal flexibility of phospholipid molecules, their molecular dynamics is somewhat more difficult to unravel. However, it has been possible to obtain a consistent description of their molecular reorientation involving rotational diffusion about the long molecular axis, the diffusion in an orienting potential of the orientation of that axis relative to the bilayer normal, *gauche-trans* isomerization, and collective order-director fluctuations. In addition, ^2H nuclear magnetic resonance has proven to be a useful means of studying the phase equilibria of binary mixtures, in particular, of mixtures of cholesterol and the lipid 1,2-dipalmitoyl-*sn*-glycero-3-phosphocholine. ^2H nuclear magnetic resonance difference spectroscopy has been used to determine the phase boundaries of the two phase region below about 37°C in this system. The nuclear magnetic resonance lineshape changes, coupled with results from differential scanning calorimetry and previous work using a number of different techniques, have resulted in a partial phase diagram of this system, covering the temperature range from about 28 to 70°C and a cholesterol concentration range from 0 to 25 mol%.

I. INTRODUCTION

Each and every living cell is enclosed within a thin membrane whose primary function is undeniably to separate the inside of the cell from its surroundings. By allowing some molecules to enter or leave the cell relatively freely while blocking the passage of all others, the membrane maintains the integrity of the cell and ensures its survival. Even if this were the only function of the cell membrane, it would still be the most important structure in biology.

The simplest cells may have only a plasma membrane. However, the cells of higher organisms have a very intricate internal membrane scaffolding, a system of subcellular organelles that compartmentalize the cell and provide convenient surfaces on which processes such as photosynthesis and oxidative phosphorylation occur. In multicellular organisms the plasma membrane has primary responsibility for cell-cell communication, adhesion, and recognition. Another important role of membranes is the organization of the molecular components of cells. Complex biochemical functions such as photosynthesis are facilitated by the incorporation of key proteins and enzymes involved into the membrane where they can be assembled in an appropriate configuration. The ability to establish and maintain gradients in chemical and electrical potentials across the membrane seems to be fundamental to the functioning of many of these complex processes (see, for example, Stryer, 1988).

A. THE MOLECULAR COMPONENTS OF THE MEMBRANE

The basic structural unit of the membrane is a bimolecular sheet formed by amphiphilic lipid molecules. The head-group regions of these lipids, some of which are drawn in Figure 1, are polar and as such can, for example, participate in hydrogen bond formation with water in the cell cytoplasm or in the surrounding aqueous medium. The long hydrocarbon lipid chains are largely hydrophobic and try to avoid contact with water. This hydrophobic effect, largely responsible for the formation of the lipid bilayer that is the structural core of the membrane, is an expression of the strong entropic "forces" that result in an increase in water structuring or order near a hydrocarbon molecule dissolved in water (Tanford, 1973). As a result of this hydrophobic effect, one finds, on mixing hydrocarbons with water, that the hydrocarbons are sequestered in regions where water is excluded. For simple hydrocarbons this leads to phase separation, while for amphiphilic molecules like lipids, a variety of structures, including the lipid bilayer, may be formed (Israelachvilli et al., 1981).

In this manner the lipids form a quasi two-dimensional matrix into which the other molecular components of the membrane are inserted. A large fraction of the dry weight of a natural membrane (typically about 50%) is protein. Because membrane proteins have a considerably higher molecular weight, averaging about 30 to 40 kDa compared to about 800 Da for lipids, the molar lipid-to-protein ratio is typically 40 or 50 to 1. While a typical cell will have hundreds of distinct membrane proteins each of which performs some vital biochemical function, there are relatively few whose functions are known and even fewer whose activities are understood at a molecular level. One reason for this is the absence of detailed structural information on membrane proteins due to the difficulty in growing crystals suitable for diffraction studies. There are a couple of notable exceptions, including the photosynthetic reaction center of *Rhodopseudomonas viridis* (Deisenhofer and Michel, 1991) and bacteriorhodopsin (Henderson et al., 1990) from *Halobacterium halobium*. The latter, whose function is to generate an H^+ (proton) gradient across the membrane when exposed to light, is shown in Figure 2. Two prominent features of membrane proteins are illustrated by this example. Firstly, bacteriorhodopsin is an example of an *intrinsic* membrane protein, i.e., one that crosses the lipid bilayer. Secondly, the protein segments that cross the bilayer are in an α-helical configuration and contain largely hydrophobic amino acid residues.

Sterols are a third important molecular component of at least the cytoplasmic membranes of plant and animal cells. Often as much as 50% of the lipid

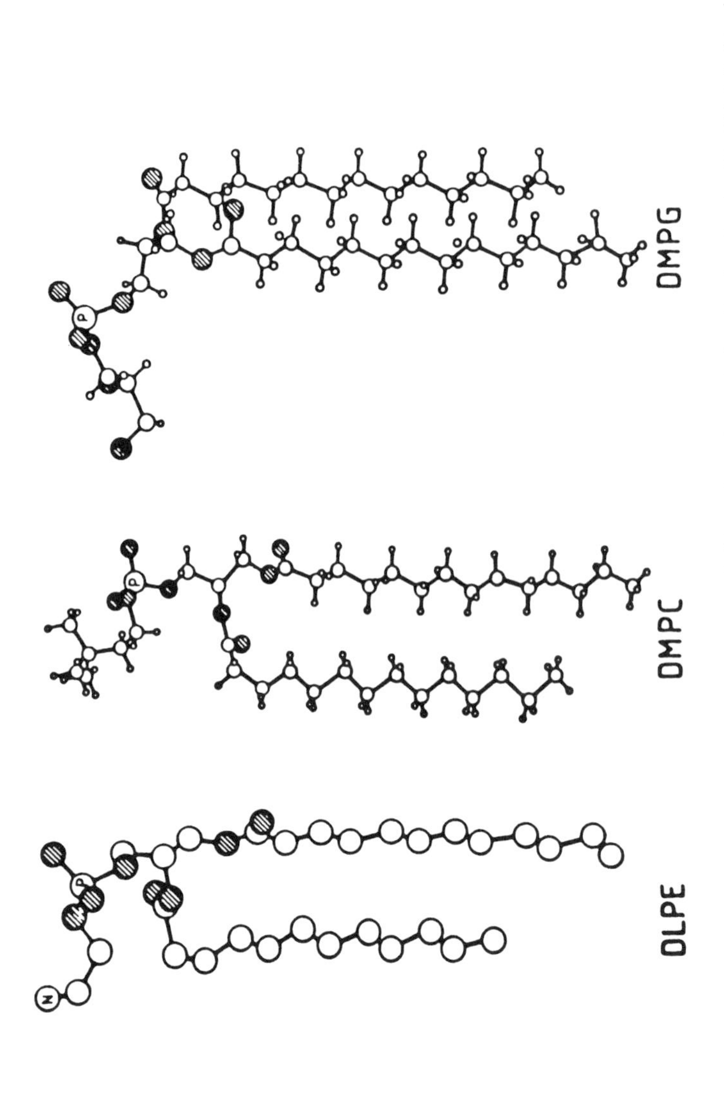

FIGURE 1. Single crystal structures of three phospholipids. The lipids shown are 1,2-dilauroyl-*sn*-glycero-3-phosphoethanolamine (DLPE); 1,2-dimyristoyl-*sn*-glycero-3-phosphocholine (DMPC); and 1,2-dimyristoyl-*sn*-glycero-3-phosphoglycerol (DMPG). (Reproduced from Seelig et al., 1987. *Biochemistry*, 26:7535–7541. With permission.)

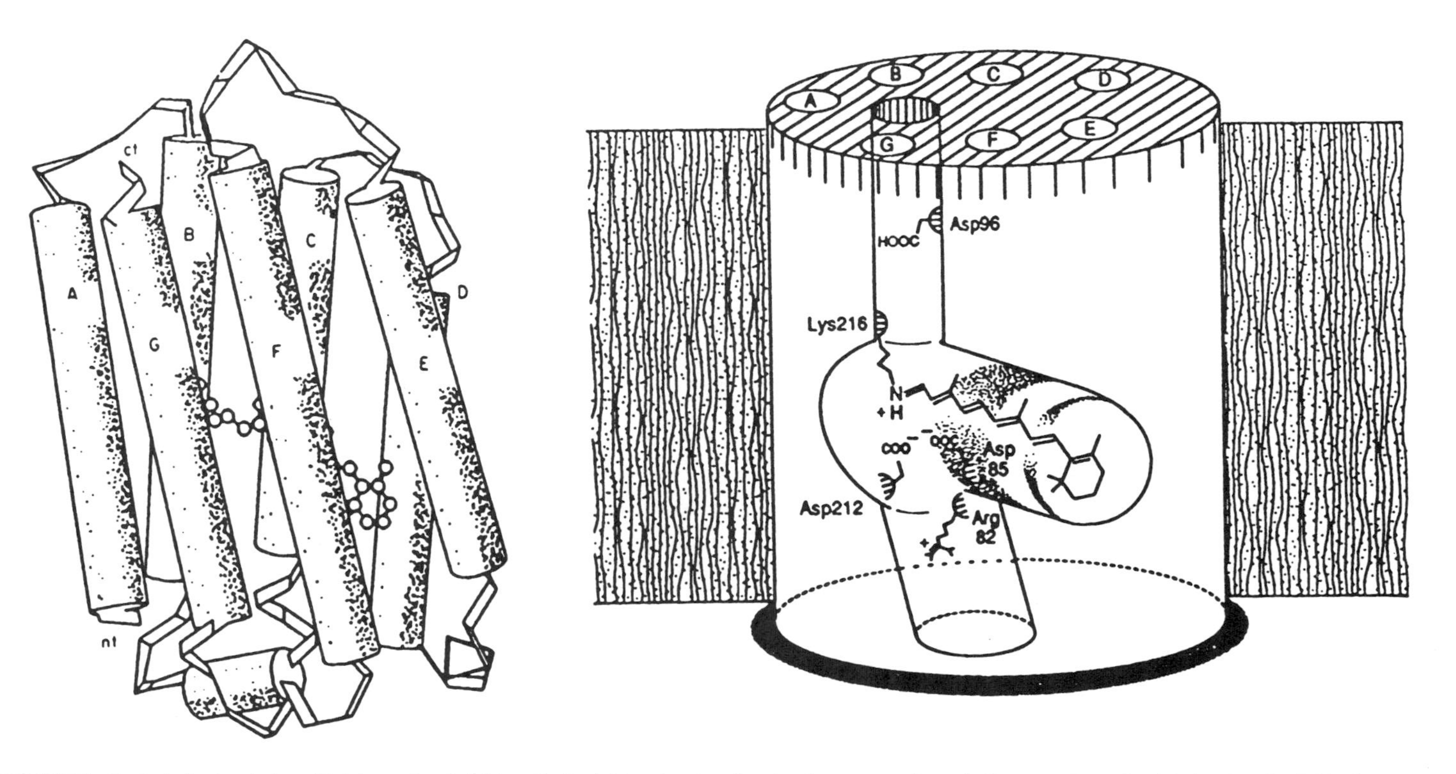

FIGURE 2. Bacteriorhodopsin from *Halobacterium halobium*. On the left is a drawing showing the transmembrane helices as rods; on the right is a drawing indicating some of the residues involved in the proton channel and the retinal binding site. The extent of the membrane bilayer is schematically indicated by the irregular shading. (Reproduced from Henderson et al., 1990. *J. Mol. Biol.*, 213:899–924. With permission.)

component of a membrane is sterol. The biological function of these sterols is still poorly understood, but their influence on the physical properties of membranes is striking, and it is intriguing, therefore, to consider whether their major biological role may be to regulate these physical properties to ensure the proper functioning of the membrane as a whole. It has even been suggested (Bloch, 1985) that cellular evolution reached an impasse until there was sufficient free oxygen in the atmosphere for the biosynthesis of cholesterol. There is no question that sterols, such as cholesterol and ergosterol, have an important role in cell biology. Their presence in virtually all eukaryotic cytoplasmic membranes and their absence from nearly all prokaryotic membranes certainly suggests that one can find a clue to their function from the differences between these types of cells (Bloom et al., 1991). Some key differences that may relate to the role of sterols in modifying membrane physical properties are the absence of a cell wall and the presence of a cytoskeletal framework (for animal cells, but not for plants) and the processes of exo- and endocytosis (Bloch, 1985; Bloom et al., 1991).

B. TECHNIQUES FOR STUDYING THE PHYSICAL PROPERTIES OF MEMBRANES

The influence of cholesterol on the physical properties of membranes and lipid bilayers has been studied by a wide spectrum of experimental techniques (for recent reviews, see Bloom et al., 1991; Vist and Davis, 1990). Changes in membrane permeability (Corvera et al., 1991; Cruzeiro-Hansson et al., 1989), molecular orientational order and dynamics (Vist and Davis, 1990; Weisz et al., 1992; Shin et al., 1990), and membrane elasticity (Needham et al., 1988; Needham and Nunn, 1990) as well as the interesting phase behavior of cholesterol/phospholipid mixtures (Vist and Davis, 1990; Mortensen et al., 1988) are a few of the important features of these systems that are revealed by physical techniques. Each of the techniques used to study these systems has its strengths and weaknesses, and to interpret the results of all of these studies, it is important to know the limitations and capabilities of each. For example, each spectroscopic technique has its own characteristic time window, i.e., what is observed is some sort of average over some finite time.

The techniques of vibrational spectroscopy, both infrared absorption and Raman scattering, have intrinsic time scales of the order of 10^{-12} s, which is short compared to the time required for molecular reorientation or rotation. The information obtained is the spectrum of vibrational frequencies in the sample and how these may change under perturbations caused by changes in sample temperature, composition, thermodynamic phase, etc. (Levin, 1984). Fluorescence polarization techniques possess time scales in the nanosecond to microsecond regime, depending on the lifetime of the fluorescence probe that is used. Thus, fluorescence can be used to study molecular reorientation and orientational order. While many protein molecules are naturally fluorescent, researchers often use simple fluorescent probe molecules with well-defined absorption and emission spectra. In addition, because of the anisotropic character of the membrane (being quasi-two dimensional), the molecular motions are anisotropic, and the shape of the probe molecule becomes important (a disk behaves very differently from a rod) (Zannoni et al., 1983; Vaz et al., 1984).

Electron paramagnetic resonance (EPR) has been an extremely popular tool for the study of membrane systems. Since there are very few naturally occurring unpaired electrons in biological molecules (transition metal ions are some exceptions), these studies have depended largely on the use of synthetic spin-label probe molecules (Devaux and Seigneuret, 1985; Marsh, 1989; Davis, 1986). The time scale of EPR is generally in the nanosecond range, but relaxation and saturation transfer techniques can extend it into the microsecond and even millisecond range (Marsh, 1989). Many molecular motions in membranes have correlation times in this range, making EPR a very sensitive tool for studies of molecular dynamics in membranes (Shin et al., 1990; Lange et al., 1985).

The nuclear magnetic resonance (NMR) spectroscopic time scale ranges from the microsecond to the second scale, depending on the nucleus studied and the magnetic field strength. This is due to the fact that the intrinsic timescales in spectroscopy are related to the inverse of the width of the spectrum. In NMR the spectrum can have a width from about a few Hz (high resolution NMR of small vesicles or in magic angle spinning experiments — even less for molecules in solution) to MHz (10^6 Hz) for quadrupolar nuclei, like deuterium, nitrogen, sodium, etc., in highly ordered systems (such as membranes). In addition, various relaxation experiments can be performed to study molecular dynamics covering time scales from 10^{-10} to 10 s, or longer (Bloom et al., 1992; Davis, 1986, 1989; Rommel et al., 1988). The great breadth of this range makes NMR, in principle at least, the most powerful tool available for studying molecular motion in partially ordered systems.

In interpreting spectroscopic measurements it is necessary to consider the time scales and how molecular motion over these time scales, both molecular reorientation and diffusion, will affect the results. Often one uses these spectroscopic and relaxation techniques to study these motions, but at times the motions themselves can interfere with the objectives of the experiment. When correlation times for molecular motions are close to the inverse of the width of the spectrum, these motions can have a dramatic effect on the shape of the spectrum. This effect can be used to advantage as a tool for studying these "slow" motions, but it can also lead to serious problems in interpretation, especially when the spectrum consists of two or more overlapping components due to distinct molecular environments or coexisting thermodynamic phases.

In studies of phase equilibria of membranes, the lateral diffusion of molecules within the membrane may lead to an exchange of molecules among the different phases in coexistence at a given temperature. If the domains of the different phases are small and the diffusion constants are sufficiently high, then appreciable exchange between domains may occur on the spectroscopic time scale. In this case one observes some average spectrum and not a spectrum characteristic of isolated nonexchanging domains.

If probe molecules are used in spectroscopic studies, they may not accurately mimic the behavior of the natural components of the system. Subtle changes in molecular configuration may have a dramatic effect on the behavior of the system. For example, the α-isomer of cholesterol has an orientation within the bilayer substantially different from that of β-cholesterol. In addition, its influence on the physical properties of the system are dramatically different (Dufourc et al., 1984). Also, in studies of phase equilibria, or in situations where phase transitions may occur, it is necessary to understand how the probe molecules partition among the various phases available. Finally, when using probes one is always studying the influence of the system on the probe molecule, thereby probing the system only indirectly.

The anisotropy of the membrane structure leads to an anisotropy in the molecular motions (cf. Gennis, 1989). For example, diffusion of molecules within the plane of the bilayer is often very rapid, while "diffusion" perpendicular to the plane (which may be considered as a "flip-flop" of molecules across the bilayer, or exchange of molecules between the two leaflets) can be extremely slow. Molecular reorientations also reflect the anisotropy of the membrane. In fluorescence the orientations of the transition moments of fluorescing molecules reflect this dynamical anisotropy. Fluorescence polarization experiments study the decay of the polarization due to molecular reorientation during the fluorescence lifetime. Because of the nonisotropic nature of these motions, the polarization anisotropy does not decay all the way to zero. The residual fluorescence anisotropy as $t \to \infty$ is a measure of the orientational order in the system.

In magnetic resonance, both EPR and NMR, the electric and magnetic interactions (for example, the dipole-dipole interaction between two magnetic moments) have strong orientation dependences. In the case of the dipole-dipole interaction between two magnetic nuclei, the strength of the interaction depends on the distance between the two nuclei and on the angle between the vector joining the two nuclei and the direction of the externally applied magnetic field (generally required for magnetic resonance experiments). If molecular reorientations are rapid and isotropic, then these orientation-dependent interactions will all average completely away. When the motions are anisotropic, there will be a residual interaction that will dominate the magnetic resonance spectrum. The strength of this residual coupling can be determined from the spectra and gives a measure of the orientational order in the system.

The existence of local (molecular) orientational order is a result of the anisotropic structure of the membrane. Another important characteristic of membranes is the absence of long-range positional order. This is due largely to the compositional heterogeneity of the system, but it is also a result of the "fluid" nature of membranes, which are most appropriately described as liquid-crystalline.

Diffraction techniques rely on long-range positional order such as is found in crystals. Membranes rarely exhibit significant long-range order, even in two dimensions (the membrane of *Halobacterium halobium*, which contains the protein bacteriorhodopsin, is a notable exception). Hence, it has been difficult to obtain high-resolution structural information with diffraction techniques. Water-soluble proteins can often be crystallized, but membrane proteins generally denature if removed from the lipidic (or hydrophobic) environment of the membrane. However, it has been possible in a number of cases to cocrystallize proteins with lipids and other amphiphiles (Deisenhofer and Michel, 1991).

Model membranes, consisting of lipid/water or lipid/protein/water mixtures can often be prepared as multilamellar dispersions or stacks of bilayers that have a well-defined repeat spacing or one-dimensional order. X-ray and neutron diffraction and neutron scattering have been successfully employed in studying these systems (Mortensen et al., 1988; Finean, 1990; Knoll et al., 1985). The influence of composition, temperature, and degree of hydration on the repeat spacing and lipid chain packing density have been widely studied in this fashion. The different phases of these mixed model membrane systems have characteristic features so that the diffraction and scattering techniques are very useful for characterizing their phase behavior. In regions of coexistence of phases, it can be difficult to be quantitative because small domains or regions of the sample that are more disordered (positionally) will not contribute as effectively to the diffraction pattern as larger, more highly ordered regions.

The phase transition from the low temperature "gel" phase to the higher temperature "fluid" or liquid crystalline phase characteristic of most membranes under physiological conditions is primarily (but probably not entirely) a hydrocarbon chain melting transition. Below the transition temperature the lipid chains are largely rigid, with little internal motion. Above the transition the chains are melted, and there is a large degree of internal rotation about C–C bonds (the melting of paraffin is a similar phase transition). There is a sizeable "heat of fusion" or enthalpy associated with this transition. Differential scanning calorimetry, where one smoothly scans the temperature of a sample cell and a reference cell and measures the difference in the heat required to keep the temperatures of both cells equal while changing both at the same rate, has become a standard tool for the study of phase transitions in membranes (McElhaney, 1982; Biltonen, 1990; Finegold and Singer, this volume). The thermograms obtained from natural membranes or from mixed model systems can appear to be complex. The shapes of the features in the thermograms reflect the nature of the phase diagram of the system (Morrow et al., 1985; Morrow and Whitehead, 1988; Ipsen et al., 1989). The phase transition of a pure lipid dispersed in excess water appears to be isothermal (very nearly first order) (Albon and Sturtevant, 1978) but there is increasing evidence that these systems are close to a critical point (Morrow et al., 1985; Morrow and Whitehead, 1988; Mouritsen, 1991).

It is possible to prepare large single bilayer vesicles of pure lipids and of mixed systems in excess water. If these vesicles are large enough, one can grasp one of them using gentle suction ($\sim 10^{-6}$ atm) through a micropipette (Needham et al., 1988). By adjusting the suction pressure it is possible to stretch the membrane surface and thereby measure the macroscopic mechanical properties of the membrane (Needham et al., 1988; Needham and Nunn, 1990). These fascinating experiments provide a completely different perspective on the physical properties of membranes (Bloom et al., 1991).

Each of the techniques described above has given valuable insight into the properties of membranes and, in particular, into the effect of the incorporation of cholesterol on simple lipid bilayers. The description of the physical properties of a simple two-component mixture of cholesterol in phosphatidylcholine bilayers will contribute greatly to our understanding of the role of cholesterol in the biological membrane. This article will describe in greater detail the use of nuclear magnetic resonance in the study of molecular order and dynamics and the phase equilibria of cholesterol/lipid mixtures. The next section will introduce the use of deuterium nuclear magnetic resonance to study the orientational order and dynamics of cholesterol in phospholipid bilayers. Then the influence of cholesterol on the orientational order and dynamics of phospholipids in bilayers will be described. Following that will be a discussion of the use of these physical techniques to obtain a description of the phase behavior of cholesterol/1,2-dipalmitoyl-*sn*-glycero-3-phosphocholine (DPPC) mixtures.

II. MOLECULAR ORDER AND DYNAMICS BY DEUTERIUM NUCLEAR MAGNETIC RESONANCE

In partially ordered systems such as liquid crystals, membranes, and solids, the second rank tensor interactions generally dominate the nuclear magnetic resonance spectra and the relaxation of nuclear spin coherences (which include the vector magnetization). These interactions include, among others, the dipole-dipole interactions among abundant nuclei such as protons (^{1}H); the anisotropic part of the chemical shift, important for carbon (^{13}C) and phosphorus (^{31}P); and the electric quadrupolar interaction, which is important for quadrupolar nuclei, (those with nuclear spin $I \geq 1$), such as deuterium (^{2}H). The theoretical descriptions of the second rank tensor interactions and their effects on NMR spectroscopy and relaxation have been extensively reviewed previously and will not be repeated here (Seelig, 1977; Spiess, 1978; Griffin, 1981; Davis, 1983; Davis, 1991; Bloom et al., 1992).

A. THE QUADRUPOLAR HAMILTONIAN AND ^{2}H NUCLEAR MAGNETIC RESONANCE

In ^{2}H NMR the quadrupolar interaction is far larger than the strongest internuclear dipole-dipole interaction, which would be between a deuterium nucleus and a neighboring hydrogen (^{1}H). The ^{2}H chemical shift interaction is, likewise, much smaller than the quadrupolar interaction. Thus, the theoretical treatment is particularly straightforward. One need only consider isolated spin I = 1 nuclei subject to the Zeeman interaction, with Hamiltonian

$$H_z = -\vec{\mu} \cdot \vec{H}_0 = -\gamma \hbar \omega_0 I_z \tag{1}$$

(here $\vec{\mu}$ is the nuclear magnetic moment, $\vec{H}_0$ is the external magnetic field, γ is the nuclear gyromagnetic ratio [$\gamma = 2\pi \times 6.536 \times 10^6 s^{-1} T^{-1}$ for deuterium], $\hbar$ is Planck's constant divided by 2π, ω_0 is the nuclear Larmor frequency, and I_z is the z-component of the nuclear spin angular momentum) and the quadrupolar interaction, whose Hamiltonian is

$$H_Q = \frac{eQ}{6I(2I-1)} \sum_{\alpha,\beta} V_{\alpha\beta}\left[\frac{3}{2}\left(I_\alpha I_\beta + I_\beta I_\alpha - \delta_{\alpha\beta} I^2\right)\right] \tag{2}$$

(where e is the electronic charge, Q is the nuclear quadrupolar moment, I is the nuclear spin, $V_{\alpha\beta}$ are the components of the electric field gradient tensor at the nucleus, and $\delta_{\alpha\beta}$ is the Kronecker delta that is zero unless $\alpha = \beta$, in which case it is 1).

In normal laboratory magnetic fields where $H_0 > 1$ Tesla, the quadrupolar interaction for deuterium is much smaller than the Zeeman interaction and can be treated as a small perturbation. Usually a first order approximation will be sufficient to describe the spectroscopy, and a second order approximation is sufficient for relaxation (Abragam, 1961; Davis, 1983; Davis, 1991). In an external magnetic field the Zeeman interaction removes the magnetic degeneracy of the nuclear ground state. For a spin 1 nucleus such as deuterium, there will be three equally spaced energy levels, as in Figure 3a. The spacing between them gives the deuterium nuclear Larmor frequency,

$$\omega_0 = \gamma H_0 \tag{3}$$

If the deuterium nucleus experiences an electric field gradient arising, for example, from a nonspherically symmetric electron distribution around the nucleus, then there will a shift in the nuclear energy levels due to the electric quadrupolar interaction between the nuclear quadrupolar moment and the electric field gradient. In first-order perturbation theory, this quadrupolar shift, as shown in Figure 3b, is traceless, i.e., the sum of the shifts of the three levels is zero. Since the three levels are no longer equally spaced, the NMR spectrum will now consist of two absorption maxima separated by an amount ω_Q called the quadrupolar splitting. The two members of this doublet are symmetrically located on either side of ω_0 (since H_Q is traceless in first order). As we shall see, the quadrupolar splitting is a function of the orientation of the electric field gradient relative to the magnetic field.

In its principal axis system the electric field gradient is axially symmetric; thus, $V_{\alpha\beta} = V_{\alpha\alpha}\delta_{\alpha\beta}$ and

$$H_Q = \frac{eQ}{6I(2I-1)}\left[V_{x'x'}\left(3I_{x'}^2 - I^2\right) + V_{y'y'}\left(3I_{y'}^2 - I^2\right) + V_{z'z'}\left(3I_{z'}^2 - I^2\right)\right] \tag{4}$$

We define the principal value of the electric field gradient tensor as

$$eq = V_{z'z'} \tag{5}$$

and the asymmetry parameter

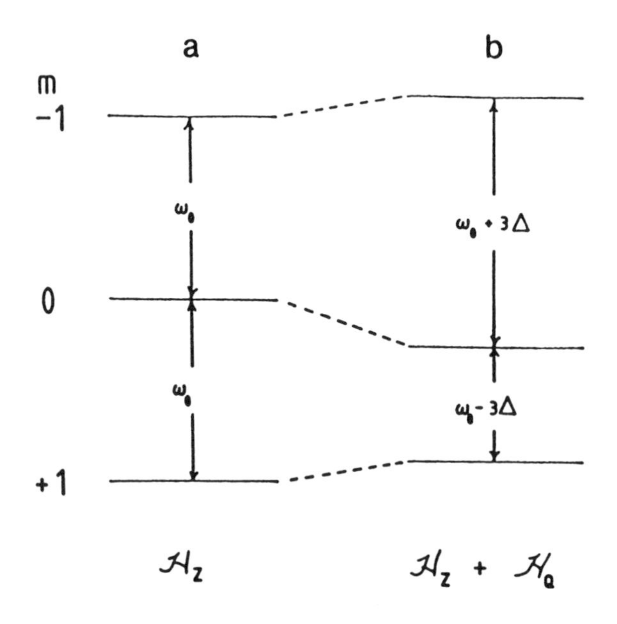

FIGURE 3. Energy levels of a spin 1 nucleus: (a) splittings due to the Zeeman Hamiltonian; (b) first order quadrupolar shifts of the energy levels. (Reproduced from Davis, 1991. In *Isotopes in the Physical and Biomedical Sciences 2*, p. 199. With permission.)

$$\eta = \left| \frac{V_{x'x'} - V_{y'y'}}{V_{z'z'}} \right| \tag{6}$$

where $0 \le \eta \le 1$. Here we have used "primed" coordinates (x',y',z') because in general the principal axis system of the electric field gradient tensor, which is fixed in the molecule, does not coincide with the laboratory reference frame where $\vec{z}$ is parallel to the external magnetic field $\vec{H}_0$. Thus, we need to make a transformation from the electric field gradient principal axis system to the laboratory frame. Before doing this we switch to a system of spherical coordinates where the principal values of the electric field gradient tensor are (Davis, 1983; Davis, 1991)

$$F^P_{2,0} = \sqrt{\frac{3}{2}} eq \tag{7}$$

$$F^P_{2,\pm 1} = 0$$

$$F^P_{2,\pm 2} = \frac{1}{2} eq\eta$$

The second rank spherical tensors for the spin operators are

$$T_{2,0} = \frac{1}{\sqrt{6}}\left[3I_z^2 - I(I+1)\right]$$

$$T_{2,\pm 1} = \mp\frac{1}{2}\left[I_z I_\pm + I_\pm I_z\right] \tag{8}$$

$$T_{2,\pm 2} = \frac{1}{2}I_\pm^2$$

(here, $I_\pm = I_x \pm iI_y$). These spherical tensors transform like the spherical harmonics (Rose, 1957; Zare, 1988), so that the electric field gradient tensor in the laboratory frame is

$$F_{2,m}^{L} = \sum_{m'=-2}^{2} F_{2,m'}^{P} D_{m',m}^{(2)}(\alpha\beta\gamma) \tag{9}$$

where $D^{(2)}(\alpha\beta\gamma)$ is the Wigner rotation matrix and the set of angles $(\alpha\beta\gamma)$ are the Euler angles (Rose, 1957; Zare, 1988) describing the successive rotations required to bring the principal axis reference frame into coincidence with the laboratory frame.

In the laboratory frame the quadrupolar Hamiltonian is

$$\begin{aligned} H_Q &= \frac{eQ}{2}\sum_{m=-2}^{2}(-1)^m T_{2,m} F_{2,-m}^{L} \\ &= \frac{eQ}{2}\sum_{m=-2}^{2}(-1)^m T_{2,m} \sum_{m'=-2}^{2} F_{2,m'}^{P} D_{m',-m}^{(2)}(\alpha\beta\gamma) \end{aligned} \tag{10}$$

Since $\vec{H}_0$ defines only the z-direction in our laboratory frame, we can choose the laboratory y-axis to lie perpendicular to the plane formed by $\vec{z}$ and $\vec{z}'$. Then the first Euler angle, γ, will be zero (see Zare (1988), pp. 77–80). Further, in first order perturbation theory we need only keep that part of the perturbing Hamiltonian, H_Q, which commutes with the Zeeman Hamiltonian, H_z; thus, we keep only the terms in $T_{2,0}$ from Equation 10 (Cohen-Tannoudji, 1977).

$$H_Q = \frac{e^2qQ}{2}T_{2,0}\left\{\sqrt{\frac{3}{2}}D_{00}^{(2)}(\alpha\beta 0) + \frac{1}{2}\eta\left[D_{20}^{(2)}(\alpha\beta 0) + D_{-20}^{(2)}(\alpha\beta 0)\right]\right\} \tag{11}$$

or, substituting for the $D_{m0}^{(2)}(\alpha\beta 0)$, (Seelig, 1977; Zare, 1988; Rose, 1957)

$$H_Q = \frac{e^2 qQ}{8}\left[3I_z^2 - I(I+1)\right]\left[(3\cos^2\beta - 1) + \eta\sin^2\beta\cos 2\alpha\right] \tag{12}$$

Then, to first order in H_Q, the Zeeman energy levels are shifted by the amounts

$$E_{Q,m} = \frac{e^2 qQ}{8}(3m^2 - 2)\left[(3\cos^2\beta - 1) + \eta\sin^2\beta\cos 2\alpha\right] \tag{13}$$

The quadrupolar splitting corresponds to the difference in frequency between the m = −1 to m = 0 and the m = 0 to m = 1 transitions. Thus,

$$\omega_Q = \frac{3e^2 qQ}{4\hbar}\left[(3\cos^2\beta - 1) + \eta\sin^2\beta\cos 2\alpha\right] \tag{14}$$

where α and β define the orientation of the electric field gradient tensor principal axis system in the laboratory frame.

For ^2H in C–^2H bonds, the electric field gradient tensor is very nearly axially symmetric ($\eta \cong 0.01$–0.02) (Millar et al., 1986). In this case we often neglect η and can take one other Euler angle, α, to be zero. This leaves only β, the angle between the electric field gradient principal axis and the applied magnetic field, so that Equation 14 becomes

$$\omega_Q = \frac{3e^2 qQ}{4\hbar}(3\cos^2\beta - 1) \tag{15}$$

In this case there is a simple, direct relation between the quadrupolar splitting and the orientation of the C–^2H bond and the magnetic field $\vec{H}_0$.

In the presence of motion the situation may become considerably more complicated. It will then often be useful to introduce another coordinate transformation from the electric field gradient principal axis system to some intermediate "crystal fixed" axis system, which may be defined by the symmetry of the crystal or by the symmetry of the motion. In this case the electric field gradient tensor becomes

$$F^D_{2,m'} = \sum_{m''} F^P_{2,m''} D^{(2)}_{m''m'}(\alpha\beta\gamma) \tag{16}$$

where $F^D_{2,m'}$ is the tensor in the new crystallographic coordinate system, and now the Euler angles (αβγ) define this new transformation. Proceding to the laboratory frame, we obtain

$$F^L_{2,m} = \sum_{m'} F^D_{2,m'} D^{(2)}_{m'm}(\alpha'\beta'\gamma') \tag{17}$$

where now $(\alpha'\beta'\gamma')$ define the transformation from the intermediate to the laboratory frame.

Earlier we had noted that the laboratory frame was axially symmetric and that in first order perturbation theory we need only retain that part of the perturbing Hamiltonian that commutes (the secular part) with the unperturbed Hamiltonian. Then,

$$H_Q = \frac{eQ}{2} T_{2,0} F^L_{2,0} = \frac{eQ}{2} T_{2,0} \sum_{m'} F^D_{2,m'} D^{(2)}_{m'0}(\alpha'\beta'\gamma') \tag{18}$$

The $D^{(2)}_{m'0}(\alpha'\beta'\gamma')$ are independent of γ', so, taking $\gamma' = 0$,

$$H_Q = \frac{eQ}{2} T_{2,0} \sum_{m'} F^D_{2,m'} D^{(2)}_{m'0}(\alpha'\beta'0) \tag{19}$$

For membrane samples the local bilayer normal defines a convenient "crystallographic" reference frame. An "oriented sample" is commonly prepared by layering the lipid (or lipid/lipid, lipid/cholesterol, lipid/peptide, lipid/protein, etc., mixture) onto glass plates. In well-aligned samples the bilayer normal is aligned perpendicular to the plane of the glass plates. The D-frame has its z_D-axis perpendicular to the glass plates, so the angle β' is the angle between the magnetic field and the normal to the plates, independent of position in the sample; or if a sample consists of many parallel plates, β' will be the same for all plates. The requirement that H_Q be independent of α' is equivalent to taking $m' = 0$ in Equation 19. Then,

$$H_Q = \frac{eQ}{2} T_{2,0} F^D_{2,0} D^{(2)}_{00}(0\beta'0) \tag{20}$$

where we have taken $\alpha' = 0$ for convenience. One might imagine, however, that different plates, or even different regions on a single plate, might have different values of α'. Thus, the NMR spectrum would, in general, be a "powder" average over the angle α' if all values of α' are equally likely, since the quadrupolar splitting is a function of α' (Auger et al., 1990). If the D-frame exhibits axial symmetry, then the quadrupolar splitting will be independent of α'. Now, in going to the axially symmetric crystallographic frame from the electric field gradient principal axis system, we have the transformation

$$F^D_{2,0} = \sum_{m''} F^P_{2,m''} D^{(2)}_{m''0}(\alpha\beta\gamma) \tag{21}$$

so that

$$H_Q = \frac{eQ}{2} T_{2,0} \sum_{m''} F^P_{2,m''} D^{(2)}_{m''0}(\alpha\beta\gamma) D^{(2)}_{00}(0\beta'0) \tag{22}$$

which, after substituting for $D_{00}^{(2)}(0\beta'0)$ and $F_{2,m''}^{P}$, becomes

$$H_Q = \frac{e^2qQ}{2}T_{2,0}\cdot\frac{1}{2}(3\cos^2\beta' - 1)\times \left\{\sqrt{\frac{3}{2}}D_{00}^{(2)}(\alpha\beta 0) + \frac{1}{2}\eta\left[D_{2,0}^{(2)}(\alpha\beta 0) + D_{-2,0}^{(2)}(\alpha\beta 0)\right]\right\} \tag{23}$$

where we have set $\gamma = 0$ because $D_{m'',0}^{(2)}(\alpha\beta\gamma)$ is independent of γ. Then, substituting for $D_{\pm 2,0}^{(2)}(\alpha\beta 0)$ and $T_{2,0}$, we obtain finally

$$H_Q = \frac{e^2qQ}{8}(3\cos^2\beta' - 1)\times \left[\frac{1}{2}(3\cos^2\beta - 1) + \frac{1}{2}\eta\sin^2\beta\cos 2\alpha\right]\left[3I_z^2 - I(I+1)\right] \tag{24}$$

For lipid bilayers or membranes the angle between the local bilayer normal $\vec{H}_0$ may often be static, but the molecular reorientations modulate the angles $\alpha(t)$ and $\beta(t)$, which are often expressed as explicit functions of time. There are collective motions, however, such as director fluctuations (Brown, 1982; Weisz et al., 1992), surface undulations (Bloom and Evans, 1990; Bloom et al., 1991), and even simple lateral diffusion over curved membrane surfaces (Bloom et al., 1992), which can lead to a modulation of β' with time.

Since NMR spectroscopy has a finite time scale associated with it, we need to perform an average over the motions that are rapid compared to this time scale. Then the quadrupolar splitting will be

$$\begin{aligned} <\omega_Q> &= \frac{3e^2qQ}{4\hbar}(3\cos^2\beta' - 1)\cdot\frac{1}{2} <(3\cos^2\beta - 1) + \eta\sin^2\beta\cos 2\alpha> \\ &= \frac{3e^2qQ}{4\hbar}(3\cos^2\beta' - 1)\left[S_{zz} + \frac{1}{3}\eta(S_{xx} - S_{yy})\right] \end{aligned} \tag{25}$$

where S_{xx}, S_{yy}, and S_{zz} are the principal elements of the order parameter tensor (DeGennes, 1974). If $\eta = 0$, then

$$<\omega_Q> = \frac{3e^2qQ}{4\hbar}(3\cos^2\beta' - 1)\cdot\frac{1}{2} <(3\cos^2\beta - 1)> \tag{26}$$

With a single crystal, where all molecules have the same orientation with respect to the magnetic field, the ^{2}H NMR spectrum is a simple doublet with quadrupolar splitting ω_Q. In the presence of rapid axially symmetric reorientation about the bilayer normal, one has effective axial symmetry ($S_{xx} = S_{yy}$ in Equation 25).

Hence, for a lipid sample oriented on a glass plate, the doublet spacing depends on the angle β between the principal axis of the electric field gradient tensor (typically the C–^{2}H bond) and the axis of motional symmetry (the bilayer normal, parallel to the normal to the glass plate), as well as on the angle β' between this symmetry axis and the magnetic field (Equation 26). In the spectrum of an oriented sample of fully hydrated, chain perdeuterated dilauroyl-*sn*-glycero-3-phosphocholine (DLPC), shown in Figure 4, the quadrupolar splittings of the different deuterons reflect the gradient in hydrocarbon chain flexibility. Positions near the center of the bilayer, for example the terminal methyl, will have splittings that are smaller than positions closer to the lipid-water interface, due to increased motional averaging. This flexibility gradient is characteristic of fluid phase lipid bilayers.

If the sample is a simple lipid/water dispersion or other unoriented membrane preparation where all orientations of the local bilayer normal are equally likely, then the ^{2}H NMR spectrum will be a superposition of doublets from all orientations. From Equation 25 we see that the maximum quadrupolar splitting occurs when $\beta' = 0$ and that all values from 0 to this maximum value also occur. The resulting powder spectra are illustrated in Figure 5. If the system has a non-zero asymmetry parameter η, spectra of the form of Figure 5a are obtained, while if $\eta = 0$ or if the rapid molecular motions are such that $S_{xx} = S_{yy}$ in Equation 25, then the characteristic axially symmetric powder pattern of Figure 5b is obtained.

B. THE ORIENTATION OF CHOLESTEROL IN PHOSPHOLIPID BILAYERS

When $\eta = 0$ there is a simple relationship between the quadrupolar splitting, ω_Q, and the orientation of the C–^{2}H bond (since the z-axis of the electric field gradient tensor's principal axis system often lies along the C–^2H bond), as shown in Equation 15 for the static case, or in Equation 26 for the dynamic case. Thus, in favorable situations the quadrupolar splittings can be used to determine bond orientations. Deuterium-labeled cholesterol incorporated into fluid phase phospholipid bilayers provides a nice example of how this can be done.

Figure 6 shows a drawing of cholesterol and another of 1,2-dimyristoyl-*sn*-glycero-3-phosphocholine (DMPC) for comparison. The cholesterol molecule-fixed coordinate system, with its origin at carbon C_3, is shown in the figure. A deuteron on C_i of the rigid ring moiety of cholesterol in the fluid phase lipid bilayer will have a quadrupolar splitting given by

$$< \omega_{Q,i} > = \frac{3e^2qQ}{4\hbar}\frac{1}{2} < (3\cos^2\beta_i - 1) > \tag{27}$$

where we have taken $\beta' = 90°$ in Equation 25 so that we are measuring the quadrupolar splitting for the 90° peaks in the powder pattern spectrum (see Figure 5). The angle β_i in Equation 27 is the angle between the C_i–^{2}H bond and the bilayer normal.

In the fluid phase of membranes, the bilayer normal is an axis of symmetry for rapid axial reorientations. Cholesterol is expected to reorient rapidly about

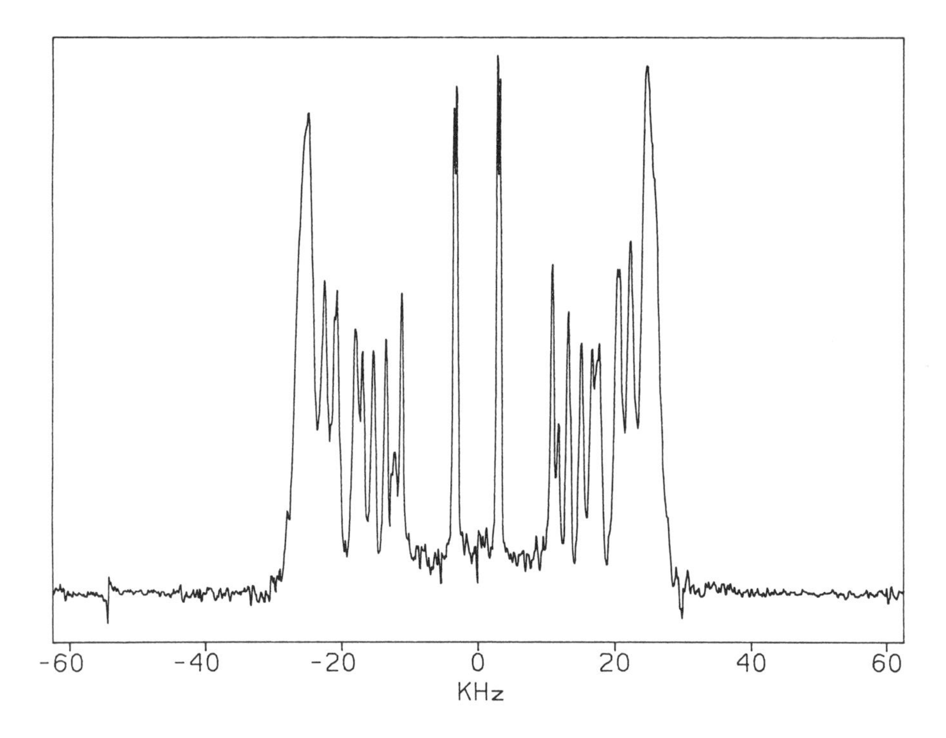

FIGURE 4. ^{2}H NMR spectrum of an oriented sample of 1,2-dilauroyl-*sn*-glycero-3-phosphocholine (DLPC). From a single glass plate coated with about 6 mg of lipid. The spectrum was taken at 35°C at 55.26 MHz, and the sample is oriented with the bilayer normal parallel to the direction of the external magnetic field. (Courtesy of R. S. Prosser.)

this axis. This motion is probably either axial diffusion through small steps or discrete jumps, having at least three sites (Bonmatin et al., 1990). In addition, one expects some reorientation of the long molecular axis in the membrane orienting potential (which lies along the bilayer normal). To properly formulate the problem, we introduce a molecule fixed coordinate system whose z-axis is the axis about which the molecule performs axial diffusion, with diffusion constant $D_{\parallel}$. In addition, we will allow this axis to reorient relative to the bilayer normal with diffusion constant $D_{\perp}$. This defines the principal axis system of the molecular rotational diffusion tensor. For a long, irregularly shaped rigid molecule we might use the moment of inertia to obtain an estimate of the orientation of this axis system within the molecule. For a flexible molecule the situation is much more complex, since we would need to perform the calculation for each possible conformation and then perform a statistical average. Cholesterol has a large 5-ring section that is essentially rigid and a short hydrocarbon tail that has at least some flexibility. Thus, it will be difficult to guess at the direction of the reorientation axis within the molecular structure.

For deuterons on the rigid ring structure of cholesterol (and for rigid molecules in general), the molecular reorientations will affect all splittings in the

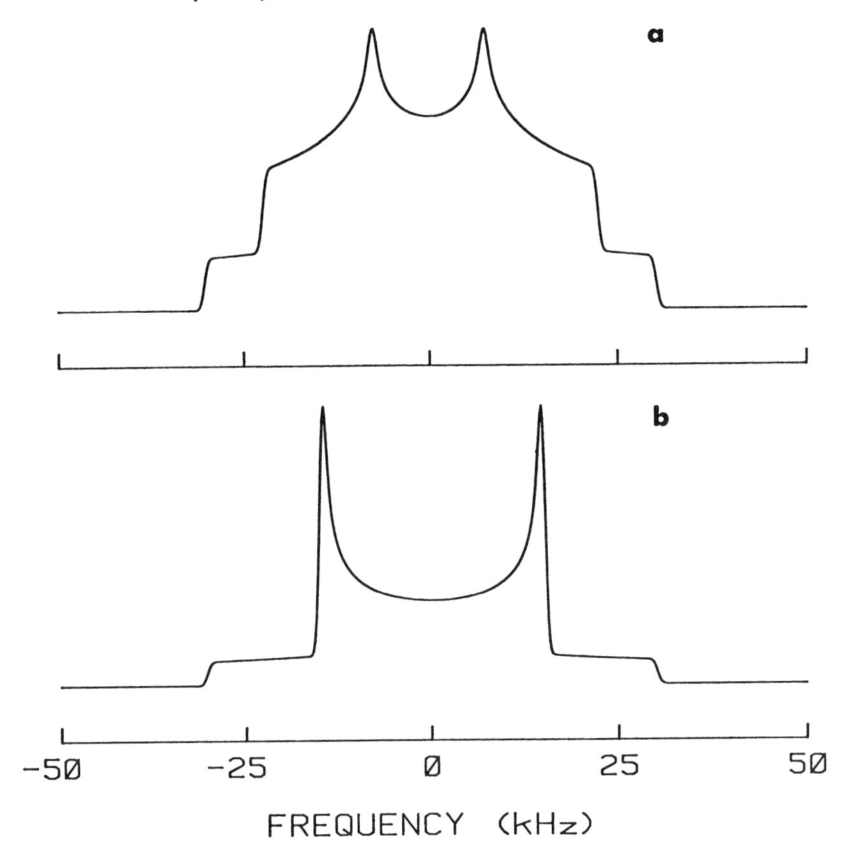

FIGURE 5. First order spin 1 powder pattern lineshapes: (a) nonaxially symmetric case with $\eta \neq 0$; (b) axially symmetric case, $\eta = 0$. (Reproduced from Davis, 1983. *Biochim. Biophys. Acta,* 737:117. With permission.)

same way. Introducing the molecular fixed frame (X_M, Y_M, Z_M) and assuming rapid axially symmetric reorientation about the Z_M axis, the quadrupolar splitting (for the 90° peaks of the powder pattern spectrum) is, in the notation of Taylor et al. (1981) and Dufourc et al. (1984),

$$< \omega_{Q,i} > = \frac{3e^2qQ}{4\hbar}\frac{1}{2}(3\cos^2\alpha_i - 1)\frac{1}{2} < 3\cos^2\gamma - 1 > \qquad (28)$$

where γ is the angle between the bilayer normal and the molecular fixed axis of reorientation, and α_i is the angle between the C_i–^2H bond and this reorientation axis. The α_i are fixed by molecular geometry and are not affected by motion for a rigid structure. The quantity $^1/_2 < 3\cos^2\gamma - 1 >$ is a molecular "order parameter" that scales all of the splittings by the same factor. Thus, if we put ^2H labels at two different positions, i and j, on the ring structure, then the ratio

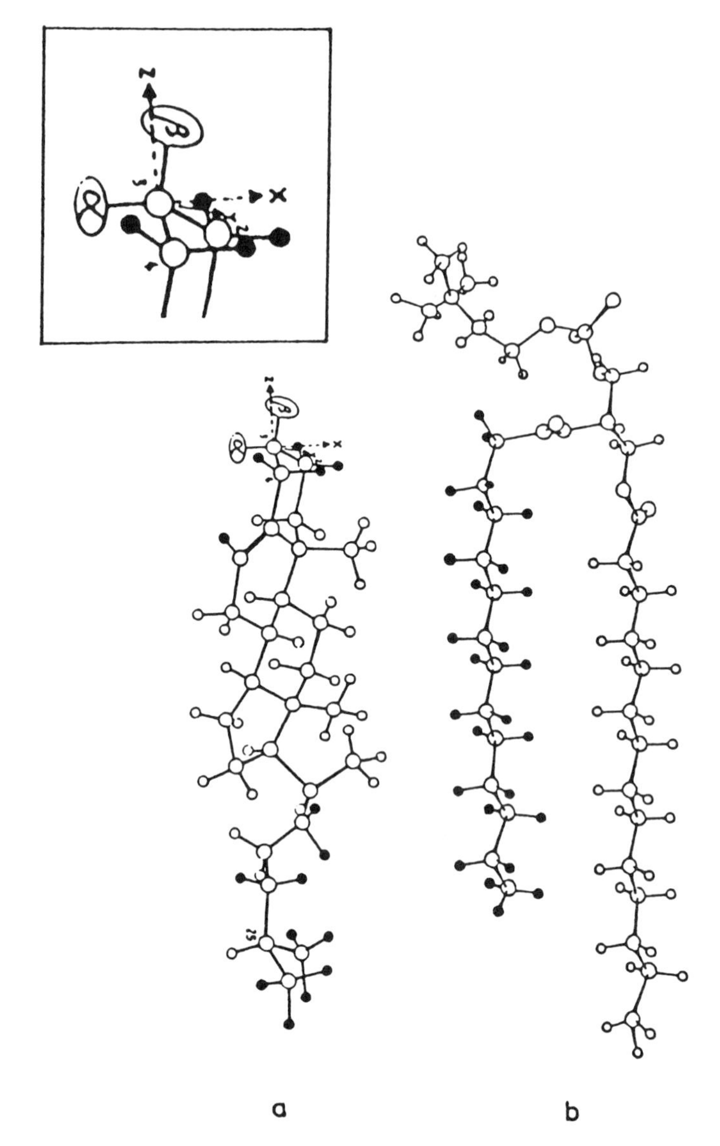

FIGURE 6. Structures of cholesterol (a) and 1,2-dimyristoyl-*sn*-glycero-3-phosphocholine (DMPC); (b): α = *H*, β = *OH*, β-cholesterol; α = *OH*, β = *H*, α-cholesterol. The filled circles represent the available deuterium labels. The sterol fixed axis system has its origin at C_3. (Reproduced from Dufourc et al., 1984. *Biochemistry,* 23:6062. With permission.)

$$R_{ij} = \frac{\omega_{Qi}}{\omega_{Qi}} = \frac{(3\cos^2 \alpha_i - 1)}{(3\cos^2 \alpha_j - 1)} \tag{29}$$

is independent of the molecular reorientation. The experimental values for R_{ij} can be used to determine the orientation of the molecular reorientation axis (Taylor et al., 1981).

As described in Taylor et al. (1981), the direction cosines defining the orientation of the diffusion axis, (l_r, m_r, n_r), in a spherical coordinate system, are $(\sin\theta\cos\phi, \sin\theta\sin\phi, \cos\theta)$. The direction of the C_i–^2H bond in the same coordinate system is given by

$$(l_i, m_i, n_i) = \frac{1}{b}\left(X_{d_i} - X_{c_i}, Y_{d_i} - Y_{c_i}, Z_{d_i} - Z_{c_i}\right) \tag{30}$$

where

$$b = \sqrt{\left(X_{d_i} - X_{c_i}\right)^2 + \left(Y_{d_i} - Y_{c_i}\right)^2 + \left(Z_{d_i} - Z_{c_i}\right)^2}$$

is the C_i–^2H bond length, and the X_{d_i}, X_{c_i}, etc. are the atomic coordinates of the i'th ^2H and the i'th C (Craven, 1979). Then the angle α_i, between the C_i–^2H bond and the diffusion axis is, by the addition theorem,

$$\begin{aligned}\cos\alpha_i &= l_i l_r + m_i m_r + n_i n_r \\ &= l_i \sin\theta\cos\phi + m_i \sin\theta\sin\phi + n_i \cos\phi\end{aligned} \tag{31}$$

Thus, there are two unknown quantities: the angles (θ, ϕ) that specify the orientation of the diffusion axis in the molecular coordinate system. In determining these two angles from the experimental ratios R_{ij} of Equation 29, it must be kept in mind that there are always two values of α_i that give the same ω_{Qi}, specifically, α'_i and $\pi - \alpha'_i$. In addition, for values of $^1/_2(3\cos^2\alpha - 1) < {}^1/_2$ there are four different angles that will give the same splitting (except when $\alpha_i = 54.7°$, the so-called "magic angle" where $^1/_2(3\cos^2\alpha - 1) = 0$).

Using cholesterol that was deuterium labeled at five different positions on the rigid steroid ring structure, Taylor et al. (1981) were able to determine the orientation of the reorientation axis, using reasonable assumptions to eliminate the extraneous solutions (e.g., one expects the reorientation axis to lie roughly along the "long axis" of the molecule). Their best values for the angles (θ, ϕ) in the molecular coordinate system defined in Figure 7 are $\theta = 79°$ and $\phi = 93°$. From this they also could deduce the value of the "molecular order parameter" defined by $S_{mol} = {}^1/_2 < 3\cos^2\gamma - 1 >$. For an equimolar mixture of cholesterol in egg yolk phosphatidylcholine (egg yolk PC) at 25°C, the value of $S_{mol} \approx 0.87$, which suggests immediately that the angular excursions of the diffusion axis relative to the bilayer normal are relatively small under these conditions (the molecule is highly ordered).

Similar approaches can be used to study the head group orientations of lipids in the presence (Brown and Seelig, 1978) or absence (Seelig et al., 1977; Seelig et al., 1987) of cholesterol. Because of their tremendous internal reorientational freedom, it is not feasible to determine the orientations of the C–^2H bonds for the lipid chains, except for the first, or α, position (Seelig and Seelig, 1975). At other positions one can speak of average orientations combined with significant

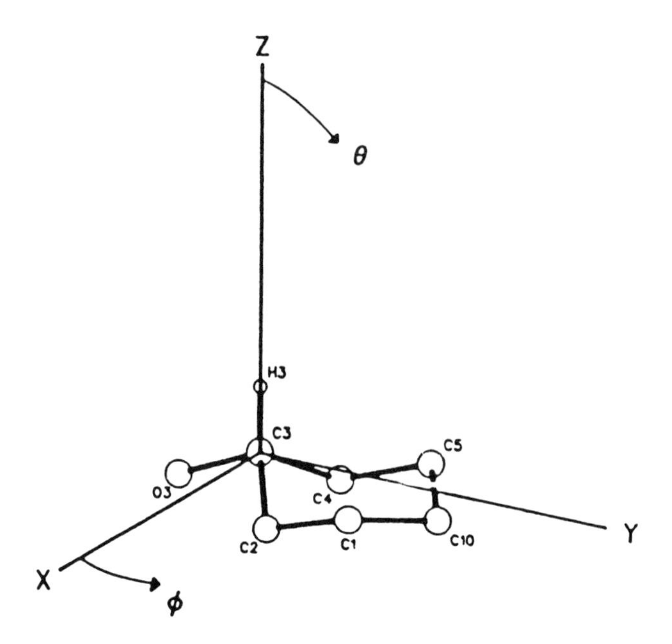

FIGURE 7. Sterol fixed axis system with origin at C_3. The angles θ and φ give the orientation of the diffusion axis within this coordinate system. (Reproduced from Taylor et al., 1981. *Chem. Phys. Lipids*, 29:327. With permission.)

dynamic orientational averaging, so that it will not, in general, be possible to separate the two.

C. RELAXATION IN DEUTERIUM NUCLEAR MAGNETIC RESONANCE

All pulsed NMR experiments involve the preparation and observation of a nonequilibrium state of the nuclear spin system. Relaxation experiments are designed to study the return of the system to its equilibrium state. The interest, of course, is in the physical mechanisms that bring the system back to equilibrium. These are the molecular motions. Because the spin interactions are orientation dependent, molecular reorientation in a magnetic field leads to fluctuations in these interactions. It is these fluctuations that cause relaxation. In order to see how this works, we need to develop a means of describing the spin system and its evolution in time. Fortunately, isolated spin 1 nuclei, whose spectroscopy and relaxation behavior are dominated by the quadrupolar interaction, provide one of the simplest theoretical examples of spin dynamics.

The statistical nature of spin systems is conveniently described using the density operator (or density matrix) formalism (Abragam, 1961; Cohen-Tannoudji, 1977; Davis, 1983; Davis, 1991). The Liouville equation describes the time evolution of the density operator ρ,

$$\frac{d\rho}{dt} = \frac{i}{\hbar}(\rho H - H\rho) = \frac{i}{\hbar}[\rho, H] \tag{32}$$

(where $[A,B] = AB - BA = -[B, A]$ is the familiar "commutator" notation employed in quantum mechanics). This equation has the formal solution

$$\rho(t) = \exp\left(-\frac{i}{\hbar}Ht\right)\rho(0)\exp\left(\frac{i}{\hbar}Ht\right) \tag{33}$$

which may be verified by substitution back into Equation 32. If we know the form of the Hamiltonian and it is not time dependent (or at least if we can divide time into finite "windows" wherein H is constant, though it may be different during different "windows"), then Equation 33 will allow us to calculate $\rho(t)$ for all times. More importantly, we can calculate the expectation value of any physical observable at any time, since the expectation value of an operator O is given by (Cohen-Tannoudji, 1977; Davis, 1991)

$$< O >= Tr\{\rho O\} = \sum_n \sum_m < n|\rho|m >< m|O|n > \tag{34}$$

For a spin system in equilibrium at a lattice temperature T, the off-diagonal elements of the density matrix are zero (Cohen-Tannoudji, 1977; Davis, 1991), while the diagonal elements are given by

$$(\rho_0)_{nn} = \frac{1}{Z}\exp\left\{-\frac{H_0}{kT}\right\}$$

where (35)

$$Z = \sum_m \exp\left\{-\frac{E_m}{kT}\right\}$$

Thus, $(\rho_0)_{nn}$ is the fractional population of the nth energy level with energy E_n. These levels are populated according to the Boltzmann factor at temperature T.

As an illustration, if H_0 is the Zeeman Hamiltonian, then the z-component of the magnetization, M_z, in a magnetic field H_0 is

$$\begin{aligned} < M_z > &= Tr\{\rho_0 M_z\} \\ &= \gamma\hbar Tr\{\rho_0 I_z\} \\ &= \gamma\hbar < I_z > \end{aligned} \tag{36}$$

In the high temperature approximation ($T \geq 1$ K) we need to keep only the term linear in 1/T in the expansion of the exponential of Equation 35. Then, since $Tr\{H_z\} = 0$ and $Tr\{1\} = 3$ (for spin 1),

$$< I_z > = \frac{\gamma\hbar H_0}{3kT} Tr\{I_z^2\} = \frac{2\hbar\omega_0}{3kT} \tag{37}$$

For deuterium, relaxation is caused predominantly by fluctuations in the quadrupolar interaction due to molecular reorientation in the static magnetic field $\vec{H}_0$. We can separate the quadrupolar Hamiltonian into its static and fluctuating parts,

$$H_Q(t) = < H_Q >_T + [H_Q(t) - < H_Q >_T] \tag{38}$$

where $< H_Q >_T$ is the residual quadrupolar interaction averaged over the ^{2}H NMR spectroscopic time scale. The response of the density operator to fluctuations in the quadrupolar interaction is readily calculated [see appendix and references: Abragam (1961); Davis (1983, 1991) and Slichter (1990)]. The calculations are considerably simplified if one introduces an operator space formalism wherein one expands the density operator in terms of a complete orthonormal set of basis operators $\{Q_n\}$ which span the $N = 9$ dimensional spin-space of this spin 1 system (Vega and Pines, 1977; Wokaun and Ernst, 1977; Bloom et al., 1980; Davis, 1983; Goldman, 1988; Bloom, 1988; and Davis, 1991).

In this formalism the density operator is written as

$$\rho(t) = \sum_q c_q(t) Q_q \tag{39}$$

where all of the time dependence is in the coefficients $c_q(t)$, the operators being fixed in operator space. Now the Liouville equation takes the form

$$\begin{aligned} \frac{d\rho}{dt} &= \sum_q \frac{dc_q}{dt} Q_q \\ &= \frac{i}{\hbar}\left[\sum_q c_q(t) Q_q, H\right] \\ &= \frac{i}{\hbar}\sum_q c_q(t)[Q_q, H] \end{aligned} \tag{40}$$

If we multiply both sides of this equation by Q_p and take the trace, using the orthonormality of the basis operators with respect to the trace ($Tr\{Q_pQ_q\} = \delta_{pq}$, where $\delta_{pq} = 1$ if $p = q$ and $\delta_{pq} = 0$ if $p \neq q$), we obtain the system of coupled linear first-order differential equations for the coefficients

$$\frac{dc_p}{dt} = \frac{i}{\hbar}\sum_q c_q(t) Tr\{Q_p[Q_q, H]\} \tag{41}$$

for $p = 1$ to N. To evaluate this equation we need to calculate the commutator of each of the nine basis operators Q_q with the Hamiltonian. This presents little problem for a system of isolated spin 1 nuclei (Bloom, 1988), but may become cumbersome for more complex systems. Through use of the cyclic property of the trace ($Tr\{ABC\} = Tr\{BCA\} = Tr\{CAB\}$), this fundamental equation can be rewritten as

$$\frac{dc_p}{dt} = \frac{i}{\hbar}\sum_q c_q(t) Tr\{Q_q[H, Q_p]\} \tag{42}$$

where now, to find the equation for a given coefficient $c_p(t)$, we need only evaluate the commutator of the single operator Q_p with H.

The system of equations represented by Equation 42 can be solved analytically in many simple cases (Bloom, Davis, and Valic, 1980; Davis, 1991). Of particular interest here are the equations describing the return of the spin system to equilibrium following some pulse excitation. For example, in an inversion recovery T_{1z} experiment, one first inverts the z-component of the magnetization by applying a 180° pulse (or rotation) about either the x or y axis. One then waits for a time τ before applying a second 90° pulse to convert the remaining z-magnetization into observable transverse magnetization. The measurement of T_{1z} consists of measuring the amplitude of this remnant magnetization as a function of the variable time τ. In many cases this recovery follows a simple single exponential with characteristic time constant T_{1z}.

Using the basis operator system discussed by Davis (1991) and in the appendix, the evolution of the x component of magnetization is given by

$$\frac{dc_1(t)}{dt} = \frac{i}{\hbar}\sum_q c_q(t) Tr\{Q_q[H_Q, Q_1]\} \tag{43}$$

where H_Q is the quadrupolar Hamiltonian of Equation 38 and $Q_1 = \frac{1}{\sqrt{2}} I_x$ is proportional to M_x (Davis, 1991; also, $Q_2 = \frac{1}{\sqrt{2}} I_y$ and $Q_3 = \frac{1}{\sqrt{2}} I_z$, etc.). In this manner we can find equations describing the recovery of all of the different elements of the density operator to their equilibrium values:

$$\frac{dc_1}{dt} = -\omega_Q c_6 - c_1 / T_{2e}$$
$$\frac{dc_6}{dt} = \omega_Q c_1 - c_6 / T_{2e} \tag{44}$$

and

$$\frac{dc_2}{dt} = \omega_Q c_5 - c_2 / T_{2e} \tag{45}$$

$$\frac{dc_5}{dt} = -\omega_Q c_2 - c_5 / T_{2e}$$

describe the precession of the x and y components of magnetization under the static part of the quadrupolar Hamiltonian and their relaxation due to the fluctuation part. The transverse relaxation rate $1/T_{2e}$ is given by the expression (Jeffrey, 1981; Davis, 1983)

$$\frac{1}{T_{2e}} = \frac{1}{8}\left[\frac{eQ}{\hbar}\right]^2 \left[\frac{3}{2} J_0(0) + \frac{3}{2} J_1(\omega_0) + J_2(2\omega_0)\right] \tag{46}$$

where the spectral densities, $J_m(m\omega)$, $m = 0, 1, 2$ are defined in the appendix. These spectral densities describe the intensity of fluctuations in the Hamiltonian at frequency $m\omega$ [see, e.g., Slichter (1990)].

The recovery of the z-magnetization, or Zeeman polarization, governed by the spin-lattice relaxation rate $1/T_{1z}$, is described by the equation

$$\frac{dc_3}{dt} = -\left[c_3 - c_3^0\right] / T_{1z} \tag{47}$$

where c_3^0 is the equilibrium value of Zeeman polarization. The spin lattice relaxation rate, in terms of spectral densities, is

$$\frac{1}{T_{1z}} = \frac{1}{8}\left[\frac{eQ}{\hbar}\right]^2 \left[J_1(\omega_0) + 4J_2(2\omega_0)\right] \tag{48}$$

A third quantity, the quadrupolar polarization, which is proportional to $Q_4 = \frac{1}{\sqrt{6}}(3I_z^2 - 2)$ (Davis, 1991), follows the equation

$$\frac{dc_4}{dt} = -c_4 / T_{1Q} \tag{49}$$

with

$$\frac{1}{T_{1Q}} = \frac{1}{8}\left[\frac{eQ}{\hbar}\right]^2 3J_1(\omega_0) \tag{50}$$

Finally, there are the two equations describing double quantum coherence (see Vega and Pines, 1977 and Davis, 1991).

$$\frac{dc_7}{dt} = -c_7 / T_{DQ}$$

$$\frac{dc_8}{dt} = -c_8 / T_{DQ} \tag{51}$$

where the double quantum coherence relaxation rate is

$$\frac{1}{T_{DQ}} = \frac{1}{8}\left[\frac{eQ}{\hbar}\right]^2 \left[J_1(\omega_0) + 2J_2(2\omega_0)\right] \tag{52}$$

Using different preparation pulse sequences, it is possible to prepare any of these "coherences" — transverse magnetization or single quantum coherences (and the associated zero-quantum coherences with which they precess under the influence of the quadrupolar Hamiltonian [Equations 44 and 45]), Zeeman polarization or longitudinal magnetization, quadrupolar polarization, and double quantum coherence — and to measure their relaxation rates (Davis, 1991). In doing so one measures the spectral densities $J_0(0)$, $J_1(\omega_0)$, and $J_2(2\omega_0)$. By performing these experiments as a function of temperature and frequency and as a function of the orientation of the motional symmetry axis (director or bilayer normal) relative to the external magnetic field, one can sometimes deduce the character and time scales of the molecular reorientations responsible for the fluctuations in the Hamiltonian, as will be discussed below.

D. THE DYNAMICS OF CHOLESTEROL

The motion of the rigid moiety of cholesterol in the "fluid" phase phospholipid bilayer reflects the axial symmetry of the system. The bilayer normal provides an axis of symmetry for the molecular reorientations. One anticipates that the cholesterol molecule will undergo some sort of rapid axially symmetric reorientation about its long axis and that the orientation of this axis may fluctuate relative to the bilayer normal. ^{2}H NMR spin lattice relaxation measurements should be able to test and quantify this dynamical model. Figure 8 shows partially relaxed ^{2}H NMR spectra from both powder and oriented samples of cholesterol in DMPC bilayers at 30°C from an inversion recovery T_{1z} experiment (Bonmatin et al., 1990).

The temperature dependences of the spin lattice relaxation time T_{1z} of specifically deuterium labeled cholesterol at high concentrations in DMPC (Weisz et al., 1992), in DPPC (Bonmatin et al., 1990), in egg yolk PC (Taylor et al., 1982), and in a magnetically oriented nematic liquid crystal (Davis, 1988) all exhibit a minimum relaxation time near room temperature (at approximately 30°C for cholesterol in DPPC, DMPC, and egg PC, and at about 10°C for cholesterol in the liquid crystal ZLI1132 [E. Merck, Darmstadt]). Figure 9 compares the T_{1z} data for three of these systems. The similarity among all of these data sets is striking. It is clear that the orientational order and dynamics of cholesterol are similar in all three cases (Davis, 1988), leading one to speculate that the most important physical factor determining the character of the molecu-

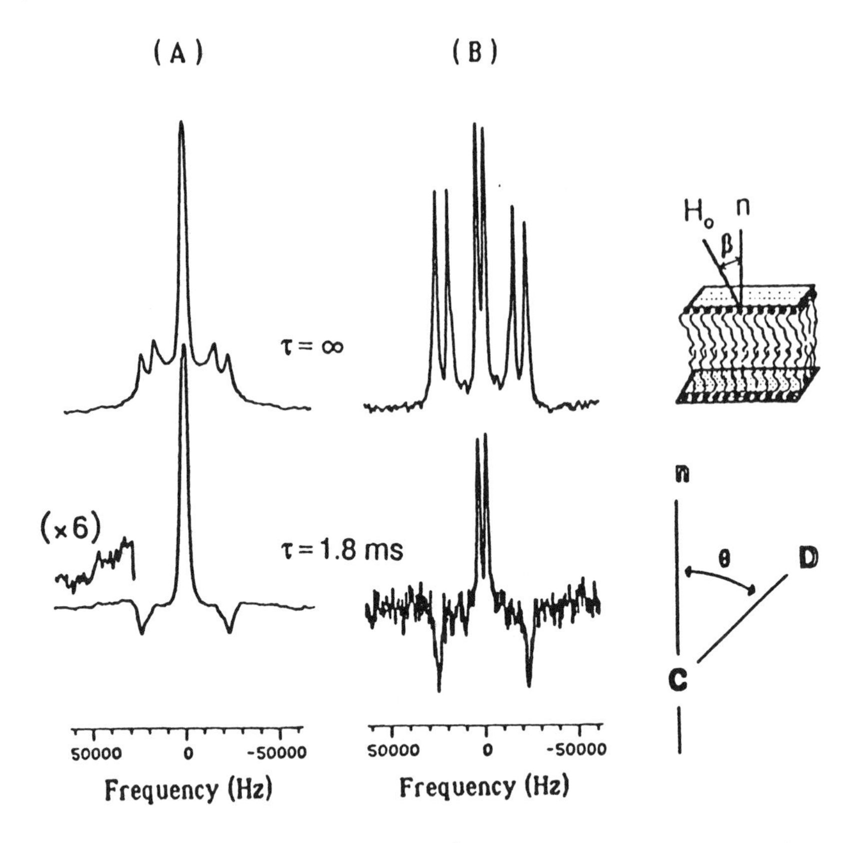

FIGURE 8. Partially relaxed spectra of [2,2,4,4,6-$^2H_5$] cholesterol in 1,2-dimyristoyl-*sn*-glycero-3-phosphocholine (DMPC) at 30°C, from inversion recovery T_{1z} experiments: (A) unoriented multilamellar dispersions of 30 mol% cholesterol in DMPC; (B) multibilayers oriented at $\beta = 90°$ relative to the external magnetic field, H_0, 50 mol% cholesterol in DMPC. The sketch of the lipid bilayer on the right defines the direction of bilayer normal relative to the magnetic field and defines the angles β and θ used in the original paper. (Reproduced from Bonmatin et al., 1990. *J. Am. Chem. Soc.*, 112:1697. With permission.)

lar dynamics seems to be the presence of an axially symmetric orienting potential.

1. Model for Diffusion about the Long Axis

In our simple dynamical model the fluctuations in orientation described by the spectral densities defined in Equation A16 of the appendix are due to fluctuations in the angles specifying the orientation of the C–^{2}H bond relative to the motional symmetry axis. The lattice functions $F_{2,m}$ in Equation A14 are the components of the electric field gradient tensor in the laboratory frame. In terms of the electric field gradient tensor's principal axis components, $F^P_{2,m}$, we have, in the laboratory frame,

$$F^L_{2,m} = \sqrt{\frac{4\pi}{5}} F^P_{2,0} \sum_{m'} Y^*_{2,m'}(\beta,\alpha) d^{(2)}_{m'm}(\beta') \tag{53}$$

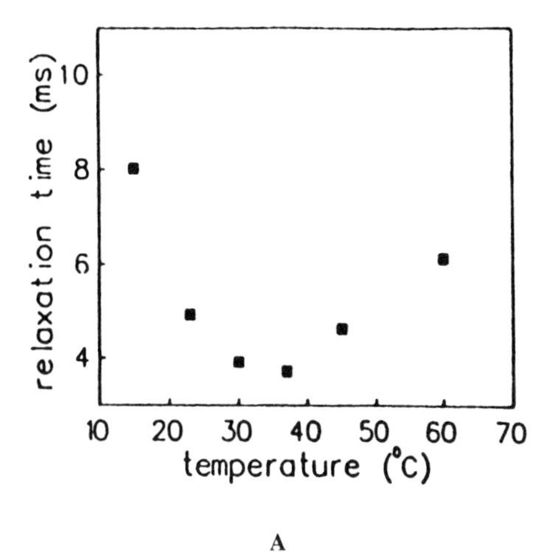

A

FIGURE 9. The temperature dependence of the spin-lattice relaxation time T_{1z} of (A) 50 mol% [α-7-$^2H_1$]cholesterol in 1,2-dipalmitoyl-*sn*-glycero-3-phosphocholine (DPPC) at 30.7 MHz (reproduced from Bonmatin et al., 1990. *J. Am. Chem. Soc.*, 112:1697. With permission.) The temperature dependence of T_{12} of (B) [7,7-$^2H_2$]cholesterol in egg yolk PC at 46 MHz (reproduced from Taylor et al., 1982. *Chem. Phys. Lipids*, 31:359. With permission.); and (C) the 6-position of [2,2,3,4,4,6-$^2H_6$]cholesterol in an oriented nematic liquid crystal ZLI1132, at 41 MHz. (Reproduced from Davis, 1988. In *Physics of NMR Spectroscopy in Biology and Medicine*, p. 302. With permission.)

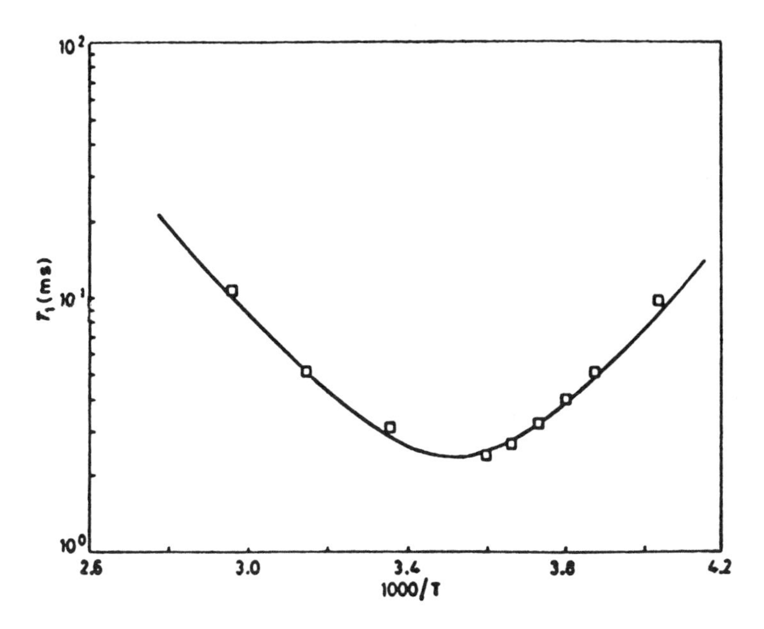

FIGURE 9B.

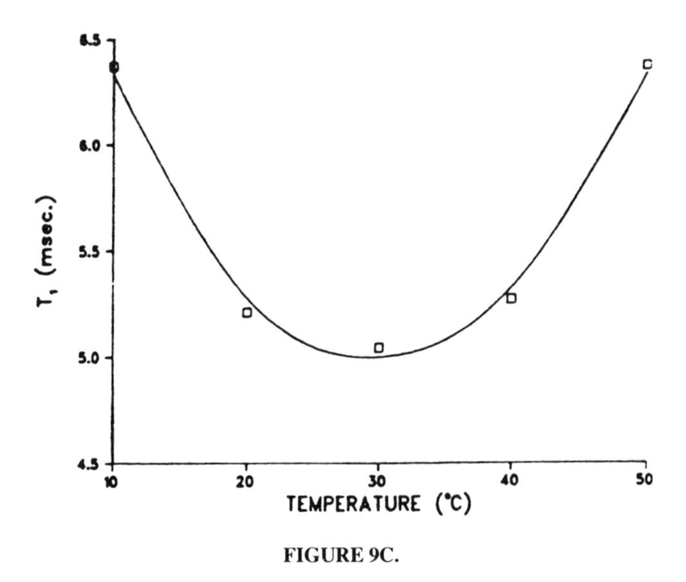

FIGURE 9C.

where we have assumed that the electric field gradient tensor is axially symmetric so that only $F^P_{2,0}$ is nonzero in the principal axis system. The $d^{(2)}_{ij}$ are the reduced Wigner rotation matrices (see e.g., Zare, 1988), and the Y^*_{2m} are the complex conjugates of the spherical harmonic functions. The Euler angles $(0\beta'0)$ specify the orientation of the axially symmetric director with respect to the external magnetic field, and the angles (β, α) give the orientation of the electric field gradient tensor's principal axis system relative to the director. The time dependence of the lattice functions is due to the fluctuations in the angles $\beta(t)$ and $\alpha(t)$. In this case, the spectral densities are (Davis, 1983)

$$J_m(m\omega) = (-1)^m \frac{4\pi}{5}\left(F^P_{2,0}\right)^2 \times \sum_{m'}\sum_{m''} d^{(2)}_{m'm}(\beta')d^{(2)}_{m''-m}(\beta') \times \int_{-\infty}^{\infty} d\tau e^{im\omega\tau} \\ \{\overline{Y^*_{2m''}(\beta(t),\alpha(t))Y^*_{2m'}(\beta(t-\tau),\alpha(t-\tau))} - \overline{(Y^*_{2m''}(\beta,\alpha))}^2\} \tag{54}$$

If we assume for simplicity that the fluctuations are only in the angle α, i.e., that β remains fixed, then the correlation function in Equation 54 is greatly simplified, and the spectral densities become

$$J_m(m\omega) = (-1)^m \frac{4\pi}{5}\left(F^P_{2,0}\right)^2 \sum_{m'} d^{(2)}_{m'm}(\beta')d^{(2)}_{-m'-m}(\beta') \\ \times Y^*_{2-m'}(\beta,0)Y^*_{2m'}(\beta,0)\int_{-\infty}^{\infty} d\tau e^{im\omega\tau} g_{m'}(\tau) \tag{55}$$

where the correlation function is

$$g_{m'}(\tau) = \overline{e^{im'\alpha(t)}e^{-im'\alpha(t-\tau)}} \tag{56}$$

In the cases presented in Figure 9, the director or axis of motional symmetry is aligned perpendicular to the magnetic field. Thus, the angle $\beta' = 90°$. Using this value in the reduced Wigner rotation matrices of Equation 55 we obtain

$$J_1(\omega) = \frac{3}{4}\left(F_{20}^{P}\right)^2\left[\sin^2\beta\cos^2\beta\left(\frac{2\tau_1}{1+\omega^2\tau_1^2}\right)+\frac{1}{4}\sin^4\beta\left(\frac{2\tau_2}{1+\omega^2\tau_2^2}\right)\right] \tag{57}$$

and

$$J_2(2\omega) = \frac{3}{4}\left(F_{20}^{P}\right)^2\left[\sin^2\beta\cos^2\beta\left(\frac{2\tau_1}{1+4\omega^2\tau_1^2}\right)+\frac{1}{16}\sin^4\beta\left(\frac{2\tau_2}{1+4\omega^2\tau_2^2}\right)\right] \tag{58}$$

where we have explicitly performed the Fourier transformation of the correlation functions for axial diffusion

$$g_{m'}(\tau) = e^{-(m')^2 D\tau} \tag{59}$$

(Woessner, 1962; Wallach, 1967; Wittebort and Szabo, 1978; Brown, 1979; Torchia and Szabo, 1982; Davis, 1983). Then, with

$$j\left(\tau_{m'}, m\omega\right) = \int_{-\infty}^{\infty} d\tau e^{im\omega\tau} g_{m'}(\tau) = \frac{2\tau_{m'}}{1+m^2\omega^2\tau_{m'}^2} \tag{60}$$

where $\tau_{m'} = 1/[(m')^2 D]$ and D is the coefficient for axial diffusion, this leads us to the expression for $1/T_{1z}$ for the special case of the diffusion axis aligned perpendicular to the external magnetic field

$$\begin{aligned} \frac{1}{T_{1z}} = \frac{1}{8}\left[\frac{3e^2qQ}{4\hbar}\right]^2 \Big[& 2\sin^2\beta\cos^2\beta \cdot j(\tau_1,\omega_0) \\ & + \frac{1}{2}\sin^4\beta \cdot j(\tau_2,\omega_0) + 8\sin^2\beta\cos^2\beta \cdot j(\tau_1, 2\omega_0) \\ & + \frac{1}{2}\sin^4\beta \cdot j(\tau_2, 2\omega_0)\Big] \end{aligned} \tag{61}$$

as given in Torchia and Szabo (1982).

The temperature dependence of the spin lattice relaxation rate in this simple model arises from the temperature dependence of the correlation time $\tau_c(T) = \tau_c^0 \exp(E_A/RT)$ where R is the gas constant and E_A is the activation energy for axial diffusion. The location of the T_{1z} minimum, and the value at the minimum, are functions of the orientation of the C—^{2}H bond relative to the diffusion axis. If $\beta = 90°$, then the T_{1z} minimum occurs when the correlation time $\tau_2 = (4D)^{-1} = 0.812/\omega_0$ (or $\tau_1 = D^{-1} = 3.248/\omega_0$), while for an isotropically reorienting molecule the minimum would be at τ_c=0.62/ω_0 (Abragam, 1961; Davis, 1983). For a C–^{2}H bond oriented at an angle $\beta = 90°$ relative to the diffusion axis (which itself is perpendicular to the magnetic field), the minimum value of $T_{1z}{}^{\min}$ is given by

$$\frac{1}{T_{1z}^{\min}} = \frac{1}{8}\left[\frac{3e^2qQ}{4\hbar}\right]^2 \frac{0.692}{\omega_0} \tag{62}$$

At a frequency $\omega_0 = 2\pi \times 30.7$ MHz, as in Bonmatin et al. (1990), simple axial diffusion gives $T_{1z}^{\min} \approx 3.46$ ms, which compares favorably with the experimental minimum value of $T_{1z}^{\exp} \approx 3.7$ ms reported for C_7-labeled cholesterol [the C_7–^{2}H bond is oriented at $\beta = 77°$ relative to the diffusion axis (Bonmatin et al., 1990)]. While it appears that there are no simple expressions for either the orientation dependence of the position of the minimum nor that of the value of T_{1z} at the minimum, it is a straightforward matter to calculate these numerically for any value of β and as a function of the orientation (β') of the diffusion axis relative to the magnetic field, using Equations 48, 55, and 59. In the diffusion model, for $\beta = 77°$, the T_{1z} minimum occurs at $\tau_1 = 2.583/\omega_0$, and the value of T_{1z} at the minimum is $T_{1z}^{\min} \approx 3.25$ ms.

2. Model for Three-Fold Jumps about the Long Axis

It is equally straightforward to consider a model for the motion of cholesterol where the molecule jumps quickly between three equally probable sites whose orientations differ only in the angle α, which takes values of 0, 120, and 240°. The expression for the spectral densities, Equation 54, still holds, but the correlation function, Equation 56, is different. As described in Torchia and Szabo (1982), the correlation function for this case is given by

$$g_{m',m''}(\tau) = \begin{cases} 1 & \text{if } m'' = m' = 0 \\ e^{-\tau/\tau_c} & \text{if } m'' = m' \neq 0, \text{ or} \\ 0 & \text{otherwise} \end{cases} \tag{63}$$

Here we have kept both indices on the correlation function because of the threefold nature of the motion. The correlation time $\tau_c = (3k)^{-1}$, where k is the rate constant for jumping between neighboring sites. Models having more than three sites give completely analogous results (Torchia and Szabo, 1982; Bonmatin et al., 1990).

For the case treated above, where the motional symmetry axis is aligned perpendicular to the magnetic field ($\beta' = 90°$), the spectral densities are

$$J_1(\omega) = \frac{3}{4}\left(F_{2,0}^P\right)^2\left[\sin^2\beta\cos^2\beta + \frac{1}{4}\sin^4\beta\right]\left(\frac{2\tau_c}{1+\omega^2\tau_c^2}\right) \tag{64}$$

and

$$J_2(2\omega)=\frac{3}{4}\left(F^{P}_{2,0}\right)^2\left[\sin^2\beta\cos^2\beta+\frac{1}{16}\sin^4\beta\right]\left(\frac{2\tau_c}{1+4\omega^2\tau_c^2}\right) \tag{65}$$

so that the spin lattice relaxation rate for threefold jumps is, from Equation 48,

$$\frac{1}{T_{1z}}=\frac{1}{8}\left[\frac{3e^2qQ}{4\hbar}\right]^2\left\{\left(2\sin^2\beta\cos^2\beta+\frac{1}{2}\sin^4\beta\right)\cdot j(\omega_0)\right.$$
$$\left.+\left(8\sin^2\beta\cos^2\beta+\frac{1}{2}\sin^4\beta\right)\cdot j(2\omega_0)\right\} \tag{66}$$

where

$$j(\tau_c,\omega)=\int_{-\infty}^{\infty}d\tau e^{-iw\tau}e^{-\tau/\tau_c}=\frac{2\tau_c}{1+\omega^2\tau_c^2} \tag{67}$$

3. Comparison of the Two Models near the T_{1z} Minimum

Figure 10a compares the predicted values of $\omega_0\tau_c$ at the T_{1z} minimum as a function of β, the angle between the C–^{2}H bond and the motional symmetry axis (bilayer normal) for the two models. The solid curve is for the diffusion model with $\tau_c = \tau_1 = D^{-1}$, and the dashed curve is for the threefold jump model with $\tau_c=(3_k)^{-1}$. Our choice of whether to choose τ_1 or τ_2 from the diffusion model for comparison with τ_c from the threefold jump model may seem arbitrary; however, τ_1 and τ_2 are directly related to each other through the axial diffusion constant D. The relative changes of $\omega_0\tau_c$ with β are significant. The change in $\omega_0\tau_1$ by a factor somewhat greater than 4 as β is changed from 90° to 60° for the diffusion model implies a change in the expected position of the T_{1z} minimum from about 305 K for β = 90° to about 345 K for β = 60° [this comparison assumes that $\tau_c = \tau_c^0\exp(E_A/RT)$ and uses the activation energy, $E_A = 30 \pm 4$ kJ/mol, and the temperature of the T_{1z} minimum of 310 K for the C_7-labeled cholesterol, as reported by Bonmatin et al. (1990)]. The same sort of analysis for the threefold jump model results in a change of temperature for the T_{1z} minimum from 308 K for β = 90° to 315 K for β = 60°. The large change that occurs for the diffusion model on changing bond orientation is due to the shifting of the predominant relaxation mechanism from those that modulate the terms in $m' = \pm 2$ in Equation 55, corresponding to the shorter correlation time τ_2 (near β = 90°, cf. Equation 57), to those that modulate the terms in $m' = \pm 1$, corresponding to τ_1 (at smaller values of β). Simply speaking, since both correlation times in the diffusion model are due to the same motional mechanism, axial diffusion, the shorter correlation time (τ_2) will have its maximum effectiveness (when $\omega_0\tau c$ is optimal) at lower temperatures than will the longer correlation time (τ_1). Thus, if the molecules are essentially undergoing axial diffusion about the bilayer normal, we expect to find a large variation in the temperature of the T_{1z} minimum as we look at bonds with different orientations relative to the diffusion axis. On the other hand, if the

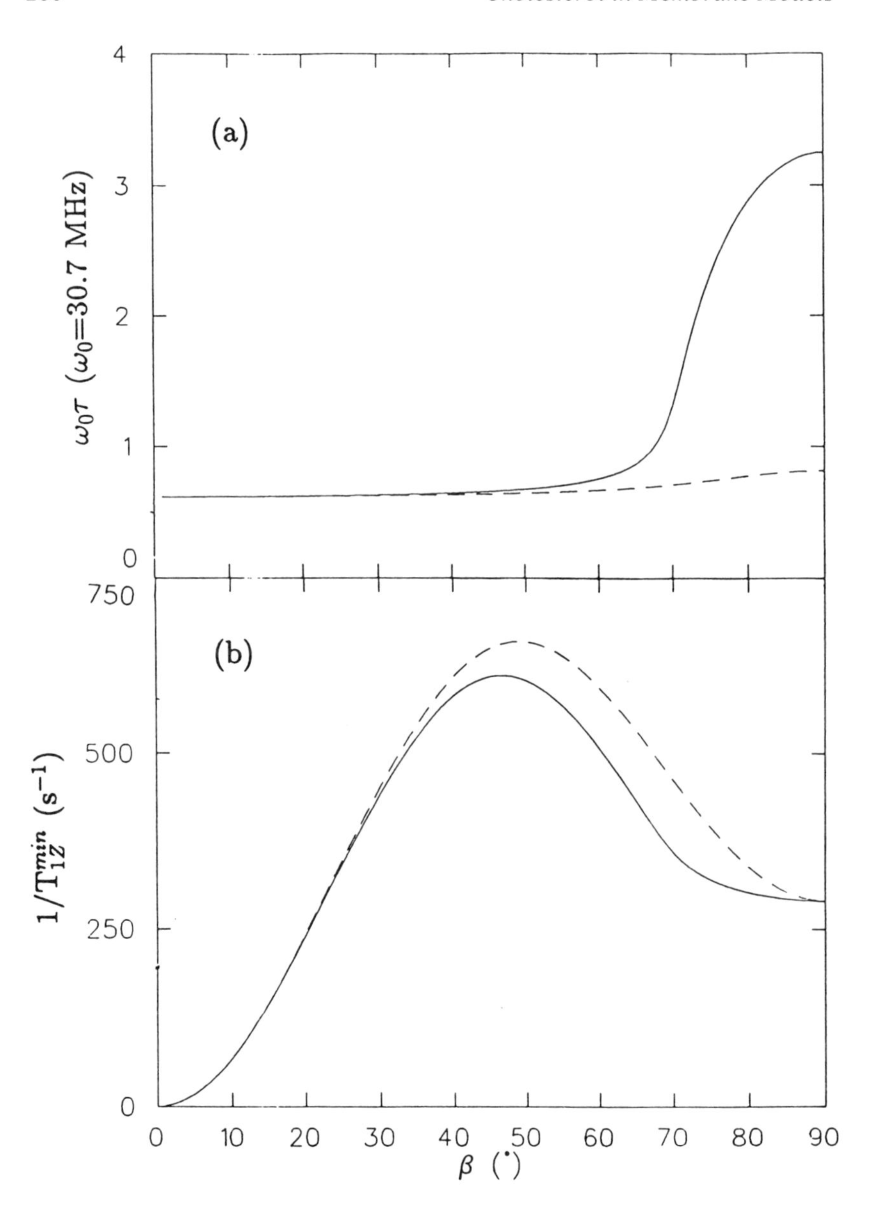

FIGURE 10. A comparison of the bond orientation dependence of the T_{1z} minimum for diffusion and three-site jump models: (a) The variation of the T_{1z} minimum, given as a factor times $\omega_0 \cdot \tau$ and (b) the value of $1/T_{1z}$ at the minimum, as functions of the angle β between the C–^{2}H bond and the axis of motional symmetry. The solid curves are calculated assuming axial diffusion; the dashed curves are for three-site jumps.

molecule undergoes threefold jumps, the variation in temperature of the minimum is much less sensitive to bond orientation.

The predicted values of $(1/T_{1z}^{min})$ are graphed as a function of β in Figure 10b. Both models exhibit roughly the same sort of dependence on β, with some differences in $(1/T_{1z}^{min})$ for values of β near 50°. From these calculations it appears that the bond orientation dependence of the temperature at the minimum is a better means of distinguishing between these two motional models than is a comparison of the actual values of $(1/T_{1z}^{min})$. Unfortunately, sufficient data on the location of the minimum for different bond orientations are not yet available.

Fortunately, T_{1z} (and, indeed, T_{1Q}, see Equation 50) are themselves strongly orientation dependent, and this dependence is different, in general, for the two different models. However, for the simple case of $\beta = 90°$, the two different motional models give the identical orientation (β') dependence, namely,

$$\frac{1}{T_{1z}} = \frac{1}{16}\left[\frac{3e^2qQ}{4\hbar}\right]^2 \left[(1-\cos^4\beta')\frac{2\tau}{1+\omega^2\tau^2} + (\cos^4\beta' + 6\cos^2\beta' + 1)\frac{2\tau}{1+4\omega^2\tau^2}\right] \tag{68}$$

(Torchia and Szabo, 1982; Davis, 1983), where $\tau = (4D)^{-1}$ for diffusion and $\tau = (3k)^{-1}$ for the threefold jump. For other values of β, where fortunately the predicted orientation dependences are different, it is more convenient to perform the calculations numerically, as was done by Bonmatin et al. (1990). The results of these simulations for the two models are compared to the experimental orientation dependences in Figure 11. They reported T_{1z} data for three different cholesterol labels, with bond angles of $\beta = 77$, 68, and 56° (for the C_7, C_2 [C_4], and C_6 positions. Bonmatin et al. (1990) used θ instead of β for this angle; their "β" corresponds to our "β'"). The bond orientation angles are indicated on the figure. The solid lines are the predictions of the threefold jump model, while the dotted lines are the predicted orientation dependences for the diffusion model using two different values of the axial diffusion constant. The jump model gives a somewhat better fit to the experimental data; but if only these data are considered, either model could satisfactorily explain the data. The values of the diffusion constant and jump rate used in these simulations all give correlation times close to the value at the minimum: with $D = 3.2 \times 10^8\ s^{-1}$, $\tau_1 = 3.125 \times 10^{-9}\ s$ and $\tau_2 = 7.8 \times 10^{-10}\ s$, for $D = 3.0 \times 10^7\ s^{-1}$, $\tau_1 = 3.33 \times 10^{-8}\ s$ and $\tau_2 = 8.33 \times 10^{-9}\ s$; for $k = 3.2 \times 10^7\ s^{-1}$, $\tau_c = 1.04 \times 10^{-8}\ s$. These values should be compared to the value of $\omega_0\tau$ at the T_{1z} minimum for $\beta = 77°$, namely: for diffusion, $\tau_2 = 3.35 \times 10^{-9}\ s = \tau_1/4$ and for 3-jump, $\tau_c = 3.88 \times 10^{-9}$. At this value of β, both τ_1 and τ_2 are important in the diffusion model, but τ_2 probably dominates. Thus, the value of k used by Bonmatin et al. (1990) is on the long correlation time side of

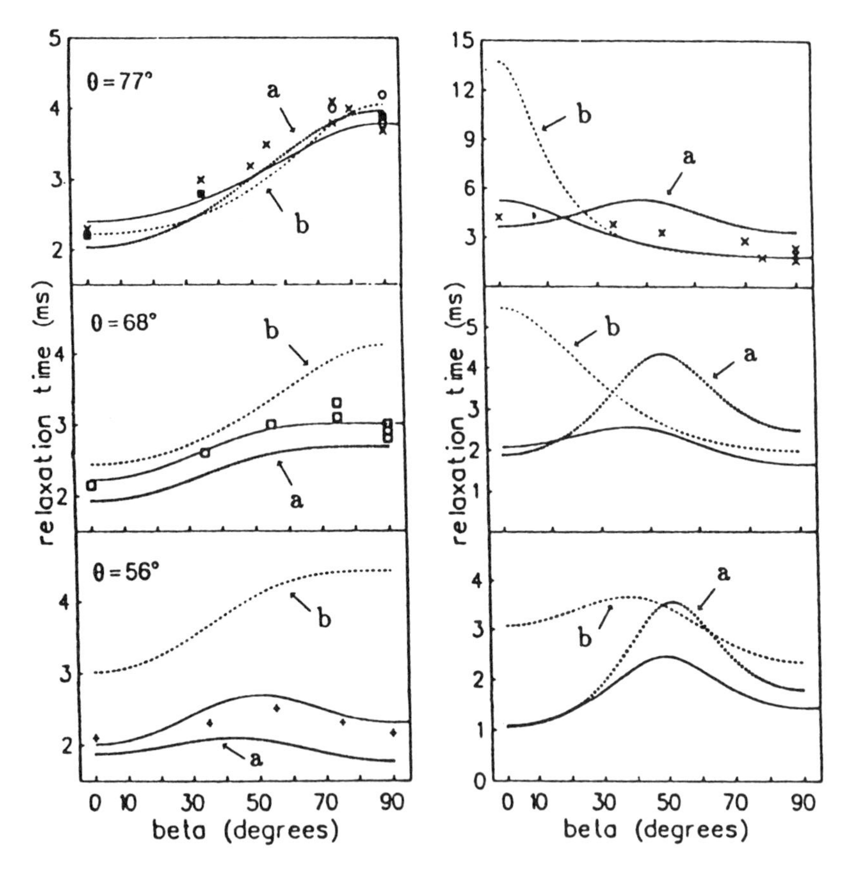

FIGURE 11. The orientation dependence of T_{1z} (left) and T_{1Q} (right) at 30.7 MHz from oriented multibilayers of 50 mol% cholesterol in DPPC at 30°C. The angle "beta" here is the angle between the bilayer normal and the external magnetic field (as defined by Bonmatin et al., 1990.) Cholesterol is labeled at the C_7 position (θ = 75°): "×"; at the C_2 and C_4 axial positions (θ = 75–77°) "O"; at the C_2 and C_4 equatorial positions (θ = 68°) "❑"; and at the C_6 position (θ = 56°): "+". The solid curves are calculated relaxation times assuming a three-site jump model with jump rate $k = 3.2 \times 10^7\ s^{-1}$. The other curves are for diffusion with D = 3.2 × 108 s^{-1} (labeled "a") and for diffusion with $D = 3.0 \times 10^7\ s^{-1}$ (labeled "b"). (Reproduced from Bonmatin et al., 1990. *J. Am. Chem. Soc.*, 112:1697. With permission.)

the predicted T_{1z} minimum for the jump model, and both the values of D used place them on the short correlation time side of the minimum predicted by the diffusion model. The data in Figure 10 are at 30°C, which is just below the reported temperature of 37°C for the minimum. Thus, since raising the temperature by 7°C decreases T_{1z}, as in Figure 9, the system must be on the long correlation time side of the minimum at 30°C, and it appears that the threefold jump model is a slightly better candidate for describing the motion of cholesterol in lipid bilayers. Clearly more experimentation would be helpful. In particular, one should look for the bond orientation dependence of the temperature of the T_{1z} minimum.

III. THE EFFECT OF CHOLESTEROL ON PHOSPHOLIPIDS

A. PHOSPHOLIPID ORIENTATIONAL ORDER

Perhaps the most striking effect of cholesterol on membrane physical properties is the large increase in phospholipid chain order that occurs when cholesterol is added. This effect has been widely observed with a variety of physical techniques, including, in particular, electron paramagnetic resonance, time-resolved fluorescence depolarization, and nuclear magnetic resonance [see Vist and Davis (1990) for a summary of the earlier work in this area]. ^{2}H NMR has provided probably the most detailed and quantitative description of phospholipid chain orientational order and of the influence of cholesterol on the order* (Oldfield, 1971; Stockton et al., 1974, 1976; Haberkorn et al., 1977; Jacobs and Oldfield, 1979; Davis et al., 1980; Blume and Griffin, 1982; Maraviglia et al., 1982; Vist and Davis, 1990; Sankaram and Thompson, 1990; Thewalt and Bloom, 1992). Figure 12 shows the influence of 0, 10, and 15 mol% cholesterol on the ^{2}H NMR spectra of oriented bilayers of di-(2,2-$^2H_2$)-myristoyl PC in the "fluid" or liquid crystalline phase. The observed increase in quadrupolar splitting is due primarily to a reduction in the extent of *gauche-trans* isomerization along the phospholipid hydrocarbon chain (see below). This effect occurs for all chain positions, as shown by the ^{2}H powder pattern spectra of chain perdeuterated DPPC/cholesterol mixtures in Figure 13 (Vist and Davis, 1990). These spectra were all taken at a temperature of 38.0°C, just above the hydrocarbon chain melting transition that occurs at 37.75°C for chain perdeuterated DPPC in excess water. [The transition occurs a few degrees lower in temperature for perdeuterated chain lipids than for unlabeled lipids, e.g., for DPPC the phase transition occurs at about 41.2°C (Davis, 1979).] Studies of the membranes of *Acholeplasma laidlawii* grown on specifically deuterated fatty acids and varying amounts of cholesterol have shown that the entire chain is affected, with perhaps slightly larger fractional increases in quadrupolar splittings for positions near the terminal methyl groups (Rance et al., 1982).

The spectra in Figure 13, arising from chain perdeuterated lipids, consist of a large number of overlapping axially symmetric powder pattern spectra like those in Figure 5b: one for each deuteron on the lipid. Since the two deuterons at each chain position are usually equivalent (except for the 2-position of the *sn*-2 chain (Seelig and Seelig, 1974), as demonstrated in Figures 12a and 12b), there are essentially 31 separate powder patterns in each of the spectra of Figure 13. The flexibility gradient of the hydrocarbon chain results in different quadrupolar splittings for different chain positions; those nearer the center of the bilayer in general having smaller quadrupolar splittings. In addition, many of the positions near the lipid/water interface have nearly the same quadrupolar splitting (see also, Seelig and Seelig, 1974; and Davis, 1979, 1983). Careful examination of

* See the references cited here and Davis (1983, 1986, 1991) and Griffin (1981) for descriptions of the technical aspects of the application of ^{2}H NMR to studies of molecular orientational order and dynamics.

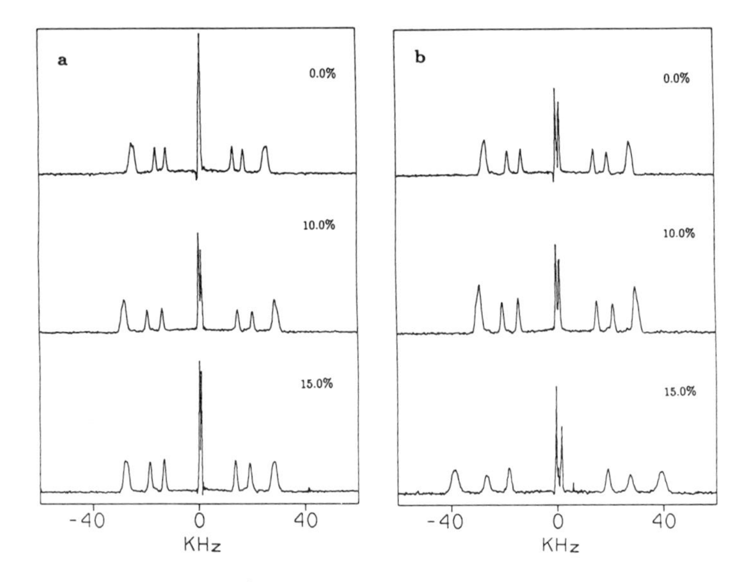

FIGURE 12. ^2H NMR spectra at 55°C of di-[2,2-$^2H_2$]myristoyl PC/cholesterol/buffer (the water used in the buffer is 5 mol% deuterated) mixtures oriented with the bilayer normal parallel to the external magnetic field. (a) At high hydration levels, the 0 mol% cholesterol sample contains 62 waters per DMPC; the 10 mol% cholesterol sample has 23 waters per DMPC; and the 15 mol% cholesterol sample has 57 waters per DMPC. (b) At lower hydration levels, the 0 mol% cholesterol sample has 18 waters per DMPC; the 10 mol% sample has 14 waters per DMPC and the 15 mol% sample has 19 waters per DMPC. Water concentrations were determined from the relative areas of the water peaks (those with the smallest quadrupolar splittings) and the DMPC peaks.

the spectra of Figure 13 reveals that all of the quadrupolar splittings in these spectra increase monotonically with cholesterol concentration. Recall from Equation 25 that the quadrupolar splitting is proportional to the order parameter for the C–^2H bond. Thus, there is an increase in hydrocarbon chain orientational order with increasing concentration of cholesterol up to at least 25 mol% cholesterol.

For a more quantitative picture of the effect of cholesterol on chain order, it is useful to introduce the moments of the ^2H powder pattern spectrum (Davis, 1979).

$$M_n = \frac{1}{A}\int_{-\infty}^{\infty} d\omega(\omega - \omega_0)^n g(\omega - \omega_0) \tag{69}$$

where $g(\omega - \omega_0)$ is the spectrum, A is the total area of the spectrum, and the integral is over all frequencies. Since for ^2H in C–^2H bonds we can usually safely neglect the chemical shift and dipole-dipole contributions to the line shape, these spectra are symmetric about ω_0 (Davis, 1979). In this case we can define the *odd* moments of the "half" spectrum, in particular, the first moment,

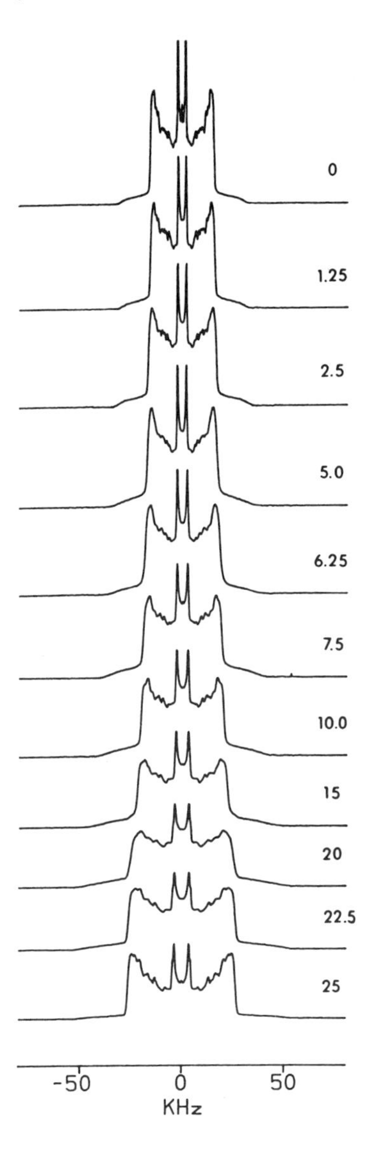

FIGURE 13. ^{2}H NMR spectra of chain perdeuterated DPPC/cholesterol mixtures at 38°C, at 12 different cholesterol concentrations (molar concentrations in mol% indicated on the figure). (Reproduced from Vist and Davis, 1990. *Biochemistry,* 29:451. With permission.)

$$M_1 = \frac{1}{A/2} \int_0^{\infty} d\omega' \omega' g(\omega') \tag{70}$$

where we have used $\omega' = \omega - \omega_0$. For spectra such as those shown in Figure 13, which consist of a superposition of axially symmetric powder patterns, M_1 is proportional to the average quadrupolar splitting

$$M_1 = \frac{2}{3\sqrt{3}} < \delta \upsilon_Q > = \frac{2}{3\sqrt{3}} < S_{C-{}^2H} > \tag{71}$$

as well as to the average C–^{2}H bond order parameter (Davis, 1979).

Figure 14 shows a histogram of M_1 as a function of temperature and cholesterol concentration. In the fluid phase, above about 37.75°C, the value of M_1 increases with cholesterol concentration by approximately a factor of 2 as cholesterol concentration is increased from 0 to 25 mol%. It is also clear that, in the fluid phase, the chain order decreases as temperature increases — for all cholesterol concentrations. The discontinuous increase in M_1 observed near 37°C at low cholesterol concentration is due to the hydrocarbon chain phase transition. Below this temperature the pure lipid/water mixture is in the "gel" phase where the *gauche-trans* isomerization rate of the chains is dramatically reduced from what it was in the fluid phase (see, e.g., Gaber and Peticolas, 1977; Mendelsohn et al., 1989).

The ^{2}H NMR spectra are very sensitive to this phase transition, as illustrated in Figure 15. In part (a) of the figure, the spectra of the pure DPPC/water mixture are shown for a narrow range of temperatures near the phase transition. The spectrum at 38°C is characteristic of the lipid bilayer fluid phase, as discussed above. At 37.5°C there has been a dramatic broadening of the spectrum, and much of the detail due to the fluid phase hydrocarbon chain flexibility gradient has been washed out. This is largely a result of two effects: the strong decrease in the rate of *gauche-trans* isomerization, and the dramatic increase in the time scales of the other motions in the gel phase. These spectra are characteristic of ^{2}H NMR spectra in the intermediate time scale regime where motions have correlation times of the same order of magnitude as the spectroscopic time scale (roughly $1/\omega_Q$) (Davis, 1979, 1983; Griffin, 1981).

At 37.75°C, in Figure 15a, the spectrum consists of a superposition of two components: one characteristic of the fluid phase and another characteristic of the gel phase. The large difference between the two spectra, characteristic of the two different phases in coexistence at this one temperature, suggests the possibility of using ^{2}H NMR to study the phase equilibria of lipid bilayers (Bienvenue et al., 1982; Huschilt et al., 1985; Morrow et al., 1986; Morrow and Davis, 1988; Vist and Davis, 1990; Morrow et al., 1992). Figure 15b shows a similar series of spectra for DPPC with 2.5 mol% cholesterol. At this low concentration of cholesterol we observe only a slight reduction of the phase transition temperature. However, as we shall see later, cholesterol has a profound influence on the phase equilibria of this model membrane system.

At high cholesterol concentrations the gel-to-liquid crystalline phase transition is strongly suppressed (see below), and the lipid bilayer retains many of the properties of the fluid phase well below the temperature of the normal transition to the gel phase. These properties include the rapid lateral diffusion within the bilayer plane and the rapid reorientation of the lipid molecules about the bilayer normal. This highly ordered fluid phase is characteristic of, for example, the

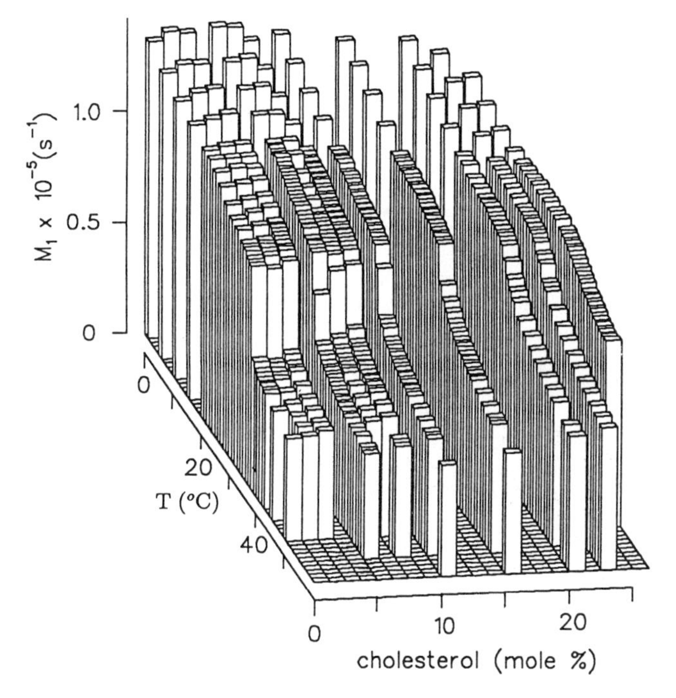

FIGURE 14. Histogram of the first moment of the ^2H NMR spectra of chain perdeuterated DPPC/cholesterol mixtures as a function of temperature (°C) and cholesterol concentration (mol%). (Reproduced from Vist and Davis, 1990. *Biochemistry,* 29:451. With permission.)

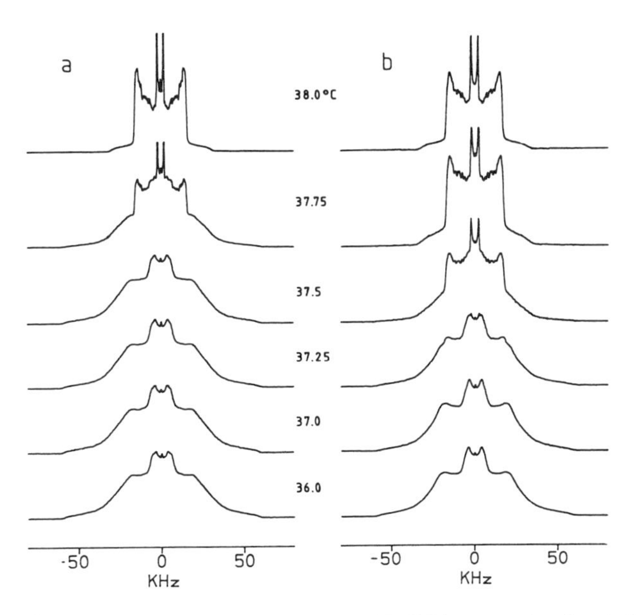

FIGURE 15. ^2H NMR spectra of (a) chain perdeuterated DPPC in excess phosphate buffer near the hydrocarbon chain melting transition, and (b) with 2.5 mol% cholesterol added. (Reproduced from Vist and Davis, 1990. *Biochemistry,* 29:451. With permission.)

human erythrocyte membrane under normal physiological conditions (Davis et al., 1979; Maraviglia et al., 1982).

B. PHOSPHOLIPID DYNAMICS

The description of phospholipid dynamics is considerably more difficult than that of the rigid moiety of cholesterol presented earlier. This is due largely to the large number of conformational degrees of freedom available to both the phospholipid head group and the hydrocarbon chains. The internal "flexibility" was illustrated in the previous section by the strong variation in quadrupolar splitting with chain position and is a reflection of the dynamical gradient that exists along the chain. The correlation times and the amplitudes of the motions also vary with position. Additionally, the motions are not independent; the *gauche-trans* isomerization occurring about the C_i–C_j bond influences the orientation of neighboring C–^{2}H bonds as well as those of deuterium atoms directly bonded to the *ith* and *jth* carbons.

Models for the motion of the lipid chains include, typically, the internal *gauche-trans* isomerization about each C_i–C_j bond (with correlation time τ_j) and whole molecule reorientation such as axial diffusion (with correlation time $\tau_{||}$) fluctuations in the orientation of the diffusion axis (with correlation time $\tau_{\perp}$) with respect to the director (the bilayer normal) and collective motions involving fluctuations in the orientation of the local director relative to its average orientation (Meier et al., 1983, 1986; Weisz et al., 1992). Alternative models replace the whole molecule diffusive motions ($\tau_{||}$ and $\tau_{\perp}$) by large angle jumps (e.g., threefold jumps) (Wittebort et al., 1987).

The application of NMR to the study of lipid dynamics has been described elsewhere (Brown, 1979, 1982; Meier et al., 1983, 1986; Davis, 1986; Bloom et al., 1991). Here we shall restrict ourselves largely to an illustration of some of the effects of cholesterol on lipid motions, mentioning only a few of the more recent general results.

From a study of the frequency dependence of ^{1}H spin-lattice relaxation (T_{1z}) in DMPC using field cycling techniques, Rommel et al. (1988) have concluded that collective motions are important only at very low frequencies (near or below 30 kHz) and do not significantly affect spin lattice relaxation in the MHz range. They reach similar conclusions from the analysis of transverse relaxation (which is sensitive to slow motions through $J_0(0)$, (see Equation 46) from ^{2}H T_{2e}^{CP} experiments on oriented samples of specifically deuterated DMPC (Stohrer et al., 1991). [The T_{2e}^{CP} experiment is a modification (Bloom and Sternin, 1987) of the Carr-Purcell-Meiboom-Gill (Meiboom and Gill, 1958) experiment for measuring transverse relaxation.] However, Pastor et al. (1988), in their analysis of the ^{13}C T_{1z} data of Brown et al.(1983), concluded that collective motions are important in the frequency range of from 10^6 to 10^8 s^{-1}. Bloom et al. (1991) proposed that surface undulations, such as those observed optically in vesicles (Evans and Rawicz, 1990) are the fundamental slow motions in membranes, and they developed a theory to describe transverse relaxation due to these undulations (Bloom and Evans, 1991). We will have to abandon for now the question of the origin of the frequency dependence of relaxation in lipid bilayers and concentrate on the changes in dynamics induced by cholesterol.

The ^{2}H NMR line shape becomes sensitive to the detailed nature of the molecular motions when these motions have correlation times comparable to the spectroscopic time scale. This time scale may be defined to be the inverse of the frequency difference between identifiable features in the spectrum. For example, a ^{2}H-labeled molecule may have two distinct, stable conformations at some temperature, which give rise to an NMR spectrum with different quadrupolar splittings in the absence of any interconversion between conformations. If raising the temperature results in interconversion, say at a rate k, then we begin to see a broadening and a shift of the two splittings towards each other when k becomes comparable to the difference between the splittings ($\delta\nu_Q$). In the limit of fast exchange [$k >> \Delta(\delta\nu_Q)$], we would observe only a single, sharp average splitting.

The formalism developed above assumes that the fluctuations in the Hamiltonian are small and that changes in the density operator occur over a time scale much longer than the correlation time (τ_c was assumed to be short compared to T_{2e}, T_{1z}, T_{1Q}, T_{DQ}, etc., in fact, short compared to even $1/\omega_Q$; see appendix). This is basically the Redfield limit (Redfield, 1965; Slichter, 1990). For motions in the intermediate time scale regime, this assumption breaks down, and we can no longer use this formalism.

Studies of pure lipid systems (Davis, 1979; Huang et al., 1980; Griffin, 1981; Blume et al., 1982; Wittebort et al., 1982; Meier et al., 1983; Davis, 1983) and of lipid mixtures (Blume and Griffin, 1982; Wittebort et al., 1982) showed that intermediate time scale motions were responsible for the dramatic changes in the shapes of the spectra at temperatures just below the gel-to-liquid crystalline phase transition (see, e.g., the spectra in Figure 15). It was also found that these lineshapes were strong functions of the time delay between the two pulses in the quadrupolar echo sequence (Griffin, 1981; Meier et al., 1983; Griffin et al., 1988). The sensitivity of the line shape to the details of the motions provides a sensitive means of testing models of molecular reorientation. However, a different approach is needed in order to calculate these intermediate time scale line shapes. The numerical simulation models mentioned above have been quite successful at extracting dynamical parameters from the temperature, orientation, and "time delay" dependences of these spectra (Meier et al., 1983, 1986; Wittebort et al., 1987).

These numerical models are based essentially on the equation

$$\frac{d\rho_\alpha(t)}{dt} = \frac{i}{\hbar}[\rho_\alpha, H_\alpha] + \frac{\partial}{\partial t}(\rho_\alpha) \tag{72}$$

where

$$\frac{\partial}{\partial t}(\rho_\alpha) = \sum_{\alpha'}[k_{\alpha'\alpha}\rho_{\alpha'} - k_{\alpha\alpha'}\rho_\alpha] \tag{73}$$

and α, α' label different sites or molecular orientations, H_α is the Hamiltonian when the molecule is in site α, and $k_{\alpha'\alpha}$ is the rate for going from site α' to site α (Meier et al., 1983, 1986). The different models used by different research groups really amount to the use of a different set of possible sites that the molecule can occupy. The diffusion and threefold jump models discussed earlier, with reference to the dynamics of cholesterol in lipid bilayers, are simple applications of this approach. For axial diffusion one would use a large number of closely spaced sites (in azimuthal angle ϕ), while the threefold jump model would have only three orientations corresponding to $\phi = 0$, 120, and 240°. This "stochastic Liouville equation" approach allows us to explicitly calculate the correlation functions for these simple models, which we can then use to calculate the spectral densities and relaxation times in the Redfield limit. In the intermediate time scale regime we may still use Equation 72, but we need to evaluate numerically the evolution of the elements of the density operator. Formally, the procedure consists of constructing the matrix

$$(\omega_{\alpha'\alpha})_{ijkl} = \frac{1}{\hbar}\delta_{\alpha'\alpha}(\delta_{jl}H_{ik} - \delta_{ik}H_{lj}) \tag{74}$$

where $H_{ik} = (H_\alpha)_{ik}$ are the matrix elements of the Hamiltonian in site α, and the matrix

$$(\boldsymbol{k})_{\alpha'\alpha} = \begin{cases} \delta_{ik}\delta_{jl}k_{\alpha'\alpha} & \text{if } \alpha' \neq \alpha \\ -\delta_{ik}\delta_{jl}\sum_{\alpha}k_{\alpha'\alpha} & \text{if } \alpha' = \alpha \end{cases} \tag{75}$$

(Meier et al., 1983, 1986; see also, Abragam, 1961 and Mehring, 1983). With these definitions, the stochastic Liouville equation becomes

$$\frac{d\rho(t)}{dt} = (i\omega + \mathbf{k})\cdot\rho(t) \tag{76}$$

which has the formal solution

$$\rho(\mathrm{t}) = \exp[(i\boldsymbol{\omega} + \boldsymbol{k})t]\cdot\boldsymbol{\rho}(0) \tag{77}$$

(we have used boldface notation to emphasize that $\boldsymbol{\rho}, \boldsymbol{\omega}, \boldsymbol{k}$ are matrices). The key to this problem is in finding the matrix S that diagonalizes $(i\omega + \boldsymbol{k})$, i.e., $S^{-1}\cdot(i\omega + \boldsymbol{k})\cdot S = \Lambda$ so that Equation 77 becomes

$$\rho(\mathrm{t}) = \boldsymbol{S}\cdot\exp(\boldsymbol{\Lambda}t)\cdot\boldsymbol{S}^{-1}\cdot\boldsymbol{\rho}(0) \tag{78}$$

since, if $\boldsymbol{\Lambda}$ is diagonal, with elements $\delta_{ij}\lambda_{ii}$, then exp $(\boldsymbol{\Lambda}t)$ is also diagonal, with elements $e^{\delta_{ij}\lambda_{ii}t}$

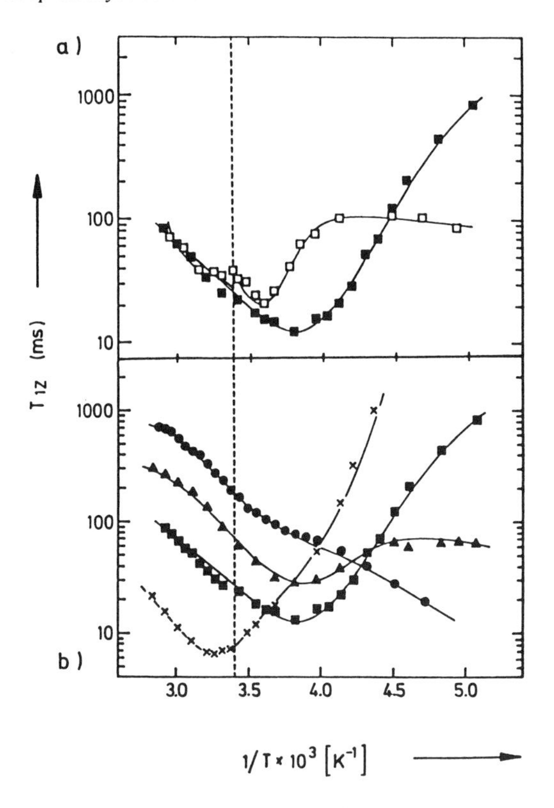

FIGURE 16. The temperature dependence of the spin-lattice relaxation time of (a) 2[6,6-$^2H_2$]-DMPC; ❑: with no cholesterol; ■: with 40 mol% cholesterol. (b) ×: cholesterol-3α-$^2H_1$; ■: 2[6,6-$^2H_2$]-DMPC/cholesterol; ▲: 2[13,13-$^2H_2$]-DMPC/cholesterol; and ●: 2[14,14,14-$^2H_3$]-DMPC/cholesterol; all samples are DMPC with 40 mol% cholesterol. (Reproduced from Weisz et al., 1992. *Biochemistry.* With permission.)

Weisz et al. (1992) separate the lipid motions into two classes that they call inter- and intramolecular. The intra-molecular motions are the *gauche-trans* isomerizations and are represented as random jumps between the different conformations of the labeled segment, with correlation time τ_J (see Meier et al., 1986, for a detailed description of the model and its implementation). The intermolecular motions are represented by restricted diffusion in an orienting potential, with correlation times $\tau_{\|}$, for axial diffusion about the long molecular axis, and $\tau_{\perp}$, describing fluctuations of the orientation of the long axis within the orienting potential. In addition to these single molecule motions, collective motions are introduced, modeled as instantaneous fluctuations of the local director axis with respect to its time-averaged orientation. Their experiments were performed on multilamellar dispersions (powders) and oriented samples of di-[6,6-$^2H_2$]myristoyl-PC with and without 40 mol% cholesterol, di-[13,13-$^2H^2$]myristoyl-PC, di-[14,14,14-$^2H_3$]myristoyl-PC and cholesterol-3α-$^2H_1$ in DMPC. They studied the temperature and orientation dependence of T_{1z}, T_{2e}, and T_{2e}^{CP}. Figure 16 presents the results of their measurements of the temperature

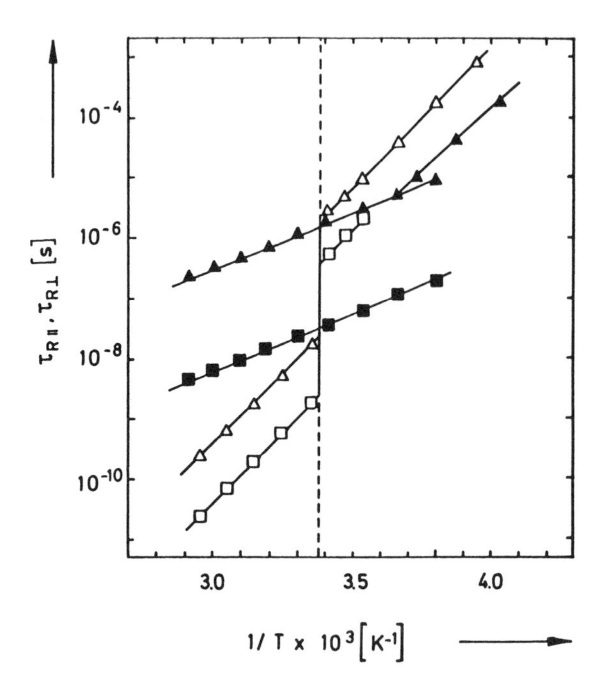

FIGURE 17. The temperature dependence of the intermolecular correlation times for DMPC: ($\tau_{\parallel}$ – ❑: no cholesterol; ■: with 40 mol% cholesterol; $\tau_{\perp}$ – Δ; no cholesterol; and ▲: with 40 mol% cholesterol. (Reproduced from Weisz et al., 1992. *Biochemistry*, With permission.)

dependence of T_{1z}. In agreement with the work of Bonmatin et al. (1990) discussed above, cholesterol exhibits a T_{1z} minimum at about 35°C, albeit with a minimum value of T_{1z} of about 7 ms. There is also a minimum in the spin-lattice relaxation time for di-[6,6-$^2H_2$]myristoyl-PC both with and without the 40 mol% cholesterol and a shallower minimum for the di-[13,13-$^2H_2$]myristoyl-PC without cholesterol.

The analysis of their data, using the dynamical model outlined above, is summarized in Figure 17. Here the correlation times $\tau_{\parallel}$ and $\tau_{\perp}$ are plotted vs. inverse temperature. In the absence of cholesterol, the liquid crystalline-to-gel phase transition results in an abrupt increase in both correlation times by roughly a factor of 10. This is accompanied by a similar increase in the *gauche-trans* correlation time τ_J [see Figure 11; Weisz et al. (1992)]. The addition of 40 mol% cholesterol has eliminated both the phase transition and the discontinuous changes in these correlation times. It has also lowered the activation energies (from the slopes of the plots) for all of these motions. The values of the correlation times $\tau_{\parallel}$ and $\tau_{\perp}$ are larger than those for the pure lipid fluid phase, but shorter than those in the pure lipid gel phase. As indicated by the lineshapes of the perdeuterated lipids (Figure 13) in the presence of large concentrations of cholesterol, the phospholipid molecules still undergo rapid reorientation about their long axes. There is a much more gradual slowing down of this motion as temperature is lowered than is the case for pure lipid bilayers. Additionally, the frequency dependence of T_{2e}^{CP} (not shown here) reveals a similar slowing down in the "collective motions" in the presence of cholesterol. Finally, this model indicates that there is a large decrease in *gauche* populations in the presence of cholesterol (Weisz et al., 1992).

While many of the principal conclusions of this study have already been well established by other techniques, it provides an excellent example of the richness of the ^{2}H NMR spectrum and of the usefullness of numerical modeling of the molecular dynamics of these complex molecules.

IV. THE PHASE EQUILIBRIA OF CHOLESTEROL/DPPC MIXTURES

Cholesterol has some remarkable effects on the physical properties of phospholipid bilayers. At temperatures where the pure phospholipid would be in the fluid phase, it can dramatically decrease the rates of molecular reorientations of the phospholipids and the population of *gauche* conformers, without changing substantially the rate of lateral diffusion (Rubenstein et al., 1979; Alecio et al., 1982). The decrease in *gauche-trans* isomerization of the hydrocarbon chains results in large increases in the chain order and in a thickening of the hydrophobic region of the lipid bilayer (Hui and He, 1983; McIntosh et al., 1989; Nezil and Bloom, 1992). In addition, cholesterol both reduces the surface compressibility by a factor of 3 to 4 and increases the cohesive strength of the bilayer (Bloom et al., 1991).

The observations made above are based on comparisons of the physical properties of phospholipid bilayers with no cholesterol and those with perhaps 30 or 40 mol% cholesterol. Clearly a basic understanding of how cholesterol induces these changes will require a careful study of these properties as a function of cholesterol concentration, starting from 0% and increasing the cholesterol concentration far enough and in small enough steps so that all important effects are observed. One of the most basic requirements of such a study is the determination of an accurate and reliable phase diagram for this two-component mixture. Interestingly enough, there is no shortage of experimental phase diagrams for cholesterol/PC mixtures (see, e.g., Ladbrooke et al., 1968; Shimshick and McConnell, 1973; Gershfeld, 1978; Lentz et al., 1980; Rand et al., 1980; Recktenwald and McConnell, 1981; Mortensen et al., 1988; Vist and Davis, 1990; Pasenkiewicz-Gierula et al., 1990; Sankaram and Thompson, 1991; Thewalt and Bloom, 1992). It should be mentioned that although there are many important differences among these phase diagrams, they all have certain important features in common and that none of them provides a complete description. Complications due to the occurrence of metastable states, to the formation of molecular complexes, or to critical phenomena may make it more difficult to interpret some experimental results in terms of a phase diagram.

The construction of a phase diagram for even a simple two-component system requires information from a number of sources. For example, in the case of phospholipid/cholesterol/water mixtures*, it is necessary to determine whether, over the range of compositions studied, the system forms a bilayer structure. In the case of PC/cholesterol mixtures in excess water, this has been clearly demonstrated by X-ray diffraction (see Finean, 1990, for a recent discussion). Another important question to consider in these efforts is whether or not the system truly contains an excess of water. Gibbs' phase rule states that $f = c - p + 2 \geq 0$ for a system where temperature, pressure, and composition are the only independent variables. Here, f is the number of degrees of freedom available to the system, c is the number of components, and p is the number of phases in equilibrium. Thus, for a three component system (like PC/cholesterol/water) one can have $p = 3$ phases in coexistence in a $f = 2$ dimensional region of the phase diagram. If we are careful to always have a bulk water phase, then effectively one of the degrees of freedom is taken by the excess water phase. The system then behaves as a pseudobinary mixture where we can have three phases in coexistence only on a line ($f = 1$). If we display our phase diagram as a temperature/composition plot, taking the vertical or y-axis as the temperature axis, then the line of three-phase coexistence will be horizontal (an isotherm).

There is a heat of transition associated with a first-order phase transition. Thus, differential scanning calorimetry (DSC) has become a standard means of identifying the temperatures at which a system crosses a phase boundary (see, e.g., Mabrey et al., 1978; Mabrey and Sturtevant, 1979; and Morrow et al., 1985; for a discussion of the interpretation of DSC experiments on two component mixtures). DSC scans of mixtures of chain perdeuterated DPPC/cholesterol at ten different concentrations from 0 to 22.5 mol% cholesterol are shown in Figure 18 (Vist and Davis, 1990). The scan for 0 mol% is typical of pure saturated PC/water mixtures, showing the main transition at about 38°C; the "pretransition" — common to saturated PC/water dispersions, occurring at about 32°C in this case — at the ripple ($P_{\beta'}$) to $L_{\beta'}$ phase boundary (notation due to Tardieu et al.,

* The procedure used in preparing cholesterol/DPPC/water samples can have an important effect on the homogeneity of the samples. We investigated several solvent systems (Vist, 1984; Vist and Davis, 1990) and found that methanol gave the most reproducible results and the most homogeneous samples over the composition range of from 0 to 25 mol% cholesterol. Because the ^{2}H NMR quadrupolar splittings are so sensitive to cholesterol concentration, the spectra themselves provide a critical test of sample homogeneity. Our protocol for testing sample homogeneity involved dissolving the two components (cholesterol and DPPC) in a particular solvent, removing the solvent by rotary evaporation at temperatures less than 40°C, and, finally, pumping on the mixture (now spread over the surface of a 50 mL round bottom flask) for at least 6 h. The samples were then mixed with 50 mM phosphate buffer at pH 7.0 at a 3:4 weight ratio (3 parts cholesterol/DPPC to 4 parts water, by weight). The samples were then stirred by hand in the NMR tube using a small glass stirring rod. At high temperatures, the ^{2}H NMR powder pattern spectrum of a homogeneously mixed sample exhibits very sharp 90° edges. If a sample is poorly mixed there is a spread in the quadrupolar splittings for each ^{2}H-labeled position, and the peaks are significantly broadened. We found that we were unable to prepare sufficiently homogeneous samples using chloroform as solvent (although it is probably the most commonly used solvent for preparing cholesterol/lipid samples), but that methanol gave us homogeneous mixtures up to cholesterol concentrations as high as 30 mol%. At higher concentrations we were unable to prepare samples that met our criteria for homogeneity.

1973); and the "subtransition" at about 12°C. The subtransition involves a rearrangement of the hydrocarbon chain packing within the plane of the bilayer (Chen et al., 1980; Ruocco and Shipley, 1982). In saturated PCs, the subtransition normally requires long incubation times at temperatures near 0°C before it can be detected (Chen et al., 1980; Finegold and Singer, 1986). This is a good illustration of how metastable states can make it difficult to construct accurate phase diagrams. The addition of 1.25 mol% cholesterol has little effect on the DSC scan, as shown in the figure. The main transition at about 38°C is still sharp, and the pretransition and subtransition still occur. The apparent displacement in temperature of the pretransition is a well-known characteristic of this peak. The temperature where it occurs, as well as its area, are sensitive to the thermal history of the sample. For these reasons, we will concentrate on the main hydrocarbon chain melting (gel-to-fluid phase) transition. At 2.5 mol% cholesterol we see already, in Figure 18, a slight depression in the temperature of the main phase transition and a slight broadening of the DSC peak. At 5 and 6.25 mol% the thermograms are similar, although the main transition of the 6.25 mol% sample seems to have narrowed slightly and the peak due to the pretransition has vanished (and is also absent at higher cholesterol concentrations). At 7.5 mol%, and more clearly at 10 mol%, we see the appearance of a broad shoulder at the high temperature side of the sharp transition peak. The relative area of this shoulder steadily increases with increasing cholesterol concentration, at the expense of the sharp peak until, somewhere between 20 and 22.5 mol% cholesterol, the sharp peak has disappeared. [It should be noted that the DSC scan at 22.5 mol% in this figure was made under slightly different conditions from the others in order to enhance the sensitivity. The scan rate for this trace was considerably higher, and a larger sample was used (Vist and Davis, 1990)].

These calorimetry scans and those of Mabrey et al. (1978) provide the first real clue to the nature of the DPPC/cholesterol phase diagram. The sharp, temperature invariant peak at 37°C that decreases in area with increasing cholesterol concentration, vanishing just above 20 mol%, is a three-phase coexistence line beginning somewhere near 6.25 or 7.5 mol% and ending near 20 mol%. The high temperature shoulder, which grows in size between about 7.5 and 20 mol% cholesterol, and the broad maximum seen for samples at 22 mol% and above will also give us some indications of the nature of the system above the threephase line.

Unfortunately, calorimetry alone is not sufficient either for the determination of the phase diagram or for the description of the nature of this two-component mixture in its different phases. We will need to resort to other experimental techniques in order to establish firmly the character and phase behavior of the system.

In Figure 15 we have already seen that the ^{2}H NMR spectrum changes dramatically at the gel-to-liquid crystalline phase transition and, at least in this case, that it is possible to recognize two-phase coexistence by the superposition of two distinct spectra characteristic of the two coexisting phases. Figure 19 shows a hypothetical phase diagram for a two-component mixture that is in a

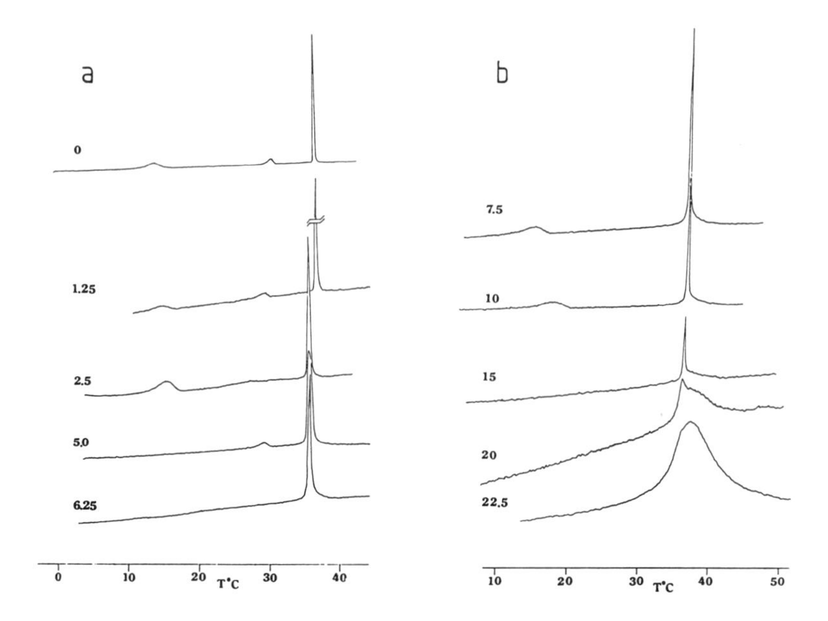

FIGURE 18. Differential scanning calorimetry traces for chain perdeuterated DPPC/cholesterol mixtures at ten different cholesterol concentrations (given on the figure). Scan rates varied from 9 to 12°C/h from sample to sample, except that for the 22.5 mol% sample whose scan rate was increased to 44°C/h to increase the sensitivity. (Reproduced from Vist and Davis, 1990. *Biochemistry*, 29:451. With permission.)

fluid phase at high temperatures and in a solid (or "gel") phase at lower temperatures. For the pure one-component system, at $x = 0$ there is an isothermal phase transition from the gel-to-fluid phase at a temperature T_m. For $x > 0$, the fluid and gel single phase regions are separated by a two-phase region, in accordance with Gibbs' phase rule. If we prepare an NMR sample whose second component (cholesterol in this case) has a concentration x_A, then at the highest temperatures we expect to have a fluid phase with composition x_A, and at the lowest temperatures a gel phase at that composition (the two components are miscible in both the fluid and gel phases). However, at temperatures between the upper and lower two-phase lines separating the two phase coexistence region from the one-phase regions above and below, the system will consist of domains of two different compositions. One type of domain is fluid, the other is gel. The compositions at a temperature T of these two types of domains are obtained from the intersection of the isotherm at T with the two phase lines. Thus, the fluid phase domains have a composition x_F, and the gel phase domains a composition x_G.

The ^{2}H NMR spectrum of such a two-phase system will consist of a superposition of two components: one for each of the two types of domains in the sample. The fluid domain component will be similar to the spectrum of Figure 15b at 37.75°C, while the gel phase component will be similar to the spectrum

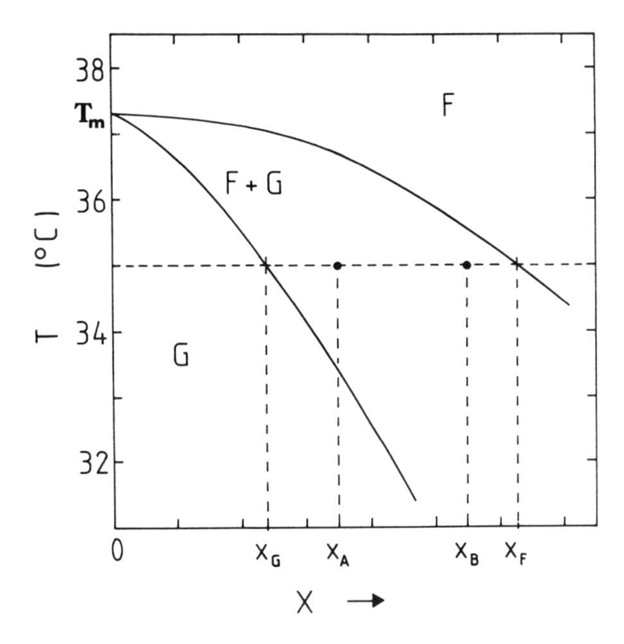

FIGURE 19. A hypothetical phase diagram showing a fluid (F) plus gel (G) two-phase coexistence region for a two component mixture. (Reproduced from Davis, 1989. *Adv. Magn. Res.*, 13:195. With permission.)

at 37.25°C. The relative amounts of each of these component spectra will depend on the relative amounts of the two types of domains. The lever rule,

$$\alpha_A(T) = \frac{x_A - x_G(T)}{x_F(T) - x_G(T)} \tag{79}$$

gives the fraction of the sample A which is in the fluid phase at temperature T. The gel-phase fraction is then simply $1 - \alpha_A(T)$. This does not give directly the amount of phospholipid in the two phases, however, because the lipid concentrations are different, (the areas of the two NMR components are proportional to the amount of phospholipid in the two phases). The concentration of phospholipid, for a sample with a cholesterol concentration x_A, is $1 - x_A = \alpha_A \cdot (1 - x_F) + (1 - \alpha_A) \cdot (1 - x_G)$, where $(1 - x_F)$ is the phospholipid concentration in the fluid-phase domains, and $(1 - x_G)$ is its concentration in the gel-phase domains. Thus, the fraction of phospholipid in the fluid phase at temperature T is

$$f_A(T) = \alpha_A(T) \cdot \frac{1 - x_F(T)}{1 - x_A} \tag{80}$$

and the gel phospholipid fraction is $1 - f_A(T)$.

The ^{2}H NMR spectrum of a sample of composition x_A taken at a temperature T where the sample is in a two-phase region is given by

$$S(x_A,T) = f_A S_F(x_F,T) + (1 - f_A)S_G(x_G,T) \tag{81}$$

where $S_F(x_F,T)$ and $S_G(x_G,T)$ are the "end-point" spectra characteristic of the fluid- and gel-phase domains at the temperature T and at the end-point cholesterol concentrations x_F and x_G. We have implicitly made two important assumptions in writing down Equation 81: firstly, we have assumed that exchange of phospholipids between different types of domains is sufficiently slow that the component spectra are distinct and distinguishable; secondly, we assumed that the transverse relaxation time T_{2e} is long enough (relative to the time delay between the two pulses used to form the quadrupolar echo) that we can ignore the differential reduction in the areas of the component spectra due to their differing relaxation times. This latter assumption is potentially serious, since T_{2e} in the gel phase is generally significantly shorter than it is in the fluid phase. This can lead to systematic errors in determining the end-point concentrations x_F and x_G (Morrow et al., 1991; Thewalt and Bloom 1992). The importance of the first assumption will become apparent shortly.

Having established this description of the ^{2}H NMR spectrum of sample A in the two-phase coexistence region, we now prepare a second sample B with cholesterol concentration x_B such that at temperature T it will also be in the two-phase region, as illustrated in Figure 19. The arguments used for sample A also hold for sample B. Thus, the spectrum of sample B at temperature T within the two-phase region is

$$S(x_B,T) = f_B S_F(x_F,T) + (1 - f_B)S_G(x_G,T) \tag{82}$$

where $f_B(T)$ is the fraction of phospholipid of sample B in the fluid phase at temperature T. Thus, the NMR spectra of these two different samples consist of linear superpositions of the same two end-point component spectra. The only difference between the two experimental spectra (for A and B) are the relative fractions of the two components.

We can solve these two equations for the end-point spectra,

$$S_G(x_G,T) = \frac{1}{(f_B - f_A)} \cdot \left[f_B S(x_A,T) - f_A S(x_B,T) \right] \tag{83}$$

and

$$S_F(x_F,T) = \frac{1}{(f_A - f_B)} \cdot \left[(1 - f_B) S(x_A,T) - (1 - f_A) S(x_B,T) \right] \tag{84}$$

If the two end-point spectra are readily distinguishable by eye (or by computer), we can subtract a fraction of one of the experimental spectra from the other and obtain either of the two end-point spectra.* Thus, to obtain the gel-phase end-point spectrum, we subtract the fraction

$$K = \frac{\alpha_A(1 - x_B)}{\alpha_B(1 - x_A)} \tag{85}$$

of the experimental spectrum of sample B (the one with the larger fluid-phase component) from the experimental spectrum of sample A. (We must normalize the areas of the two experimental spectra before attempting the subtraction.) Conversely, we obtain the fluid phase end-point spectrum by subtracting the fraction

$$K' = \frac{(1 - \alpha_B)(1 - x_A)}{(1 - \alpha_A)(1 - x_B)} \tag{86}$$

of spectrum S_A from spectrum S_B.

These end-point spectra are from domains of the sample with compositions x_F and x_G, which, in general, are different from the compositions of any of the actual experimental samples. This makes these spectra interesting in their own right. Perhaps more important, however, is that from Equations 79, 80, 85, and 86, we can solve for the end-point concentrations

$$x_G = \frac{[(1 - x_B)x_A - K(1 - x_A)x_B]}{[(1 - x_B) - K(1 - x_A)]} \tag{87}$$

and

$$x_F = \frac{[(1 - x_A)x_B - K'(1 - x_B)x_A]}{[(1 - x_A) - K'(1 - x_B)]} \tag{88}$$

Thus, the technique of ^{2}H NMR difference spectroscopy can be used to determine the location of the phase boundaries of two-phase coexistence regions

* The technique of ^{2}H NMR difference spectroscopy is described in detail in Vist (1984), Huschilt et al. (1985), and Vist and Davis (1990). After normalizing the two experimental spectra to equal areas, we display the difference $S_A - K{\cdot}S_B$ on the computer screen. The value of the constant K is determined by the position of a potentiometer. By adjusting the potentiometer knob, the fraction of spectrum B that is subtracted from spectrum A can be increased or decreased. The end points are determined by the value of K where features characteristic of one of the end-point spectra have vanished from the difference spectrum. For example, for a spectrum consisting of a superposition of fluid- and gel-phase components, the disappearance of the sharp fluid-phase methyl group peaks from the center of the difference spectrum, or the disappearance of the very broad shoulders characteristic of the gel phase, are used to determine K for the two end points. Though admittedly somewhat subjective, we can often determine K to within a few percent using this technique, and it is straightforward to estimate the uncertainty by over- or undersubtracting.

(Huschilt et al., 1985; Morrow et al., 1985; Morrow and Davis, 1988; Vist and Davis, 1990; Morrow et al., 1991; Thewalt and Bloom, 1992).

The importance of our first assumption about the absence of exchange between domains is now apparent. If phospholipids can diffuse from one type of domain to another in a time short or comparable to the spectroscopic time scale (say about 100 μs), then the experimental spectra will not be superpositions of the two end-point spectra but some sort of average between them, and the subtraction procedure will not work well.

Spectra such as those at 37.75°C in Figure 15a and at 37.5°C in Figure 15b are prime candidates for this subtraction procedure. Unfortunately, in this case the fluid-gel two-phase coexistence region of DPPC/cholesterol mixtures at low cholesterol concentrations (below about 7.5 mol% cholesterol) is so narrow that it was not possible to use this method to map the phase boundaries in this region. Instead, the phase boundaries for $x_{chol} < 0.075$ were determined by inspection of the ^2H NMR spectra and of the DSC curves in Figure 18. As demonstrated in Figure 15, it is easy to identify from these spectra the temperatures where the fluid and gel phases are in coexistence. For this reason, spectra were taken at 0.25°C intervals for samples having cholesterol concentrations x_{chol} = 0.0, 0.0125, 0.025, 0.05, and 0.0625. From these spectra and the DSC scans, the two phase region was seen to broaden to only about 0.75°C between 2.5 and 6.25 mol% cholesterol, and then to narrow again at 6.25 mol%.

At higher cholesterol concentrations the spectra do not show such an obvious two-component coexistence. The spectra in Figure 20, taken with a sample containing 15 mol% cholesterol, show a dramatic increase in the total width of the spectrum on lowering the temperature from 40 to 35°C, but none of the spectra in between (taken at 1°C intervals) look like the two-component spectra of Figure 15. However, comparison of the spectra below 37°C at cholesterol concentrations of 7.5, 10.0, 15.0, 20.0, and 22.5 mol% cholesterol suggested that there may be two components to these spectra. Performing the subtractions, we found that these spectra are indeed superpositions of two distinguishable end-point spectra, as illustrated in Figure 21. The spectra in parts "a" and "b" of the figure are the normalized experimental spectra of samples with 10 and 15 mol% cholesterol at 31°C. The normalized difference spectrum in part "c" is obtained by subtracting 0.34 times the spectrum in "b" from the spectrum in "a". The difference spectrum "d" is obtained by subtracting 0.59 times the spectrum of "a" from that of "b". While the two experimental spectra in "a" and "b" look quite similar, the difference spectra of "c" and "d" are very different.

The end-point concentrations determined from the subtractions illustrated in Figure 21 were $x_F \approx 0.22$ and $x_G \approx 0.07$, both of which are very close to the compositions of two of our other samples. The end-point spectra obtained here from samples with 10 and 15 mol% cholesterol at 31°C should be characteristic of domains with those end-point compositions. Thus, these end-point spectra should be similar to those of the samples at $x_{chol} = 0.075$ and $x_{chol} = 0.225$ at 31°C. Figure 22 compares the end-point difference spectra with these experimental spectra. The 22.5 mol% spectrum "b" is then directly subtracted from the end-

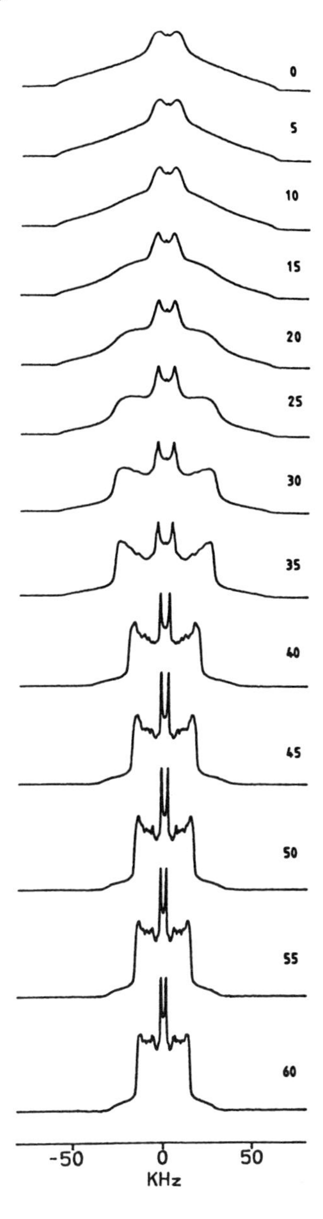

FIGURE 20. ^{2}H NMR spectra of a mixture of chain perdeuterated DPPC with 15 mol% cholesterol at 13 temperatures (indicated on the figure) in °C. (Reproduced from Vist and Davis, 1990. *Biochemistry,* 29:451. With permission.)

point difference spectrum "a" and the result plotted on the same vertical scale in "c". Similarly, the 7.5 mol% spectrum "e" is subtracted from the end-point difference spectrum "d" and plotted in "f". This remarkable agreement between four separate samples is even more firmly established by performing the same

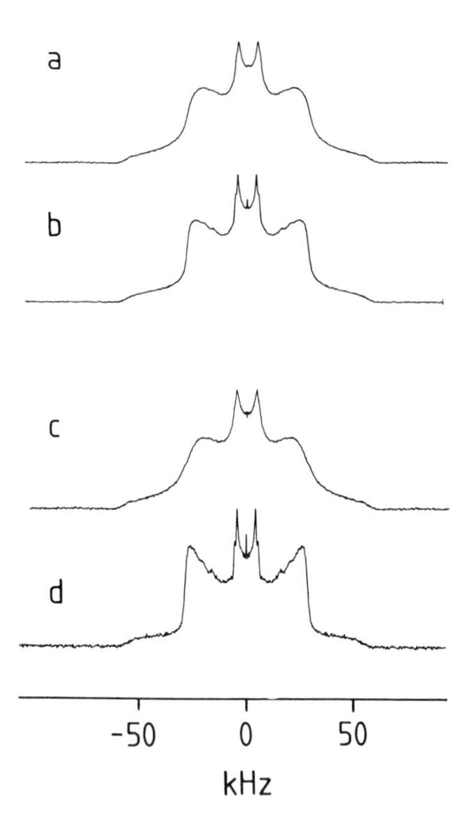

FIGURE 21. ^{2}H NMR difference spectroscopy at 31°C: (a) the spectrum of a 10 mol% cholesterol sample; (b) that of a 15 mol% sample; (c) the spectrum of "a" minus 0.34 times that of "b"; (d) the spectrum in "b" minus 0.59 times that of "a". All spectra are normalized to have the same total area. (Reproduced from Vist and Davis, 1990. *Biochemistry,* 29:451. With permission.)

comparisons with experimental spectra at 31°C for the samples with 6.25 and 20 mol% cholesterol. The large differences obtained (Vist, 1984) make it clear that the end-point concentrations are very close to 7.5 and 22.5 mol% and that the difference technique works very well in this region of the phase diagram. The end-point spectrum at $x_G = 0.07$ is typical of the phospholipid gel phase. The other end-point spectrum, at $x_F = 0.22$, has an appearance much like that of the fluid phospholipid phase, but the quadrupolar splittings are very much larger (as expected for domains with such a high cholesterol content; see above).

At 37°C we should find three phases in coexistence. The DSC results had suggested a horizontal phase boundary — which must be a three-phase line. One of these phases is the normal fluid phase; another is the gel phase. *The third must be the phase that is in coexistence with the gel phase below 37°C* (see Figure 23). This phase, which we have called the β phase, has many properties in common with the usual phospholipid bilayer fluid phase: the molecules undergo rapid axially symmetric reorientation about their long axis (relative to the spectroscopic time scale, although we noted above that the reorientational rates are

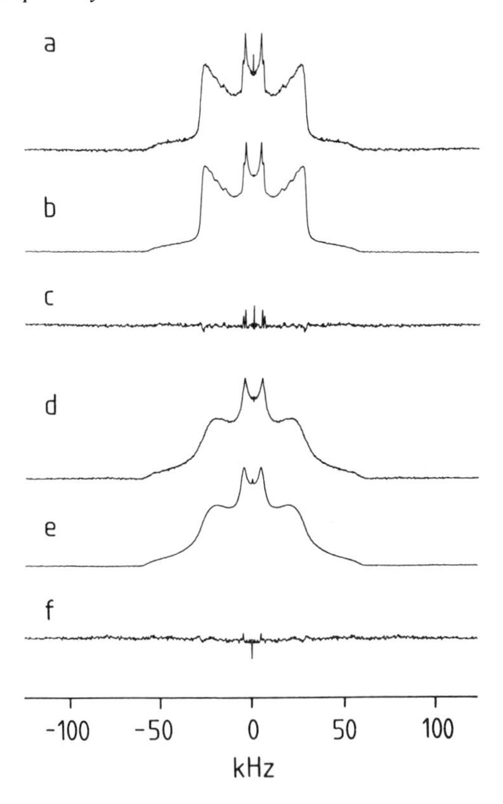

FIGURE 22. Comparison of the end-point difference spectra of Figure 21 with the spectra from samples with cholesterol compositions near the end-point concentrations (at the same temperature). (a) The end-point spectrum from Figure 21d; (b) the spectrum of a 22.5 mol% cholesterol sample; (c) the difference between the two spectra in "a" and "b", shown on the same vertical scale; (d) the end-point spectrum from Figure 21c; (e) the spectrum of a 7.5 mol% cholesterol sample; and (f) the difference between the spectrum in "d" and that in "f", on the same vertical scale. (Reproduced from Vist and Davis, 1990. *Biochemistry,* 29:451. With permission.)

lower at high cholesterol concentrations than they are for the pure lipid), and the rate of lateral diffusion is comparable to that observed in the fluid phase. The major difference seems to be that the phospholipid chains are much more highly ordered in the β phase. For this reason, Ipsen et al. (1987), in their theoretical models for the phase diagram presented by Vist (1984), called this β phase the "liquid ordered" or l_o phase, and they called the fluid phase the "liquid disordered" or *ld* phase. They also called the gel phase the "solid ordered", *so* phase. (Further theoretical models are given by Zuckermann et al. and by Scott in this volume.)

Above 37°C we should find a two phase "fluid-fluid" coexistence region where the *ld* and *lo* phases are in equilibrium. Because of the rapid rates of lateral diffusion in both the fluid and β phases, and the apparent small size of the domains, the phospholipids appear to exchange rapidly by diffusion between the two types of domains. This is evident from the ^{2}H NMR spectra in this region.

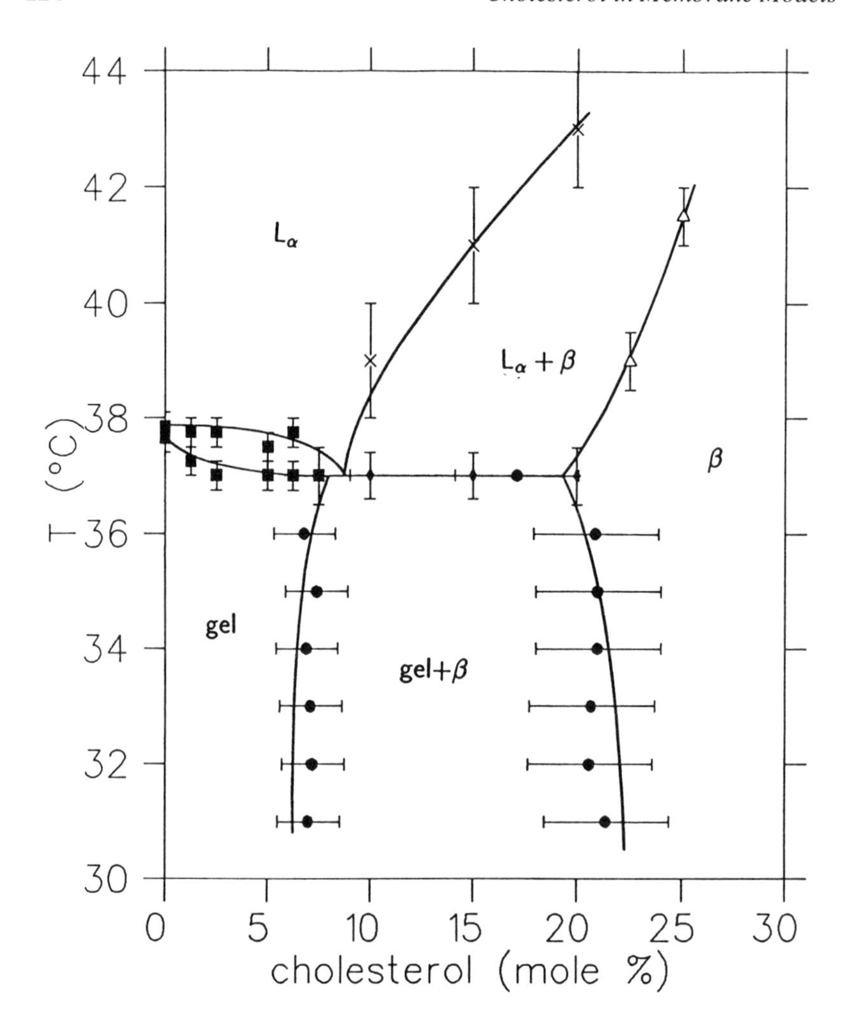

FIGURE 23. Partial phase diagram for cholesterol/DPPC mixtures. The phases are labeled as follows: L_α is the fluid or liquid-crystalline phase; gel is the gel phase; and β is the high-cholesterol-content phase characteristic of biological membranes. The squares (■) are determined by inspection of spectra such as those of Figure 15; the circles (●) are from ^{2}H NMR difference spectroscopy; the diamonds (◆) show the the three-phase line determined from the sharp peak in the DSC traces; the "x"-es (×) are from the upper limit of the broad component of the DSC traces; and the triangles (Δ) are from the abrupt sharpening of the resonances at high cholesterol concentrations. The solid lines simply delineate the phase boundaries. (Reproduced from Vist and Davis, 1990. *Biochemistry,* 29:451. With permission.)

At low cholesterol concentrations and higher temperatures, the individual peaks in these multicomponent fluid-phase powder-pattern spectra are very sharp and readily resolvable. Similarly, for 22.5 and 25 mol% cholesterol the individual peaks in the spectra are very sharp (see, for example, Figure 13). Between 10 and 22.5 mol% cholesterol the individual peaks are smeared out, broadened by exchange between domains. One can identify the temperatures at which the lines sharpen and thereby locate the phase boundary on the high-cholesterol side (shown as triangles in Figure 23), but on the other side of this two-phase region it was more difficult, so we did not attempt it. EPR has a much shorter spectroscopic time scale, so that it should be possible to identify the two components in this fluid-fluid coexistence region. In fact, the EPR data of Shimshick and McConnell (1973) and of Recktenwald and McConnell (1981), the fluorescence data of Lentz et al. (1980), and the end-point of the high temperature shoulders of the DSC thermograms all give an indication of the position of this phase boundary. Recently, Sankaram and Thompson (1991) demonstrated the existence of this two-phase coexistence region using EPR.

V. CONCLUSION

The phase diagram presented in Figure 23 is the result of a large body of experimental evidence gathered from a variety of sources using a number of different physical techniques. The PC/cholesterol mixture is still very actively studied because there remain a number of interesting questions. For example, the neutron scattering studies of Mortensen et al. (1988) on DMPC/cholesterol mixtures raise some interesting questions regarding the nature of the domains in the gel-β-phase coexistence region. The small size of the domains indicated by their work and by the freeze-fracture studies of Copeland and McConnell (1980) would suggest that there should be significant phospholipid exchange between domains due to the rapid rate of lateral diffusion in the β-phase domains. The success of the subtraction technique in this region suggests that this is not the case. The domains discussed by Copeland and McConnell (1980) and by Mortensen et al. (1988) are related to the "ripples" seen in the $P_{\beta'}$ phase, so it seems important to understand more completely the nature of the pretransition and the effect of cholesterol on this transition and the related problem of phospholipid tilt in both the $P_{\beta'}$ and $L_{\beta'}$ phases.

The ^{2}H NMR samples used in the study described above were all fully hydrated (Vist and Davis, 1990), while the samples used in the neutron scattering work of Mortensen et al. (1988) and of Copeland and McConnell (1980) were probably less than fully hydrated. To investigate this possible source of the disagreement between these results, we have performed experiments on oriented samples of specifically deuterated DMPC as a function of temperature, cholesterol concentration, and level of hydration. We found no obvious evidence in the gel-β coexistence region for a large influence of water hydration (above about 12 waters per lipid) on the spectral line shapes (X. Shan and J. H. Davis, private communication). However, the phospholipid dynamics in this region are somewhat tricky, and it will require more detailed analysis before we can reach any concrete conclusions.

While we have not directly answered the question "What is cholesterol doing in the membrane?", we have found that the β or liquid ordered (l_o) phase is the normal state of natural membranes that are rich in cholesterol. We also now have a quantitative description of the physical properties of this state. High concentrations of cholesterol increase the phospholipid hydrocarbon chain orientational order, thicken the bilayer, decrease the surface compressibility, and decrease the rates of molecular reorientations without significantly altering the rates of molecular lateral diffusion, thereby maintaining the bilayer's "liquid" character over wide temperature and broad compositional ranges. It is evident that one of the important biological roles of cholesterol is to stabilize the dynamic liquid crystallinity of the lipid bilayer.

VI. APPENDIX

A. QUADRUPOLAR RELAXATION IN PARTIALLY ORDERED SYSTEMS

We now define a transformation to an interaction representation by

$$\hat{\rho}(t) = U_R \rho(t) U_R^{-1} \tag{A1}$$

where

$$U_R = \exp[iH_z t / \hbar] \tag{A2}$$

and H_Z is the Zeeman Hamiltonian, Equation 1. Then, the Liouville equation becomes, in this interaction representation,

$$\frac{d\hat{\rho}}{dt} = -\frac{i}{\hbar}\left[\hat{H},\hat{\rho}\right] + \frac{i}{\hbar}\left[H_Z,\hat{\rho}\right] \tag{A3}$$

with

$$\hat{H} = U_R H U_R^{-1} \tag{A4}$$

If, for example, $H = H_Z + H_Q$, then

$$\hat{H} = H_Z + U_R H_Q U_R^{-1} = H_Z + \hat{H}_Q \tag{A5}$$

so that

$$\frac{d\hat{\rho}}{dt} = -\frac{i}{\hbar}\left[\hat{H}_Q,\hat{\rho}\right] \tag{A6}$$

Using Equation 38,

$$\hat{H}_Q(t) = < H_Q >_T + U_R H_q(t) U_R^{-1} = < H_Q >_T + \hat{H}_q(t) \quad \text{(A7)}$$

where $H_q(t) = H_Q(t) - < H_Q >_T$, and we have used the fact that the static part of the quadrupolar Hamiltonian, which leads to the observed residual quadrupolar splittings, commutes with the Zeeman Hamiltonian. To describe relaxation we will treat the fluctuating part of the quadrupolar Hamiltonian to second order using time-dependent perturbation theory (Abragam, 1961; Cohen-Tannoudji, 1977; Slichter, 1990). Equation A6 can be formally integrated to yield

$$\hat{\rho}(t) = \hat{\rho}(0) - \frac{i}{\hbar} \int_0^t \left[\hat{H}, \hat{\rho}\right] dt' \quad \text{(A8)}$$

where $\hat{\rho}(0)$ is the density matrix at $t = 0$ in the interaction representation.

We can substitute Equation A8 into itself iteratively and stop at the second order (Abragam, 1961; Slichter, 1990) to obtain

$$\begin{aligned} \hat{\rho}(t) = \hat{\rho}(0) &- \frac{i}{\hbar} \int_0^t dt' \left[< H_Q >_T, \hat{\rho}(0)\right] \\ &- \frac{i}{\hbar^2} \int_0^t dt' \int_0^{t'} dt'' \left[\hat{H}_q(t'), \left[\hat{H}_q(t''), \hat{\rho}(0)\right]\right] \end{aligned} \quad \text{(A9)}$$

where we have also used equation A7.

Each molecule in a homogeneous sample should experience the same time-averaged or residual quadrupolar interaction, i.e., should have the same splitting for a given orientation. However, the detailed time dependence of fluctuations on the molecular scale will vary randomly from one molecule to another. Thus, we need to average over all molecules in the sample. Replacing $\hat{\rho}$ by its average value $\bar{\hat{\rho}}$ and differentiating Equation A9 once with respect to time, we obtain

$$\frac{d\overline{\hat{\rho}(t)}}{dt} = -\frac{i}{\hbar}\left[< H_Q >_T, \bar{\hat{\rho}} - \overline{\hat{\rho}_{eq}}\right] - \frac{1}{\hbar^2} \int_0^t dt' \overline{\left[\hat{H}_q(t), \left[\hat{H}_q(t'), \hat{\rho} - \hat{\rho}_{eq}\right]\right]} \quad \text{(A10)}$$

where we have replaced $\hat{\rho}(0)$ by $\hat{\rho}(t) - \hat{\rho}_{eq}$. This assumes that the fluctuations in $\hat{H}_q(t)$ are much more rapid than the changes in $\hat{\rho}(t)$ and that the correlation between $\hat{H}_q(t)$ and $\hat{H}_q(t')$ rapidly approaches zero (this is the short correlation time regime, as will become clearer below). Also, we include $\hat{\rho}_{eq}$ so that the system approaches its equilibrium configuration at long times. This approach, basically that of Redfield (1965), will allow us to define the relaxation times describing the recovery of the spin system to equilibrium following a pulsed excitation.

The time dependence of the Hamiltonian is caused by molecular reorientation; thus, the quadrupolar Hamiltonian can be written (from Equation 10)

$$\hat{H}_Q(t) = C_Q \sum_m (-1)^m \hat{T}_{2m} F_{2-m}(t) \quad \text{(A11)}$$

so that

$$\hat{H}_q(t) = \hat{H}_Q(t) - \langle H_Q \rangle_T = C_Q \sum_m (-1)^m \hat{T}_{2m}\left[F_{2-m}(t) - \langle F_{2-m} \rangle_T\right] \quad \text{(A12)}$$

Inserting this into Equation A10 gives

$$\begin{aligned}\frac{d\overline{\hat{\rho}(t)}}{dt} &= -\frac{i}{\hbar}\left[\langle H_Q \rangle_T, \overline{\hat{\rho}(t)} - \overline{\hat{\rho}_{eq}}\right] \\ &-\frac{1}{\hbar^2} C_Q^2 \sum_m \sum_{m'} (-1)^{m+m'} \left[\hat{T}_{2m}, \left[\hat{T}_{2m'}, \overline{\hat{\rho}(t)} - \overline{\hat{\rho}_{eq}}\right]\right] \times \quad \text{(A13)} \\ &\int_0^\infty d\tau e^{-i\omega_0 t(m+m')} e^{-i\omega_0 \tau m'} \left(\overline{F_{2-m}(t) F_{2-m'}(t-\tau)} - \overline{\langle F_{2-m} \rangle_T^2}\right)\end{aligned}$$

Here we have assumed that: (i) the fluctuations in $\hat{H}_Q(t)$ are described by a stationary Markov process so that the correlation function depends only on the time difference $\tau = t - t'$ and (ii) that the correlation between $\hat{H}_q(t)$ and $\hat{H}_q(t')$ decays rapidly to zero so that we may extend the limit of integration to infinity.

Keeping only those terms where $m' = -m$ (the others oscillate rapidly) and introducing the correlation functions

$$G_m(\tau) = \overline{F_{2-m}(t) F_{2m}(t-\tau)} - \overline{\langle F_{2-m} \rangle_T^2} \quad \text{(A14)}$$

the equation for the density matrix becomes

$$\begin{aligned}\frac{d\overline{\hat{\rho}(t)}}{dt} &= -\frac{i}{\hbar}\left[\langle H_Q \rangle_T, \overline{\hat{\rho}(t)} - \overline{\hat{\rho}_{eq}}\right] \\ &-\frac{1}{\hbar^2} C_Q^2 \sum_m \left[\hat{T}_{2m}, \left[\hat{T}_{2-m}, \overline{\hat{\rho}(t)} - \overline{\hat{\rho}_{eq}}\right]\right] \int_0^\infty d\tau e^{im\omega_0 \tau} G_m(\tau)\end{aligned} \quad \text{(A15)}$$

We can define the spectral densities of the fluctuating Hamiltonian as the Fourier transforms of the correlation functions

$$J_m(m\omega) = (-1)^m \int_{-\infty}^\infty d\tau e^{im\omega\tau} G_m(\tau) \quad \text{(A16)}$$

ACKNOWLEDGMENTS

J'aimerais remercier mes amis (Michèle Auger, Michael Morrow, Scott Prosser, Xi Shan, et David Siminovitch) qui ont lu ce manuscrit et qui m'ont donné leurs commentaires critiques et dire merci au rédacteur, Len Finegold, pour sa patience et sa bonne humeur. Je voudrais aussi exprimer ma reconnaissance à Xi Shan pour m'avoir aidé avec toutes les figures. Enfin, il faut reconnaitre le Conseil de Recherches en Sciences Naturelles et en Génie du Canada pour son soutien financier.

REFERENCES

Abragam, A. 1961. *The Principles of Nuclear Magnetism*, Oxford University Press, Oxford.

Albon, N. and J. M. Sturtevant. 1978. Nature of the gel to liquid crystal transition of synthetic phosphatidylcholines, *Proc. Natl. Acad. Sci. U.S.A.*, 75:2258–2260.

Alecio, M. R., D. E. Golan, W. R. Veatch, and R. R. Rando. 1982. Use of a fluorescent cholesterol derivative to measure lateral mobility of cholesterol in membranes. *Proc. Natl. Acad. Sci. U.S.A.*, 79:5171–5174.

Auger, M., D. Carrier, I. C. P. Smith, and H. C. Jarrell. 1990. Elucidation of motional modes in glycoglycerolipid bilayers. A ^{2}H NMR relaxation and line-shape study, *Biochemistry*, 112:1373–1381.

Bienvenue, A., M. Bloom, J. H. Davis, and P. F. Devaux. 1982. Evidence of protein-associated lipids from deuterium nuclear magnetic resonance studies of rhodopsin-dimyristoylphosphatidylcholine recombinants. *J. Biol. Chem.*, 257:3032–3038.

Biltonen, R. L. 1990. A statistical-thermodynamic view of cooperative structural changes in phospholipid bilayer membranes: their potential role in biological function. *J. Chem. Thermodynam.*, 22:1–19.

Bloch, K. 1985. Cholesterol, evolution of structure and function. In *Biochemistry of Lipids and Membranes*, Vance, D. E. and Vance, J. E., Eds., Benjamin/Cummings, New York, 1–24.

Bloom, M., J. H. Davis, and M. I. Valic. 1980. Spectral distortion effects due to finite pulse widths in deuterium nuclear magnetic resonance spectroscopy. *Can. J. Phys.*, 58:1510–1517.

Bloom, M. and E. Sternin. 1987. Transverse nuclear spin relaxation in phospholipid bilayer membranes. *Biochemistry*, 26:2101–2105.

Bloom, M. 1988. NMR studies of membranes and whole cells. In *Physics of NMR Spectroscopy in Biology and Medicine*, Maraviglia, B., Ed., Soc. Italiana di Fisica, Bologna, pp. 121–157.

Bloom, M. and E. Evans. 1990. Observation of surface undulations on the mesoscopic length scale by NMR. In *NATO Adv. Workshop. Biologically Inspired Physics*, Pelliti, L., Ed., Plenum Press, New York, pp. 137–147.

Bloom, M., E. Evans, and O. G. Mouritsen. 1991. Physical properties of the fluid lipid-bilayer component of cell membranes: a perspective, *Q. Rev. Biophys.*, 24:293–397.

Bloom, M., C. Morrison, E. Sternin, and J. L. Thewalt. 1992. Spin echoes and the dynamic properties of membranes. In *Pulsed Magnetic Resonance: NMR, ESR, and Optics, a Recognition of E. L. Hahn*, Bagguley, D. M. S., Ed., Clarendon Press, Oxford.

Blume, A. and R. G. Griffin. 1982. Carbon-13 and deuterium nuclear magnetic resonance study of the interaction of cholesterol with phosphatidylethanolamine. *Biochemistry*, 21:6230–6242.

Blume, A., D. M. Rice, R. J. Wittebort, and R. G. Griffin. 1982. Molecular dynamics and conformation in the gel and liquid-crystalline phases of phosphatidylethanolamine bilayers. *Biochemistry,* 24:6220–6230.

Bonmatin, J.-M., I. C. P. Smith, H. C. Jarrell, and D. J. Siminovitch. 1990. Use of a comprehensive approach to molecular dynamics in ordered lipid systems: cholesterol reorientation in oriented lipid bilayers. A ^{2}H NMR study. *J. Am. Chem. Soc.*, 112:1697–1704.

Brown, M. F. and J. Seelig. 1978. Influence of cholesterol on the polar region of phosphatidylcholine and phosphatidylethanolamine bilayers. *Biochemistry,* 17:381–384.

Brown, M. F. 1979. Deuterium relaxation and molecular dynamics in lipid bilayers. *J. Magn. Reson.*, 35:203–215.

Brown, M. F. 1982. Theory of spin-lattice relaxation in lipid bilayers and biological membranes. ^{2}H and ^{14}N quadrupolar relaxation. *J. Chem. Phys.*, 77:1576–1599.

Brown, M. F., A. A. Ribeiro, and G. D. Williams. 1983. New view of lipid bilayer dynamics from ^{2}H and ^{13}C NMR relaxation time measurements. *Proc. Natl. Acad. Sci. U.S.A.*, 80:4325–4329.

Chen, S. C., J. M. Sturtevant, and B. J. Gaffney. 1980. Scanning calorimetric evidence for a third phase transition in phosphatidylcholine bilayers. *Proc. Natl. Acad. Sci. U.S.A.*, 77:5060–5063.

Cohen-Tannoudji, C., B. Diu, and F. Laloe. 1977. *Quantum Mechanics*, vol. 2, Hermann, Paris.

Copeland, B. R. and H. M. McConnell. 1980. The rippled structure in bilayer membranes of phosphatidylcholine and binary mixtures of phosphatidylcholine and cholesterol. *Biochim. Biophys. Acta,* 599:95–109.

Corvera, E., O. G. Mouritsen, M. A. Singer, and M. J. Zuckermann. 1991. The permeability and the effect of acyl chain length for phospholipid bilayers containing cholesterol: theory and experiment, *Biochim. Biophys. Acta,* in press.

Craven, B. M. 1979. Pseudosymmetry in cholesterol monohydrate. *Acta Crystallogr. Sect. B.*, 35:1123–1128.

Cruzeiro-Hansson, L., J. H. Ipsen, and O. G. Mouritsen. 1989. Intrinsic molecules in lipid membranes change the lipid-domain interfacial area: cholesterol at domain interfaces. *Biochim. Biophys. Acta,* 979:166–176.

Davis, J. H. 1979. Deuterium magnetic resonance study of the gel and liquid crystalline phases of dipalmitoyl phosphatidylcholine. *Biophys. J.*, 27:339–358.

Davis, J. H., M. Bloom, K. W. Butler, and I. C. P. Smith. 1980. The temperature dependence of molecular order and the influence of cholesterol in *Acholeplasma laidlawii* membranes. *Biochim. Biophys. Acta,* 597:477–491.

Davis, J. H. 1983. The description of membrane lipid conformation, order and dynamics by ^{2}H NMR. *Biochim. Biophys. Acta,* 737:117–171.

Davis, J. H. 1986. The influence of membrane proteins on lipid dynamics. *Chem. Phys. Lipids,* 40:223–258.

Davis, J. H. 1988. NMR studies of cholesterol orientational order and dynamics, and the phase equilibria of cholesterol/phospholipid mixtures. In *Physics of NMR Spectroscopy in Biology and Medicine*, Maraviglia, B., Ed., Soc. Italiana di Fisica, Bologna, pp. 302–312.

Davis, J. H. 1989. Deuterium nuclear magnetic resonance and relaxation in partially ordered systems. *Adv. Magn. Res.*, 13:195–223.

Davis, J. H. 1991. Deuterium nuclear magnetic resonance spectroscopy in partially ordered systems. In *Isotopes in the Physical and Biomedical Sciences* 2, Buncel, E. and Jones, J. R., Eds., Elsevier, Amsterdam, pp. 199–157.

De Gennes, P. 1974. *The Physics of Liquid Crystals*, Oxford University Press, Oxford, p. 27.

Deisenhofer, J. and H. Michel. 1991. High-resolution structures of photosynthetic reaction centers. *Annu. Rev. Biophys. Biophys. Chem.*, 20:247–266.

Devaux, P. F. and M. Seigneuret. 1985. Specificity of lipid-protein interactions as determined by spectroscopic techniques. *Biochim. Biophys. Acta,* 822:63–125.

Dufourc, E. J., E. J. Parish, S. Chitrakorn, and I. C. P. Smith. 1984. Structural and dynamical details of cholesterol-lipid interaction as viewed by deuterium NMR. *Biochemistry,* 23:6062–6071.

Evans, E. and W. Rawicz. 1990. Entropy-driven tension and bending elasticity in condensed-fluid phases. *Phys. Rev. Lett.*, 64:2094–2097.

Finean, J. B. 1990. Interaction between cholesterol and phospholipid in hydrated bilayers. *Chem. Phys. Lipids,* 54:147–156.

Finegold, L. and M. A. Singer. 1986. The metastability of saturated phosphatidylcholines depends on the acyl chain length. *Biochem. Biophys. Acta,* 855:417–420.

Gaber, B. P. and W. L. Peticolas. 1977. On the quantitative interpretation of biomembrane structure by Raman spectroscopy. *Biochim. Biophys. Acta,* 465:260–274.

Gennis, R. B. 1989. *Biomembranes*, Springer-Verlag, New York.

Gershfeld, N. L. 1978. Equilibrium studies of lecithin-cholesterol interactions. I. Stoichiometry of lecithin-cholesterol complexes in bulk systems. *Biophys. J.*, 22:469–488.

Griffin, R. G. 1981. Solid state nuclear magnetic resonance of lipid bilayers. *Methods Enzymol.*, 72:108–174.

Griffin, R. G., K. Beshah, R. Ebelhauser, T.-H. Huang, E. T. Olejniczak, D. M. Rice, D. J. Siminovitch, and R. J. Wittebort. 1988. Deuterium NMR studies of dynamics in solids. In *The Time Domain in Surface and Structural Dynamics*, Long, G. I. and Grandjean, F., Eds., Kluwer Academic Publishers, Norwell, MA.

Goldman, M. 1988. The fundamentals of NMR. In *Physics of NMR Spectroscopy in Biology and Medicine*, Maraviglia, B., Ed., Soc. Italiana di Fisica, Bologna, pp. 1–42.

Haberkorn, R. A., R. G. Griffin, M. D. Meadows, and E. Oldfield. 1977. Deuterium nuclear magnetic resonance investigation of the dipalmitoyl lecithin-cholesterol-water system. *J. Am. Chem. Soc.*, 99:7353–7355.

Henderson, R., J. M. Baldwin, T. A. Ceska, F. Zemlin, E. Beckmann, and K. H. Downing. 1990. Model for the structure of bacteriorhodopsin based on high-resolution electron cryo-microscopy. *J. Mol. Biol.*, 213:899–929.

Huang, T.-H., R. P. Skarjune, R. J. Wittebort, R. G. Griffin, and E. Oldfield. 1980. Restricted rotational isomerization in polymethylene chains. *J. Am. Chem. Soc.*, 102:7377–7379.

Hui, S. W. and H.-B. He. 1983. Molecular organization in cholesterol-lecithin bilayers by X-ray and electron diffraction measurements. *Biochemistry,* 22:1159–1164.

Huschilt, J. C., R. S. Hodges, and J. H. Davis. 1985. Phase equilibria in an amphiphilic peptide-phospholipid model membrane by ^{2}H nuclear magnetic resonance difference spectroscopy. *Biochemistry,* 24:1377–1385.

Ipsen, J. H., G. Karlstrom, O. G. Mouritsen, H. Wennerstrom, and M. J. Zuckermann. 1987. Phase equilibria in the phosphatidylcholine-cholesterol system. *Biochim. Biophys. Acta,* 905:162–172.

Ipsen, J. H., O. G. Mouritsen, and M. J. Zuckermann. 1989. Theory of thermal anomalies in the specific heat of lipid bilayers containing cholesterol. *Biophys. J.*, 56:661–667.

Israelachvilli, J., S. Marcelja, and R. G. Horn. 1981. Physical principles of membrane organization. *Q. Rev. Biophys.*, 13:121–200.

Jacobs, R. and E. Oldfield. 1979. Deuterium nuclear magnetic resonance investigation of dimyristoyllecithin-dipalmitoyllecithin and dimyristoyllecithin-cholesterol mixtures. *Biochemistry,* 18:3280–3285.

Jeffrey, K. R. 1981. Nuclear magnetic relaxation in a spin 1 system. *Bull. Magn. Res.*, 3:69–82.

Knoll, W., G. Schmidt, K. Ibel, and E. Sackmann. 1985. Small-angle neutron scattering study of lateral phase separation in dimyristoylphosphatidylcholine-cholesterol mixed membranes. *Biochemistry,* 24:5240–5246.

Ladbrooke, B. D., R. M. Williams, and D. Chapman. 1968. Studies on lecithin-cholesterol-water interactions by differential scanning calorimetry. *Biochim. Biophys. Acta,* 150:333–340.

Lange, A., D. Marsh, K.-H. Wassmer, P. Meier, and G. Kothe. 1985. Electron spin resonance study of phospholipid membranes employing a comprehensive line-shape model. *Biochemistry,* 24:4383–4392.

Lentz, B. R., D. A. Barrow, and M. Hoechli. 1980. Cholesterol-phosphatidylcholine interactions in multilamellar vesicles. *Biochemistry,* 19:1943–1954.

Levin, I. W. 1984. Vibrational spectroscopy of membrane assemblies. *Adv. Infrared Raman Spectr.* Clark, R. J. H. and Hester, R. E., Eds., 11:1–48.

Mabrey, S., P. L. Mateo, and J. M. Sturtevant. 1978. High-sensitivity scanning calorimetric study of mixtures of cholesterol with dimyristoyl- and dipalmitoylphosphatidylcholines. *Biochemistry,* 17:2464–2468.

Mabrey, S. and J. M. Sturtevant. 1979. High-sensitivity differential scanning calorimetry in the study of biomembranes and related model systems. *Methods Membr. Biol.*, 9:237–274.

Maraviglia, B., J. H. Davis, M. Bloom, J. Westerman, and K. W. A. Wirtz. 1982. Human erythrocyte membranes are fluid down to −5°C. *Biochim. Biophys. Acta,* 686:137–140.

Marsh, D. 1989. Experimental methods in spin-label spectral analysis. In *Biological Magnetic Resonance*, vol. 8, Berliner, L. J. and Reuben, J., Eds., Plenum Press, New York, pp. 255–304.

McElhaney, R. N. 1982. The use of differential scanning calorimetry and differential thermal analysis in studies of model and biological membranes. *Chem. Phys. Lipids,* 30:229–259.

McIntosh, T. J., A. D. Magid, and S. A. Simon. 1989. Cholesterol modifies the short-range repulsive interactions between phosphatidylcholine membranes. *Biochemistry,* 28:17–25.

Mehring, M. 1983. *High Resolution NMR in Solids*, Springer-Verlag, Berlin.

Meiboom, S. and D. Gill. 1958. Modified spin-echo method for measuring nuclear relaxation times. *Rev. Sci. Instrum.*, 29:688–691.

Meier, P., E. Ohmes, G. Kothe, A. Blume, J. Weldner, and H.-J. Eibl. 1983. Molecular order and dynamics of phospholipid membranes. A deuteron magnetic resonance study employing a comprehensive line-shape model. *J. Phys. Chem.*, 87:4904–4912.

Meier, P., E. Ohmes, and G. Kothe. 1986. Multipulse dynamic nuclear magnetic resonance of phospholipid membranes. *J. Chem. Phys.*, 85:3598–3613.

Mendelsohn, R., M. A. Davies, J. W. Grauner, H. F. Schuster, and R. A. Dluhy. 1989. Quantitative determination of conformational disorder in the acyl chains of phospholipid bilayers by infrared spectroscopy. *Biochemistry,* 28:8934–8939.

Millar, J. M., A. M. Thayer, H. Zimmerman, and A. Pines. 1986. High resolution studies of deuterium time-domain zero-field NQR. *J. Magn. Reson.*, 69:243–257.

Morrow, M. R., J. C. Huschilt, and J. H. Davis. 1985. Simultaneous modeling of phase and calorimetric behavior in an amphiphilic peptide/phospholipid model membrane. *Biochemistry,* 24:5396–5406.

Morrow, M. R., J. H. Davis, F. J. Sharom, and M. P. Lamb. 1986. Studies of the interaction of human erythrocyte band 3 with membrane lipids using deuterium nuclear magnetic resonance and differential scanning calorimetry. *Biochim. Biophys. Acta,* 858:13–20.

Morrow, M. R. and J. H. Davis. 1988. Differential scanning calorimetry and ^{2}H NMR studies of the phase behavior of gramicidin-phosphatidylcholine mixtures. *Biochemistry,* 27:2024–2032.

Morrow, M. R. and J. P. Whitehead. 1988. A phenomenological model for lipid-protein bilayers with critical mixing. *Biochim. Biophys. Acta,* 941:271–277.

Morrow, M. R., R. Srinivasan, and N. Grandal. 1991. The phase diagram of dimyristoyl phosphatidylcholine and chain-perdeuterated distearoyl phosphatidylcholine: a deuterium NMR spectral difference study. *Chem. Phys. Lipids,* 58:63–72.

Morrow, M. R., J. P. Whitehead, and D. Lu. 1992. Chain length dependence of lipid bilayer properties near the liquid crystal to gel phase transition, *Biophys. J.* 62, in press.

Mortensen, K., W. Pfeiffer, E. Sackmann, and W. Knoll. 1988. Structural properties of a phosphatidylcholine-cholesterol system as studied by small-angle neutron scattering: ripple structure and phase diagram. *Biochim. Biophys. Acta,* 945:221–245.

Mouritsen, O. G. 1991. Theoretical models of phospholipid phase transitions. *Chem. Phys. Lipids,* 57:178–194.

Needham, D., T. J. McIntosh, and E. Evans. 1988. Thermomechanical and transition properties of dimyristoylphosphatidylcholine/cholesterol bilayers. *Biochemistry,* 27:4668–4673.

Needham, D. and R. S. Nunn. 1990. Elastic deformation and failure of lipid bilayer membranes containing cholesterol. *Biophys. J.*, 58:997–1009.

Nezil, F. A. and M. Bloom. 1992. Combined influence of cholesterol and synthetic amphiphillic peptides upon bilayer thickness in model membranes, *Biophys. J.* 61:1176–1183.

Oldfield, E. and D. Chapman. 1971. Effects of cholesterol and cholesterol derivatives on hydrocarbon chain mobility in lipids. *Biochem. Biophys. Res. Commun.*, 43:610–616.

Pasenkiewicz-Gierula, M., W. K. Subczynski, and A. Kusumi. 1990. Rotational diffusion of a steroid molecule in phosphatidylcholine-cholesterol membranes: fluid-phase microimmiscibility in unsaturated phosphatidylcholine-cholesterol membranes. *Biochemistry,* 20:4059–4069.

Pastor, R. W., R. M. Venable, M. Karplus, and A. Szabo. 1988. A simulation based model of NMR T_1 relaxation in lipid bilayer vesicles. *J. Chem. Phys.*, 89:1128–1140.

Rance, M., K. R. Jeffrey, A. P. Tulloch, K. W. Butler, and I. C. P. Smith. 1982. Effects of cholesterol on the orientational order of unsaturated lipids in the membranes of *Acholeplasma laidlawii. Biochim. Biophys. Acta,* 688:191–200.

Rand, R. P., V. A. Parsegian, J. A. C. Henry, L. J. Lis, and M. McAlister. 1980. The effect of cholesterol on measured interaction and compressibility of dipalmitoylphosphatidylcholine bilayers. *Can. J. Biochem.*, 58:959–968.

Recktenwald, D. J. and H. M. McConnell. 1981. Phase equilibria in binary mixtures of phosphatidylcholine and cholesterol. *Biochemistry,* 20:4505–4510.

Redfield, A. G. 1965. The theory of relaxation processes. *Adv. Magn. Res.*, 1:1–32.

Rommel, E., F. Noack, P. Meier, and G. Kothe. 1988. Proton spin relaxation dispersion studies of phospholipid membranes. *J. Phys. Chem.*, 92:2981–2987.

Rose, M. E. 1957. *Elementary Theory of Angular Momentum,* John Wiley & Sons, New York, chap. 4.

Rubenstein, J. L. R., B. A. Smith, and H. M. McConnell. 1979. Lateral diffusion in binary mixtures of cholesterol and phosphatidylcholines. *Proc. Natl. Acad. Sci. U.S.A.*, 76:15–18.

Ruocco, M. J. and G. G. Shipley. 1982. Characterization of the sub-transition of hydrated dipalmitoylphosphatidylcholine bilayers. X-ray diffraction study. *Biochim. Biophys. Acta,* 684:59–66.

Sankaram, M. B. and T. E. Thompson. 1990. Modulation of phospholipid acyl chain order by cholesterol. A solid state ^{2}H nuclear magnetic resonance study. *Biochemistry,* 29:10676–10684.

Sankaram, M. B. and T. E. Thompson. 1991. Cholesterol-induced fluid-phase immiscibility in membranes. *Proc. Natl. Acad. Sci. U.S.A.*, 88:8686–8690.

Scott, H. L. 1992. Lipid-cholesterol phase diagrams: theoretical and numerical aspects. In *Cholesterol Membrane Models,* Finegold, L., Ed., CRC Press, Boca Raton, FL.

Seelig, A. and J. Seelig. 1974. The dynamic structure of fatty acyl chains in a phospholipid bilayer measured by deuterium nuclear magnetic resonance. *Biochemistry,* 13:4839–4845.

Seelig, A. and J. Seelig. 1975. Bilayers of dipalmitoyl-3-*sn*-phosphatidylcholine: conformational differences between the fatty acyl chains. *Biochim. Biophys. Acta,* 407:1–5.

Seelig, J. 1977. Deuterium magnetic resonance: theory and application to lipid membranes. *Q. Rev. Biophys.*, 10:353–418.

Seelig, J., H.-U. Gally, and R. Wohlgemuth. 1977. Orientation and flexibility of the choline head group in phosphatidylcholine bilayers. *Biochim. Biophys. Acta,* 467:109–119.

Seelig, J., P. M. MacDonald, and P. G. Scherer. 1987. Phospholipid head groups as sensors of electric charge in membranes. *Biochemistry,* 26:7535–7541.

Shimshick, E. J. and H. M. McConnell. 1973. Lateral phase separation in binary mixtures of cholesterol and phospholipids. *Biochem. Biophys. Res. Commun.*, 53:446–451.

Shin, Y.-K., J. K. Moscicki, and J. H. Freed. 1990. Dynamics of phosphatidylcholine-cholesterol mixed model membranes in the liquid crystalline state. *Biophys. J.*, 57:445–459.

Slichter, C. P. 1990. *Principles of Magnetic Resonance*, 3rd ed., Springer-Verlag, Berlin.

Spiess, H. W. 1978. Rotation of molecules and nuclear spin relaxation. In *NMR. Basic Principles and Progress* 11, Diehl, P., Fluck, E., and Kosfeld, R., Eds., Springer-Verlag, Berlin, pp. 55–214.

Stockton, G. W., C. F. Polnaszek, L. C. Leitch, A. Tulloch, and I. C. P. Smith. 1974. A study of mobility and order in model membranes using ^{2}H NMR relaxation rates and quadrupole splittings of specifically deuterated lipids. *Biochem. Biophys. Res. Commun.,* 60:844–850.

Stockton, G. W., C. F. Polnaszek, A. P. Tulloch, F. Hasan, and I. C. P. Smith. 1976. Molecular motion and order in single layer vesicles and multilamellar dispersions of egg-yolk lecithin and lecithin-cholesterol mixtures. A deuterium nuclear magnetic resonance study of specifically labeled lipids. *Biochemistry,* 15:954–966.

Stohrer, J., G. Grobner, D. Reimer, K. Weisz, C. Mayer, and G. Kothe. 1991. Collective lipid motions in bilayer membranes studied by transverse deuteron spin relaxation. *J. Chem. Phys.*, 95:672–678.

Stryer, L. 1988. *Biochemistry*, W. H. Freeman, New York.

Tanford, C. 1973. *The Hydrophobic Effect: Formation of Micelles and Biological Membranes*, Wiley-Interscience, New York.

Tardieu, A., V. Luzzati, and F. C. Reman. 1973. Structure and polymorphism of the hydrocarbon chains of lipids: a study of lecithin-water phases. *J. Mol. Biol.*, 75:711–733.

Taylor, M. G., T. Akiyama, and I. C. P. Smith. 1981. The molecular dynamics of cholesterol in bilayer membranes. A deuterium NMR study. *Chem. Phys. Lipids,* 29:327–339.

Taylor, M. G., T. Akiyama, H. Saito, and I. C. P. Smith. 1982. Direct observation of the properties of cholesterol in membranes by deuterium NMR. *Chem. Phys. Lipids,* 31:359–379.

Thewalt, J. L. and M. Bloom. 1992. Phosphatidylcholine:cholesterol phase diagrams, *Biophys. J.*, in press.

Torchia, D. A. and A. Szabo. 1982. Spin-lattice relaxation in solids. *J. Magn. Reson.* 49:107–121.

Vaz, W. L. C., Z. I. Derzko, and K. A. Jackobson. 1984. Photobleaching measurements of the lateral diffusion of lipids and proteins in artificial phospholipid bilayer membranes. *Cell Surf. Rev.*, 8:83–136.

Vega, S. and A. Pines. 1977. Operator formalism for double quantum NMR. *J. Chem. Phys.*, 66:5624–5644.

Vist, M. R. 1984. Partial phase behavior of perdeuteriated dipalmitoylphosphatidylcholine-cholesterol model membranes. MSc. thesis, University of Guelph, Guelph, Ontario, Canada.

Vist, M. and J. H. Davis. 1990. Phase equilibria of cholesterol/dipalmitoylphosphatidylcholine mixtures: ^{2}H nuclear magnetic resonance and differential scanning calorimetry. *Biochemistry,* 29:451–464.

Wallach, D. 1967. Effect of internal rotation on angular correlation functions. *J. Chem. Phys.*, 47:5258–5268.

Weisz, K., G. Grobner, C. Mayer, J. Stohrer, and G. Kothe. 1992. Deuteron nuclear magnetic resonance study of the dynamic organization of phospholipid/cholesterol bilayer membranes: molecular properties and viscoelastic behavior. *Biochemistry*, 31:1100–1112.

Wittebort, R. J. and A. Szabo. 1978. Theory of NMR relaxation in macromolecules: restricted diffusion and jump models for multiple internal rotations in amino acid side chains. *J. Chem. Phys.*, 69:1722–1736.

Wittebort, R. J., A. Blume, T.-H. Huang, S. K. Das Gupta, and R. G. Griffin. 1982. Carbon-13 nuclear magnetic resonance investigations of phase transitions and phase equilibria in pure and mixed phospholipid bilayers. *Biochemistry,* 21:3487–3502.

Wittebort, R. J., E. T. Olejniczak, and R. G. Griffin. 1987. Analysis of deuterium nuclear magnetic resonance line shapes in anisotropic media. *J. Chem. Phys.*, 86:5411–5420.

Woessner, D. E. 1962. Spin relaxation processes in a two-proton system undergoing anisotropic reorientation. *J. Chem. Phys.*, 36:1–4.

Wokaun, A. and R. Ernst. 1977. Selective excitation and detection in multilevel spin systems: application of single transition operators. *J. Chem. Phys.*, 67:1752–1758.

Zannoni, C., A. Arcioni, and P. Cavatorta. 1983. Fluorescence depolarization in liquid crystals and membrane bilayers. *Chem. Phys. Lipids,* 32:179–250.

Zare, R. N. 1988. *Angular Momentum. Understanding Spatial Aspects in Chemistry and Physics*, John Wiley & Sons, New York, chap. 3.

Zuckermann, M. J., J. H. Ipsen, and O. G. Mouritsen. 1992. Theoretical studies of the phase behavior of lipid bilayers containing cholesterol. In *Cholesterol in Membrane Models,* Finegold, L., Ed., CRC Press, Boca Raton, FL.

Chapter 5

CHOLESTEROL/PHOSPHOLIPID INTERACTIONS STUDIED BY DIFFERENTIAL SCANNING CALORIMETRY: EFFECT OF ACYL CHAIN LENGTH AND ROLE OF THE C(17) STEROL SIDE GROUP

Leonard Finegold and Michael A. Singer

TABLE OF CONTENTS

0-8493-4207-4/93/$0.00+$.50

ABSTRACT

A popular experimental mimetic model for the cell plasma membrane is the liposome, which is the bilayer structure formed when long-chain lipids are dispersed in water. This chapter introduces experimental liposomes and a thermal technique for their biophysical properties (differential scanning calorimetry). The power of differential scanning calorimetry is that exogenous probes are unnecessary. To study the interactions between sterol and lipids in the mimetic membrane, structural variations are made. These variations are, for the lipid, varying the head group (phosphatidyl choline or phosphatidyl ethanolamine) or varying the lipid chain length (from 10 through 20 carbons), and for the sterol, removing the tail (side group) from cholesterol (at the C17 position) to make androsten. It is found that cholesterol's interaction with the lipids is mainly through the sterol ring structure, and not its C17 side group. The results of the preceding studies enable the prediction of phase diagrams for lipids of all chain lengths (10 to 20) with cholesterol. Future theoretical models for cholesterol in lipids must include the above results. A dynamic model is outlined.

PREAMBLE

All eukaryotic membranes contain phospholipids, which make the basic bilayer structure. All eukaryotic membranes contain, inside the bilayer structure, sterols. In mammalian membranes the sterol is always cholesterol. "Why" this system is so widespread is discussed in other chapters, particularly that of P. Yeagle. In brief, membranes contain a wide variety of phospholipids, but their chain lengths are restricted typically to the range 10 to 20 carbons, consistent with the maintenance of "fluidity." A sterol, being rigid, supplies certain physical properties to the membrane. Why cholesterol should be (teleologically) chosen by membranes, out of the large number of structurally similar cholesterol analogs, seems to reside in two realms: the biochemical and the physical. The biochemical realm is discussed in P. Yeagle's chapter: biochemical pathways are of great importance in determining that sterols are a chosen molecule and why the chosen sterol should be cholesterol. We treat the physical realm in detail in this chapter.

That cholesterol is an essential component of membranes is illustrated by attempts to deprive cells of cholesterol. Only a very few mammalian cell lines can be so deprived (Esfahani and Swaney, 1990), and these cells demonstrate the importance of cholesterol by responding to cholesterol depletion with little reproduction and little viability. (Organisms at the evolutionary boundary of eukaryotes and prokaryotes, such as *Acheloplasma* sp., can be sterol deprived; from our mammalicentric viewpoint regarding cholesterol, mammalian cell lines are more relevant.)

In this chapter we discuss, in the following order, (1) the basis of a mimetic model for membranes (the liposome, from *lipos* fat + *soma* body), (2) an experimental technique for studying the biophysical properties of liposomes (differential scanning calorimetry), (3) experimental tools for varying the liposomes (variation of membrane lipid chain length n and variation of the cholesterol analogue), (4) the results of the preceding studies, (5) their refinement of our present picture of cholesterol in membranes, and, finally, (6) predictions as to how the one sterol — cholesterol — in membranes interacts with the multiplicity of lipids present in all natural membranes. Cholesterol interacts with the phospholipids mainly through its sterol ring structure.

I. INTRODUCTION

Since cholesterol is a ubiquitous — and the only sterol — component of mammalian membranes, it seems only logical to assert that this molecule has significant regulatory functions. However, the actual definition of such functions has remained elusive. Part of the difficulty of definition resides in the complexity of biological membranes themselves. Within such a complex structure, it is difficult to isolate and, hence, study the interaction of cholesterol with other specific membrane molecules. Therefore, investigators have turned to the use of model systems, and one popular system — which we like — is the phospholipid bilayer dispersed in water. The phospholipid bilayer mimics many of the properties of a real biological membrane and, in addition, allows the investigator to control precisely the bilayer composition. In this chapter we will explore the interaction between cholesterol and phospholipids, using the bilayer model system. We will examine that interaction as a function of phospholipid head group and chain length and of the structure of the sterol molecule.

The bilayer multilamellar model system that we have used is known as the multilamellar liposome system. Liposomes are closed vesicles (Figure 1), each consisting of a number of concentric bilayer membranes, separated by discrete and isolated water compartments. Topologically, liposomes can be thought of as concentric spheres. The liposome system was pioneered by A.D. Bangham (Bangham, 1983) whose original paper (Bangham et al., 1965) described the organization and permeability properties of the liposomes; the color micrographs and electron micrographs in the paper are still worth seeing. (Liposomes are also called "bangosomes.") One of the important advantages of this model system over other model systems is the ease and precision with which bilayer composition can be controlled. Another advantage for calorimetry and many other techniques is the high amount of lipid per unit volume. The liposomes of this chapter are composed of saturated phospholipids: the simplest of the membrane lipids. These lipids, upon heating, undergo a cooperative order-to-disorder phase transition (reminiscent of the melting of ice into water) at a characteristic temperature (Figure 2B). This characteristic temperature depends upon the nature of the phospholipid head group (e.g., be it a choline or an ethanolamine) and upon the acyl chain length (of n carbons long). The enthalpy change ΔH associated with this transition (i.e., the "heat of melting," which is 80 calorie/gram for ice) is measured in a calorimeter ("caloric" = heat). The incorporation of cholesterol into the bilayer membrane (Figure 2B) simulta-

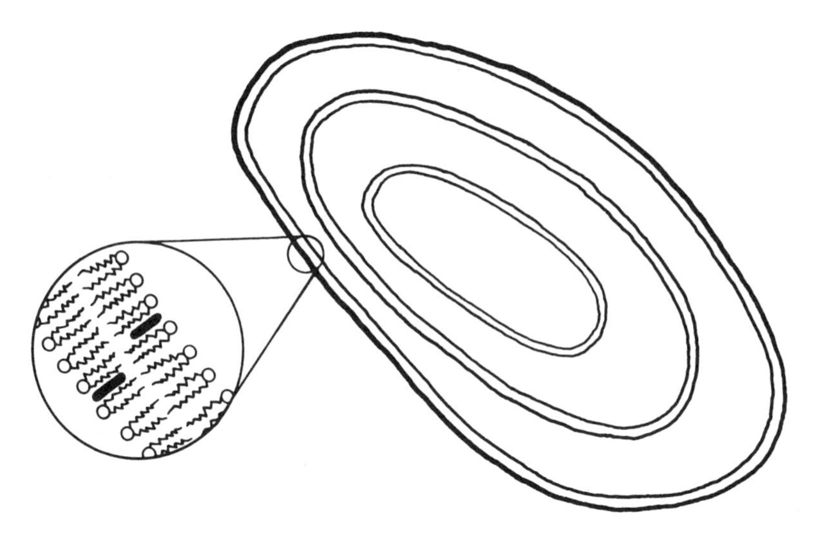

FIGURE 1. The multilamellar liposome — a schematic. A section of three concentric lipid bilayers is shown. The overall size can range from 0.1 to 5 µm. Part of one of the bilayers is shown enlarged, with two cholesterol molecules (shown darkened). Water — the molecules of which are too small to show clearly — is everywhere; water molecules would be about the size of the smallest kinks in the lipid chains.

neously blurs the sharpness of the transition (i.e., increases the temperature width of the transition) and reduces the area of the transition (i.e., reduces the enthalpy change ΔH of the transition). Both of these effects are directly dependent on the concentration of cholesterol, and from this dependence one can infer the molecular interaction between the cholesterol and the phospholipid. The details of this interaction will be discussed in later sections. It should be noted that an advantage of this calorimetric approach is that (1) calorimetry measures all the energetics of a process and (2) the "probes" used are the molecules themselves, so there is a clarity of interpretation that is lost when reporter molecules are perforce used in many other experimental approaches.

II. METHODOLOGY

Multilamellar liposomes readily form (Bangham, 1983) when dried phospholipid is simply shaken in the presence of excess water: by hand in a test tube or with a tabletop vortexing machine, provided that the temperature is above (Singer et al., 1990) that of the main transition temperature (41.5°C for C16PC). However, one must be fastidious with the procedure in order to obtain consistent results. Our procedure, which is typical, starts with the lipids dissolved in chloroform. An aliquot of the lipid is then dried in a rotary evaporator (Buchi, Brinkmann Inc., Ontario) so that a thin uniform layer of lipid is deposited in the hemispherical end of a glass tube. The lipid is then further dried overnight under vacuum to remove any traces of chloroform. Liposomes are formed as soon as

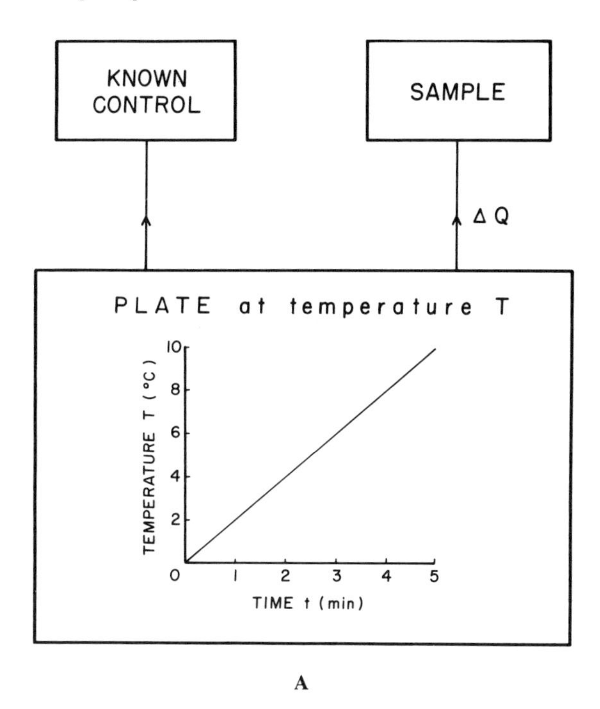

A

FIGURE 2. The essentials of the differential scanning calorimeter. (A) The sample, and a very similar known control, are on a plate at the same temperature T. The temperature of the plate is changed continuously in a time Δt (minutes) by an amount ΔT (°C), resulting in the sample absorbing an extra (compared differentially with the known control) of heat ΔQ. The temperature T is changed very linearly with time, i.e., "scanned". The example shows scan rate = 2°C/min. (B) The differential scanning calorimeter output is shown; the endotherm arrow indicates that heat Q is absorbed in the direction of the arrow. The vertical axis is $\Delta Q/\Delta T$, which is automatically plotted vs. (time t or) temperature T. The area under a curve is the Q-change of the transition, which is (with the usual experimental conditions) the enthalpy change ΔH:

$$\int (\Delta Q/\Delta T)\, dT = Q\text{-change} = \Delta H.$$

The endotherms are for liposomes of pure C16PC (dipalmitoylphosphatidylcholine), and C16PC with 20 mol% of a sterol (androsten; the endotherm for cholesterol is indistinguishable).

water (at say 50°C) is pipetted onto the lipid layer and mechanically shaken for a standard time. We rewarm the water and reshake, repeating the sequence thrice in all. A sample (say, 20 μl) of the liposome suspension is then pipetted into a pan.

III. DIFFERENTIAL SCANNING CALORIMETRY

This section is to help the reader understand differential scanning calorimetry and to evaluate the various differential scanning calorimetry techniques and results reported in the literature. Most differential scanning calorimetry workers use equipment that, historically, was developed for industry; more recently, differential scanning calorimeters have been developed for research. For com-

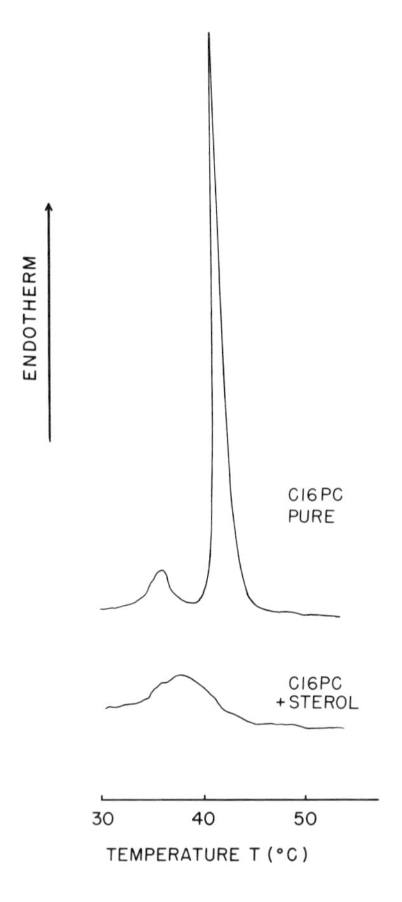

FIGURE 2B.

pleteness we only mention the existence of the most accurate adiabatic calorimeters (which are relatively simple at very low temperatures (Finegold, 1972), but which require large samples and which are very slow at room temperature) and "a.c." calorimeters, which measure tiny samples, but lose absolute accuracy.

The essentials of a differential scanning calorimeter are shown in Figure 2. To fix ideas, example values will be given in parentheses for a common type of differential scanning calorimeter. The sample (say liposomes of C16PC in water, of 20 μl) is placed in a sealed aluminum pan of mass some 50 mg (see Figure 2A and its caption). A known sample is prepared as a control (e.g., 20 μl of water), in a similar pan. Both pans are placed in a symmetrical situation on the plate of the calorimeter. The temperature of the plate is changed ("scanned") linearly in time, and so the temperatures of each of the two pans will also change linearly; if both pans contained only the same amount of water, then the difference in their heat absorption (from the plate) would be zero, i.e., no difference gives the "baseline" plot. Suppose that the sample pan now absorbs a little more heat ΔQ (because the lipid sample is going through a phase transition) than does the control pan, during a time interval Δt or an equivalent temperature interval ΔT. The calorimeter automatically plots $\Delta Q/\Delta T$ with time, as shown in Figure 2B, as "endotherms." Here we are interested in the enthalpy change ΔH of the phase transition, which is given by the area under the plotted curve (Figure 2B, caption). The effect of cholesterol is to reduce the ΔH of the transition.

Technically, differential scanning calorimeters can be divided into two broad classes, depending upon whether they measure the heat ΔQ directly (by electronic feedback) or whether they measure the temperature difference between the two pans (and arrange that this is, by thermal conduction, directly proportional to ΔQ). The latter are sometimes termed "differential thermal analyzers." In practice the difference between the two classes is unimportant, as judged by their measurements on standard samples.

There are many competing suppliers of differential scanning calorimeters; they are to be found listed in the annual guides to scientific instruments in *Science* and *Nature*. Because of their industrial genetics, they are rarely described critically in the scientific literature, and a close perusal of the manufacturer's own description and specifications may not be completely enlightening. It would be inappropriate here to list manufacturers. (Effectively, systematic and critical studies of the ills that differential scanning calorimeters are heir to were made for certain instruments by Van Dooren and Muller, 1981, and Schwarz, 1986, 1991). It appears that because of competition there may be very little difference between instruments, though in very specific applications (or even temperature ranges), one instrument might be best. Therefore, when critically examining published work or trying to decide which instrument to acquire, a prudent way is to contact someone who has recently bought one of the instruments in question and who has themselves gone through all these thought processes. In North America there are informal *Thermal Analysis Forums*, under the aegis of the North American Thermal Analysis Society (1431 St. Georges Ave. #6, Colonia, NJ 07067), which are a good source of contacts.

For lipid work there are indeed two main classes of differential scanning calorimeters, which for convenience will be termed "small" and "large" sample. Small-sample calorimeters typically handle up to 50 µl of sample volume, whereas large-sample calorimeters go up to 1000 µl (1 ml). Because of the time that heat takes to diffuse through any sample, the small-sample calorimeters can operate at faster scan rates (even 10°C/min, although 1 to 5°C/min is more usual); this speed can be valuable in exploratory investigations or where many samples are to be measured. The sample pans are small, are usually disposable, and can be rapidly (easily within a minute) loaded and unloaded. Conversely, the large-sample calorimeters must scan more slowly to enable heat to diffuse evenly. They then have advantages of larger sample mass and, hence, better sensitivity (so they are often termed "high-resolution differential scanning calorimeters"). Sometimes the slower scan rate can give better thermal equilibrium, and the higher sensitivity allows the use of more dilute samples (which is of more importance in biochemical work). However, the larger sample size can have disadvantages. The (1 ml) sample container ("cell") is typically of special construction and expensive, and filling through a narrow capillary can be time consuming for lipids, since their solutions may be quite viscous. (The equivalent filling "capillary" for a small-sample calorimeter is a short wide-bore pipette of negligible viscous resistance.) In the larger cells it is possible for the lipid to settle and stratify. Calorimetric measurements are expressed per unit mass (usually gram-mole); with electronic balances, sufficiently accurate measurements of solutions are readily accomplished, but there remain problems in filling the large cells (Schwarz, 1986, 1991). For measurements below 0°C, the small-sample pans can be frozen with impunity; the freezing of a large cell may result in a several month wait for instrument repair and recalibration. In practice, small- and large-sample calorimeters are complementary and have a wide range of overlap.

Computers are both a boon and a bane to differential scanning calorimetry: a boon in that they enable the calorimeter to be automatically controlled and the output data to be readily handled, and a bane in that the software must be exceedingly carefully checked. (We know of at least one paper submitted for publication — not ours — that violated thermodynamics because of a programming error.) In some species of calorimeter a great deal of computer work is necessary to produce what would be the "raw" data for other calorimeters, so there are even more steps between the experiment and the experimenter. A solution is to measure and report results on a standard sample; this solution is discussed later.

IV. EXPERIMENTAL RESULTS

The essential picture to bear in mind, given by the main experimental variations that we have made on the mimetic membrane system (the liposome), is (1) the variation of ΔH with cholesterol concentration probes the cholesterol-lipid interactions in the plane of the bilayer, and (2) the variation of ΔH with n (at a given cholesterol concentration) probes the cholesterol-lipid interactions perpendicular to (i.e., across) the bilayer. Other experimental variables that we have used are (3) to vary the sterol from cholesterol to an analog and (4) to vary the lipid head group.

In the following discussion, initially we focus on the results for cholesterol and phosphatidyl cholines, then we proceed to phosphatidyl ethanolamines and a sterol analog (Figure 3). (For conciseness of presentation, cholesterol analog results are presented on the same graphs as is cholesterol, and likewise for lipid head groups).

A. GENERAL OBSERVATIONS

Before proceeding to the detailed results, we make some general observations. When cholesterol is added to the phospholipid bilayer, the enthalpy change ΔH of the main transition is reduced. Calorimetric scans for pure C16PC and for a mixture of C16PC/cholesterol are illustrated in Figure 2B. The morphology of the mixture scan is complex; some authors have deconvoluted it (as described later) into a sharp and a broad component. The sharp component diminishes as

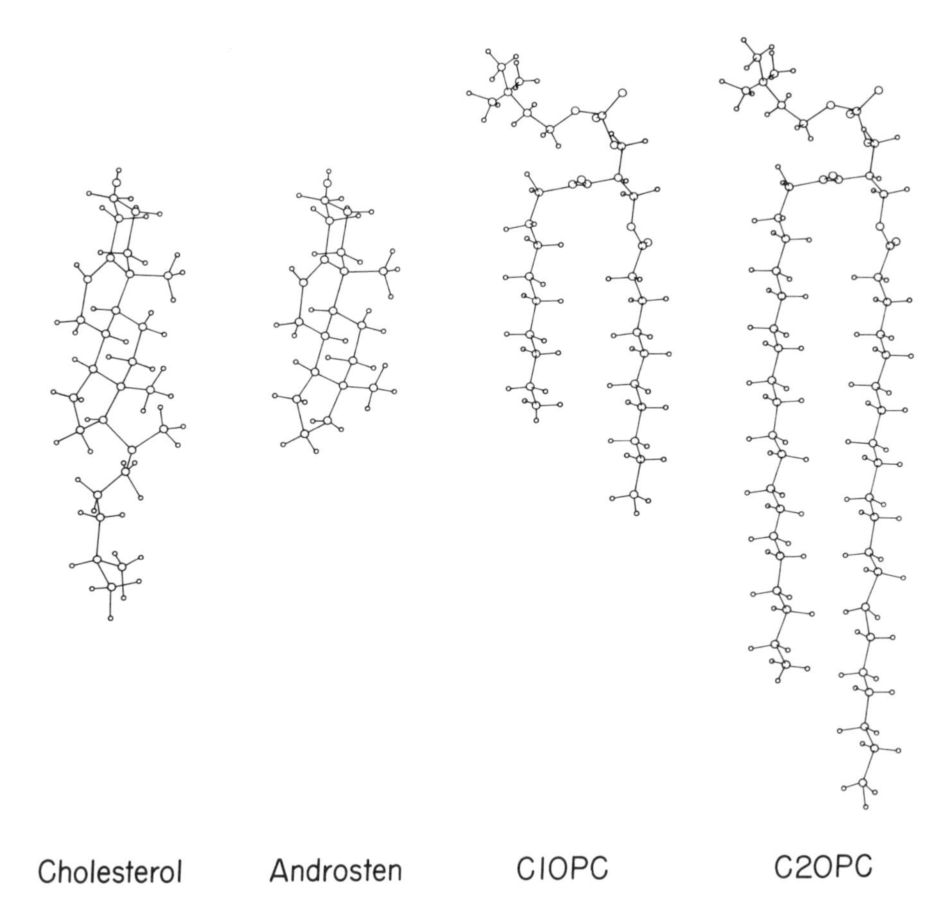

FIGURE 3. The structures of two sterols and two lipids are shown as they would be in a liposome bilayer (Singer and Finegold, 1990b). They are cholesterol (5-cholesten-3β-ol), the tailless cholesterol analog androsten (5-androsten-3β-ol), a 10-carbon phosphatidylcholine (C10PC) and a 20-carbon phosphatidylcholine (C20PC). Note that the one chain of a lipid (*sn*-1) is longer than the other (*sn*-2). The lipids span the shortest and longest lipids of these studies. (Reproduced by permission of Elsevier Scientific Publishers.)

the cholesterol concentration increases and then it disappears, leaving only the broad component. At a still higher cholesterol concentration, the broad component itself disappears.

Should the baseline extrapolation be linear (Figure 2B), then the area under the graph to be assigned to a transition's ΔH is straightforward. If the baseline after the thermal transition is not a linear extrapolation of the baseline preceding the transition (Figure 2B), then effectively there has been a change in the relative specific heats of the sample and its known control, and the ΔH assignation is unclear. Hence, the measurement of the enthalpy change, ΔH, to be associated with this broad component is subject to some indeterminacy.

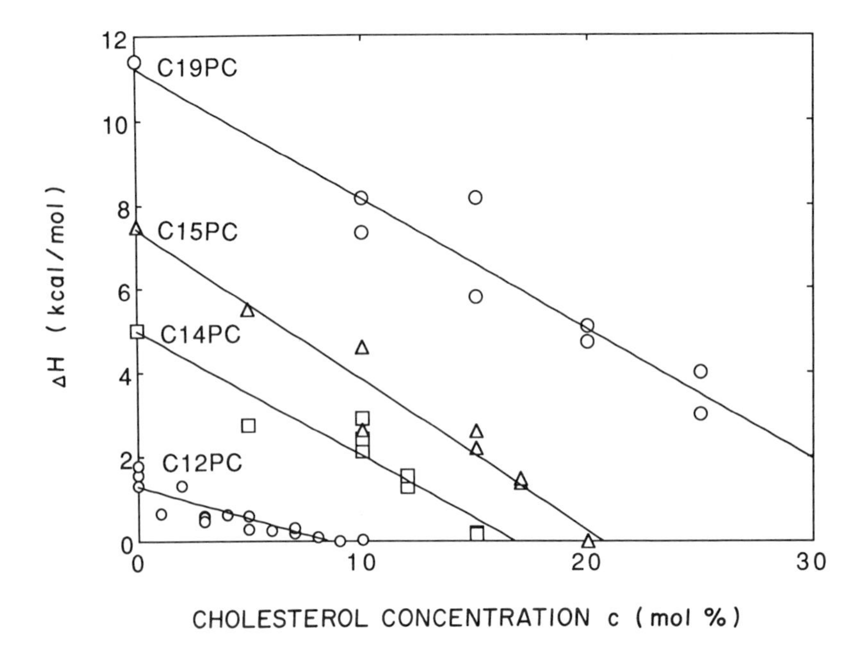

FIGURE 4. The addition of cholesterol to bilayers of phosphatidylcholines C*n*PC causes a linear decrease in the enthalpy change ΔH of the main transition (Singer and Finegold, 1990a). Here are illustrated four acyl chain lengths. (Reproduced by copyright permission of the Biophysical Society.)

The interpretation of these two components will be discussed in a later section. Here suffice it to say that the sharp component has been thought to represent a transition of the "free" unassociated lipid, whereas the broad component has been thought to represent some type of phospholipid/cholesterol association. In addition, not all phospholipid/cholesterol mixtures — ours and the literature's — give scans that can be systematically deconvoluted into one sharp and one broad component. *Hence, in the results that immediately follow, we have restricted our measurements to* (what others might interpret as) *the sharp component only.*

B. ΔH VERSUS CHOLESTEROL CONCENTRATION

Figures 4 and 5 show the enthalpy change, ΔH, of the main transition of lipids of acyl chain lengths 12, 14, 15, 19, and 20 with increasing concentration c of the sterol cholesterol. Noteworthy is the linearity of each of the ΔH vs. c plots. Curvature of the plots towards the c axis at higher concentrations might be due to limits of solubility of cholesterol in the liposome preparation; here the linearity is consistent with complete solubility.

The linearity of ΔH with concentration c shows that cholesterol molecules appear to mix uniformly with the lipid molecules and with each other, i.e., cholesterol-lipid interactions are independent of the amount of cholesterol present, in the concentration range studied. The linearity of ΔH with n also shows

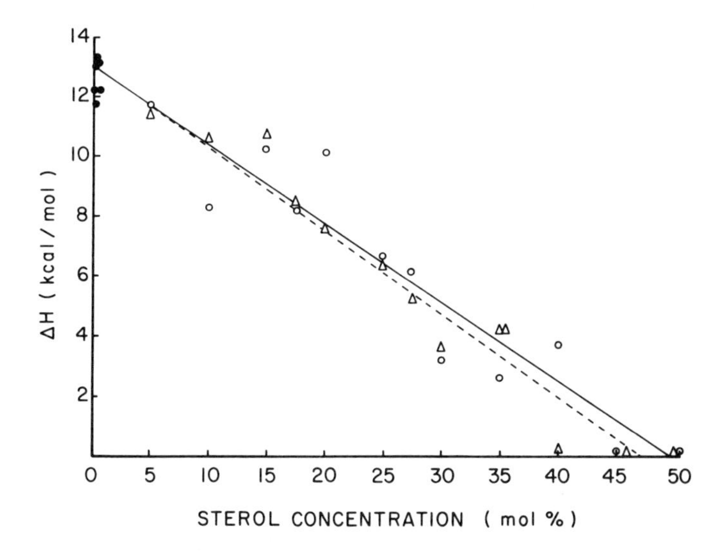

FIGURE 5. The tailless cholesterol analog has the same effect as cholesterol (Singer and Finegold, 1990b). The Figure is as Figure 4, but only one acyl chain length (C20PC) is shown. Cholesterol data points are shown as circles (o), and the cholesterol analog androsten points are shown as triangles (Δ); the closed circles are for the lipid with no sterol added. The full line is a least-squares fit for cholesterol, and the dashed line for androsten. (Reproduced by permission of Elsevier Scientific Publishers.)

that, overall, distal CH_2 groups of the lipid molecule interact with cholesterol as much as groups that are closer (proximal) to the membrane surface.

Also noteworthy is that the slope of each plot is the same for each lipid and that there is not even an indication of a systematic trend of slopes with chain length n (Figure 6).

C. ΔH VERSUS CHAIN LENGTH n

It has been long known that, in the absence of cholesterol, the enthalpy change ΔH of the main transition increases linearly with acyl chain length n (see references in Finegold and Singer, 1986); we find for phosphatidylcholines $\Delta H = -9.43 + 1.01\, n$ kcal/mol (Singer and Finegold, 1990a), and for phosphatidylcholines and phosphatidylethanolamines together (Figure 7) $\Delta H = -8.93 + 1.05\, n$ kcal/mol.

[The temperature of the main transition vs. n deviates somewhat from a straight line (Wilkinson and Nagle, 1981)]. Figures 4, 5, and 6 show that, even in the presence of cholesterol, this linearity still holds for phosphatidylcholines as the lipid chain length is changed: $\Delta H = -9.43 + 1.01\, n - 0.268\, c$ kcal/mol, where c is the cholesterol concentration (mol cholesterol/mol cholesterol + mol lipid), in mol % (Singer and Finegold, 1990a). The addition of cholesterol eliminates the main transition in a linear manner; note that each incremental mol % addition of cholesterol subtracts the same ΔH for all the lipid chain lengths [C12PC is known to be anomalous (Finegold et al., 1990; Hatta et al., submit-

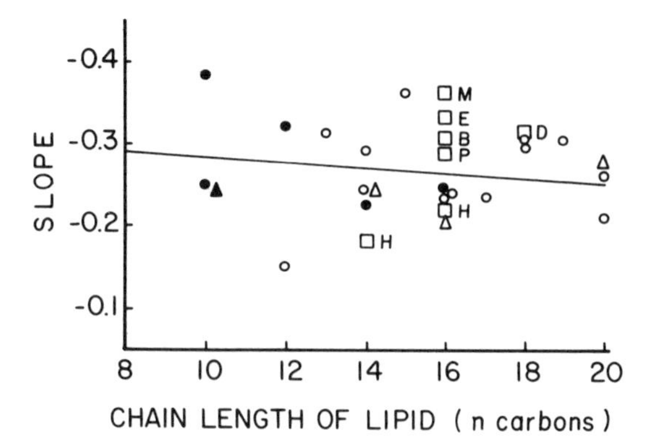

FIGURE 6. The slope of the ΔH vs. sterol concentration c plots (as in Figures 4 and 5) is independent of the chain length n of the lipid and of the head group. The slope of the line, a least-squares fit (Singer and Finegold, 1990b), is essentially zero. Again, cholesterol data points are shown as circles (o), and the androsten points are shown as triangles (Δ); open symbols are for phosphatidylcholine head groups, and closed symbols are for phosphatidylethanolamine head groups. Also shown are literature values for cholesterol for the sharp component of the endotherms: B is from Bittman (Kan and Bittman, 1991), D is Davis et al., 1983; E is Estep et al., 1981; H is Hinz and Sturtevant, 1972; M is Mabrey et al., 1978; and P is the present work (Singer and Finegold, 1990b), but restricted (see text) to data points less than 15 mol% cholesterol. (After Singer and Finegold, 1990b. Reproduced by permission of Elsevier Scientific Publishers.)

ted)]. We focus on the concentration of c that just eliminates ("shields") the ΔH of the transition, and term it the shielding concentration. The shielding concentration (at which $\Delta H = 0$) is plotted in Figure 8 as a function of n. The satisfying linearity of Figure 8 demonstrates clearly the internal consistency of the experimental measurements. The extrapolation of the cholesterol line of Figure 8, to a minimal chain length of n of about 9 carbons, is consistent with a value of 9 to 10 carbons from other workers (Figure 7). The shielding concentration for the shorter lipids (e.g., about 5% for C10PC) represents a ratio of only one cholesterol molecule for every 19 C10PC molecules, yet each cholesterol molecule is quite evidently interacting with all those molecules with which it can not contact on a static steric model. Hence, as far as those interactions that contribute to the ΔH of the main transition are concerned [predominantly *trans-gauche* isomerization of the methylene groups within the acyl chains, and the van der Waals interactions between different acyl chains (Nagle, 1980)], cholesterol molecules interact equally with all the lipid molecules. (The existence of groupings, clusters, complexes, or pools would be demonstrated by curvatures of the otherwise straight lines; so there is no evidence for groupings.) We emphasize that this equality of interaction takes place within the time scale of differential scanning calorimetry (many seconds), which provides ample time for low concentrations of cholesterol effectively to diffuse and interact.

D. HEAD GROUPS — ETHANOLAMINES

When the phosphatidyl head group of the lipids is changed from choline, $(CH_3)_3$-N-$(CH_2)_2$-, to ethanolamine, $(H_3)_3$-N-$(CH_2)_2$-, the pattern with choles-

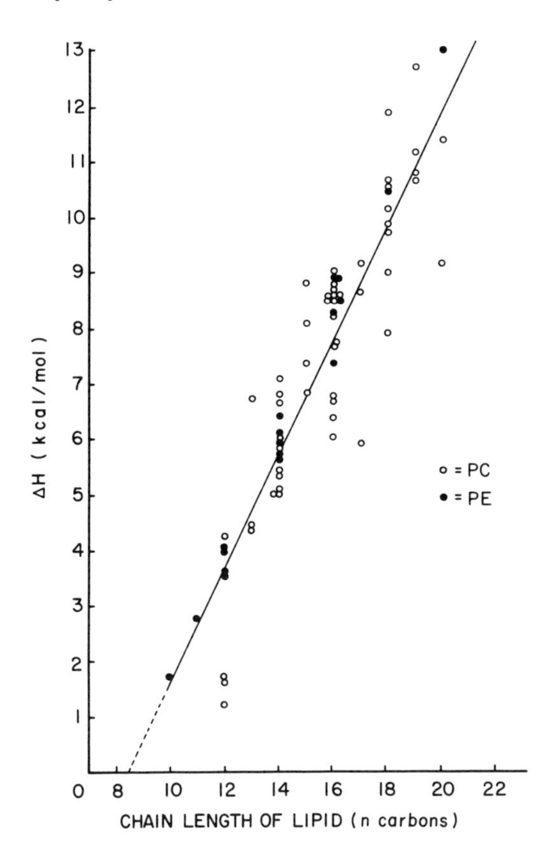

FIGURE 7. The enthalpy change ΔH for the main phase transition for two head groups — pure phosphatidyl cholines (open circles) and ethanolamines (closed circles) — with linear saturated chains of n carbons, from our and literature data (Singer and Finegold, 1990b). The least squares fit is ΔH (kcal/mol) = −8.93 + 1.05 n. (Reproduced by permission of Elsevier Scientific Publishers.)

terol described above is essentially the same. The slopes of the ΔH vs. c lines, as a function of chain length n, for the phosphatidyl ethanolamines are illustrated with those for the phosphatidyl cholines in Figure 6. The slopes are essentially independent of n for the ethanolamines as well as for the cholines. Likewise, a plot (Figure 8) of the shielding concentration (that concentration at which ΔH is zero) is the same for the ethanolamines as it is for the cholines (Singer and Finegold, 1990b). Together, these observations indicate that the (two-dimensional) stoichiometry of the shielding concentration — and, hence, that of the cholesterol/phospholipid interactions — within the plane of the bilayer is strongly determined by the acyl chain length (which represents a "vertical" dimension perpendicular to the plane of the bilayer). This stoichiometry varies from 1 cholesterol to 19 lipids (approximately) when $n = 10$, to 1 cholesterol to 1.5 lipids (approximately) when $n = 20$.

Both cholesterol and the phospholipid acyl chains lie parallel to one another and essentially perpendicular to the bilayer plane. Figure 3 illustrates lengthways

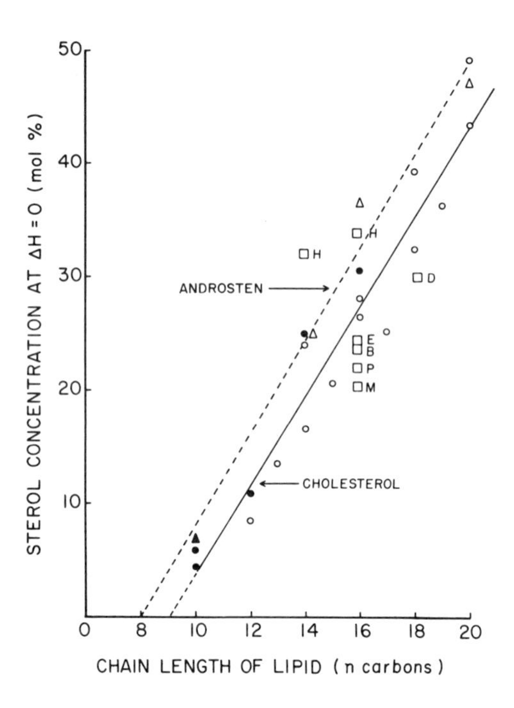

FIGURE 8. The tailless cholesterol analog has the same effect as cholesterol, for several acyl chain lengths and two head groups (Singer and Finegold, 1990b). Here the shielding concentration (that sterol concentration at which the enthalpy change ΔH of the lipids is zero) is plotted vs. the acyl chain length n. The cholesterol (full) and androsten (dashed) lines do not differ significantly. The symbols are as in Figure 6, with the same literature comparisons. (Reproduced by permission of Elsevier Scientific Publishers.)

views of cholesterol and lipids of differing chain lengths. Assuming that the cholesterol hydroxyl group lies at the level of the lipid ester bond (Yeagle, 1985), then cholesterol would have a depth in the bilayer equivalent to that of a lipid with acyl chains of 14 carbons. Clearly the cholesterol/phospholipid interaction is substantial even when both molecules differ in length (Figure 3). The static molecular model depiction of Figure 3 shows that, within the bilayer, cholesterol would project deeper than would a 10-carbon phospholipid, but significantly shallower than a 20-carbon phospholipid. However, despite these contrasting length mismatches, the ΔH reduction for each additional molecule of cholesterol is independent of the chain length n. The disparity in length can be further amplified by using an analog of cholesterol that lacks the side chain at the sterol C17 position (Figure 3). This tailless analog (androsten) is identical to cholesterol in terms of the thermal behavior in the lipid bilayer, of the ΔH vs. c slopes, and of the shielding concentrations for each n (Figures 4 through 8). The identity of behavior has been confirmed by Bittman and Kan (1991) up to 15 mol% sterol.

The thermal properties of phospholipid/cholesterol mixtures have been studied by many for years, but, unfortunately, most investigations have been restricted to lipids of n of 14, 16, or 18. Published values for ΔH vs. c slopes, and the shielding concentration at which ΔH becomes zero, are illustrated in Figures 6 and 8. The published values are consistent with ours, except for a few outliers.

V. INTERPRETATION — DISCUSSION

Before finally interpreting these observations, we must review what is known about the energetics of the phospholipid main transition. This problem is still not completely understood and is still actively studied (Finegold et al., 1990; Hatta et al., submitted). C16PC is the most-studied lipid, and its conceptual picture (Nagle, 1980) probably can be generalized to other lipids. The main transition ΔH can be virtually all accounted for by van der Waals interaction between chains and the energy cost of going from *trans* to *gauche* bonds. At temperatures below that of the main transition, the chains are ordered and essentially all *trans*. Above the transition, each acyl chain would contain about 4.2 *gauche* bonds (Nagle, 1980; Mendelsohn et al., 1989). Hence, about 60% of the ΔH is accounted for by the loosening of the van der Waal's interactions, and 35% is accounted for by conversion of *trans* to *gauche* bonds. (Changes in hydration of head groups seem negligible for this transition.)

A. THE SHAPE OF DIFFERENTIAL SCANNING CALORIMETER ENDOTHERMS

Phase transitions of pure substances have been the subject of much physics interest, yet the shapes of endotherms, such as those for pure C16PC shown in Figure 2B, still defy close theoretical interpretation. The endotherms of mixtures of sterols in lipids (Figure 2B) are still more complex, and a theoretical model for their shape is still more distant. (Even only one sterol in mixtures of only two lipids gives complex ΔH-c results (Finegold and Singer, 1991)). The current theoretical situation is discussed in this volume (in the chapters by Scott and by Zuckermann et al.). Because curve fitting computer programs are now ubiquitous, it is tempting to resolve complex endotherms, such as those of Figure 2B, into separate peaks. This procedure is valid in fields such as spectroscopy, where the line shapes have a well-known shape (i.e., equation). For lipids, where the equivalent line-shapes are simply not known, there is almost an infinity of possible fits. It is still possible that some of the fits are realistic; a favorable check would be that the same proposed peak shape gives consistent results when applied to different lipids and to the data of different workers. The numerical resolution of the C16PC-cholesterol endotherms into two peaks (Mabrey et al., 1978) results in each peak having a linear ΔH -cholesterol concentration like that of Figure 4. Unfortunately, the same workers find that three peaks result for the similar lipid C14PC-cholesterol endotherms. Estep et al. (1981), for C16PC-cholesterol, resolved the endotherm peak into two components, one of which showed a ΔH-cholesterol concentration like Figure 4, but the other peak showed a maximum in ΔH-cholesterol concentration, which result is in contradiction with Mabrey et al. (1978) and with Figure 4. Mabrey et al. and Estep et al. used large-sample calorimeters; the data of Figure 4 was obtained with a small-sample calorimeter. Hence, the favorable consistency check is not found, and the separation of differential scanning calorimeter endotherms into well-determined broad and narrow components is not well founded. Kan and Bittman (1991) also used a large-sample calorimeter for a series of sterols in C16PC, up to 15 mol% sterol (Figures 6 and 8). They found no need for multiple peaks, and their results are in good agreement with those shown as *P* (on Figures 6 and 8), which are for the same sterol concentration range, but with a small-sample calorimeter. The results shown as *P* were chosen for this comparison of large- and of small-sample calorimeters.

Actually, the numerical resolution of differential scanning calorimeter peaks involves the deconvolution of the instrument transfer function. If a phase transition were to give an input endotherm that is a perfect rectangle in shape, then the instrument (itself, including its electronics and non-zero scan rates) would round off the corners of the square to give an output endotherm looking somewhat like a bell. The instrument transfer function relates the bell to the square, enabling the square to be reconstructed from the bell. The publication of differential scanning calorimetry scans of standard samples would enable this to be done. However, it still seems that there is a relatively sharp component, together with a relatively broad component, for cholesterol with some lipid mixtures.

B. INTERLABORATORY COMPARISONS

The published enthalpies of the main transitions of even pure lipids, and even the most-studied C16PC (Figure 7), differ between laboratories by much more — one feels — than they should, even when measured with large-sample differential scanning calorimeters. These discrepancies have long been known (Nagle, 1987; Schwarz, 1986, 1991; Caffrey et al., 1991), and their origin is unclear. Thermal energies are usually more difficult to measure than temperatures; cell-filling problems were mentioned above (Schwarz, 1986, 1991), and perhaps we must be much more careful about sample thermal history (Tenchov et al., 1989). The transition temperatures of the main lipid phase transition show much less intra-laboratory variation than do the corresponding enthalpies (Schwarz, 1986, 1991; Caffrey et al., 1991). Much variation between laboratories could be reduced were authors to report enthalpies for the same standard substance. The Calorimetry Conference did sterling service in distributing standard samples for specific heat measurements using adiabatic calorimeters (Bloom et al., 1970; Finegold, 1972); Schwarz's (1986) suggestion of diphenylether is a convenient one for differential scanning calorimeters.

VI. THE PHASE DIAGRAM

The experimental results of this chapter enable us to predict (Figure 9) the phase diagram of other chain length *n* phosphatidylcholines CnPC with cholesterol, starting with the experimentally determined phase diagram for cholesterol in C16PC (Figure 12 of Vist and Davis, 1990) (see chapter by J.H. Davis in this

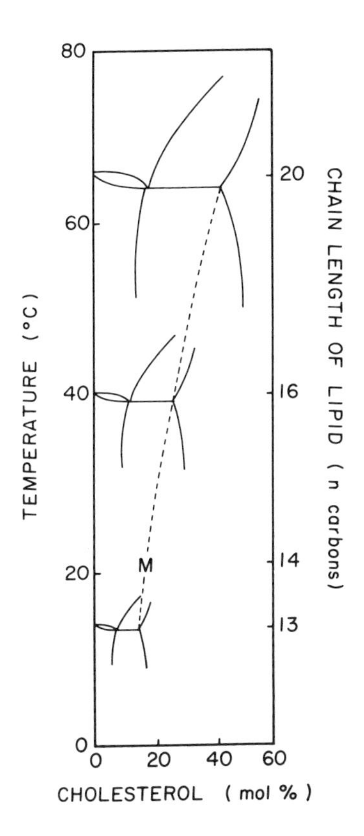

FIGURE 9. The predicted phase diagram for phosphatidylcholines with saturated acyl chains of *n* carbons (C*n*PC), of *n* = 13 through 20, with cholesterol (constructed using the data of Vist and Davis, 1990, and of this chapter; see text for details). M marks the highest cholesterol point measured by neutron scattering (Mortensen et al., 1988). The dashed line connects points common to the phase diagrams of lipids of different *n*.

volume). A phase diagram relates the temperature T of phase boundaries with composition (here, of cholesterol concentration c), but lacks direct ΔH information. Our experimental results here correlate ΔH with n and with c. We bridge the two distinct sets of results (the phase diagram at $n = 16$ only and the ΔH vs. c for many values of n) by identifying the highest cholesterol concentration of the phase diagram where three only phase lines meet [at about "c three-line" = 20 mol% for the deuterated C16PC (Vist and Davis, 1990)] with the shielding concentration (that cholesterol concentration where $\Delta H = 0$, about 25 mol% for protiated C16PC). (The difference between these numbers is considered insignificant in the light of deuteration and of inter-laboratory techniques.) It is also useful to recall that for the main transition, the temperature increases monotonically (and almost linearly) with n (Nagle, 1980). Hence, on Figure 9 we scale "c three-line" with the shielding concentration for C16PC and predict (as shown on Figure 9) that the other CnPC phase diagrams will have the same general features

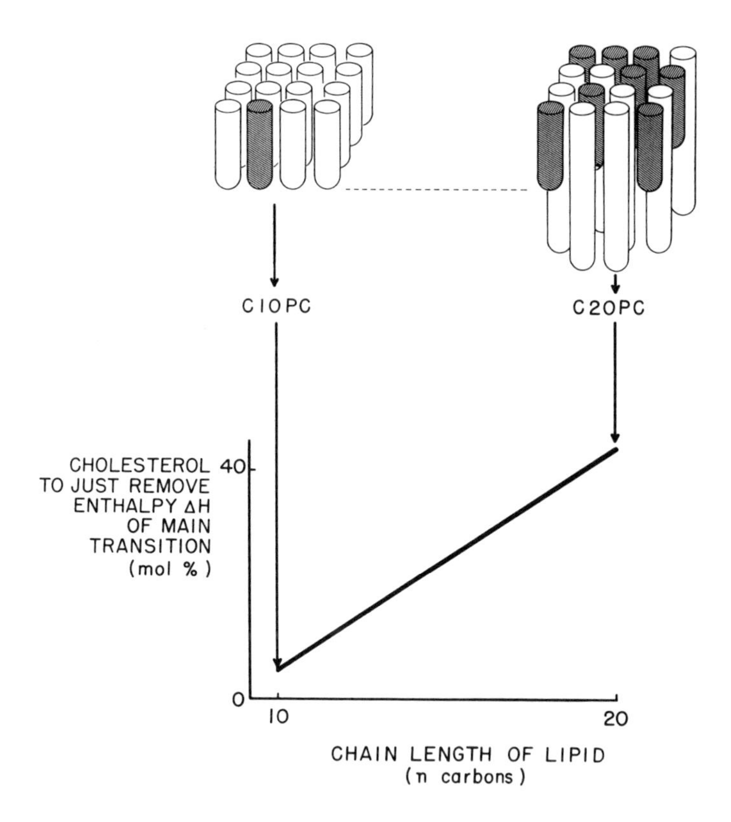

FIGURE 10. The longer-chain-length lipid C20PC (open cylinders) requires more cholesterol (shaded cylinders) than does the shorter-chain lipid (C10PC) in order to overcome the greater lipid-lipid interactions that provide the ΔH of the main transition. Cylinders are shown, instead of molecular structures, to emphasize that all the lipid molecules will interact with all the cholesterol molecules, within the time scale of the thermal measurements.

as those for C16PC, but scaled according to shielding concentration vs. cholesterol concentration c (i.e., scaled according to T (or n) vs. c). A good check on this procedure is given by the phase diagram of C14PC (deuterated)-cholesterol of Mortensen et al. (1988), by neutron scattering. The point shown in Figure 9 is taken to correspond to the highest cholesterol concentration (19 mol%) of Figure 24b of their work and fits well on the proposed phase diagram for all n.

VII. SUMMARY

One likes to have a picture or cartoon of the results (Figure 10). Such a picture must be a moving one, perhaps drawn at the tops of succeeding pages so that a rapid turning of pages gives a visual cinematic illusion. It would show the cholesterol and lipid molecules interacting completely, perhaps by spatial contact or an effective contact mediated by many-body effects, and so should be a blur, except at the shortest time scales. The longer lipid tails might even interact with the rigid rings of the sterols, since it is established that the tail of the cholesterol is unnecessary for the observed interactions. It should be stressed that static configurational modeling will not give a completed picture of cholesterol/ phospholipid interactions. A complete picture will require the inclusion of dynamical parameters such as the lateral motions of both lipid and cholesterol molecules. *It would indicate that static configurational modeling is incomplete.* Such dynamical modeling is a challenge to our theoretical friends. Simplistically, the cholesterol molecules dilute out the lipid-lipid interactions. In a tentative attempt to pin down these physical ideas, Figure 10 is offered.

ACKNOWLEDGMENTS

We thank Ms. Marlene Young and Ms. Maria Dinda for technical assistance, Camu Zdat for manuscript and graph preparation, Andrew Pedersen for drawings, and Drs. Robert Bittman and C.-C. Kan for results before publication.

Financial assistance was provided by the Medical Research Council (Canada, grant MA 4591), the North Atlantic Treaty Organization (grant 900121), and the National Institutes of Health, U.S. (B.R.S.G., Drexel University).

REFERENCES

Bangham, A. D., Ed. 1983. *Liposome Letters*, Academic Press, New York.

Bangham, A. D., M. M. Standish, and J. C. Watkins. 1965. Diffusion of univalent ions across the lamellae of swollen phospholipids. *J. Mol. Biol.*, 13:238–252.

Bittman, R. and C.-C. Kan. 1991. Private communication.

Bloom, D. W., D. H. Lowndes, Jr., and L. Finegold. 1970. Low temperature specific heat of copper: comparison of two samples of high purity. *Rev. Sci. Instrum.*, 41:690–695.

Caffrey, M., D. Moynihan, and J. Hogan. 1991. A database of lipid phase transition temperatures and enthalpy changes. *Chem. Phys. Lipids*, 57:275–291.

Davis, J. D. 1992. The molecular dynamics, orientational order and the thermodynamic phase equilibria of cholesterol/phosphatidylcholine mixtures. In *Cholesterol in Membrane Models*, Finegold, L., Ed., CRC Press, Boca Raton, FL.

Davis, P. J. and K. M. W. Keough. 1983. Differential scanning calorimetric studies of aqueous dispersions of mixtures of cholesterol with some mixed-acid and single-acid phosphatidylcholines. *Biochemistry*, 22:6334–6340.

Esfahani, M. and J. B. Swaney, Eds. 1990. *Advances in Cholesterol Research*, Telford Press, New Jersey.

Estep, T. N., D. B. Mountcastle, R. L. Biltonen, and T. E. Thompson. 1981. Studies on the anomalous thermotropic behavior of aqueous dispersions of dipalmitoylphosphatidylcholine-cholesterol mixtures. *Biochemistry*, 20:7115–7118.

Finegold, L. 1972. Electronic density of states and low temperature specific heat of metals. In *Techniques in Metals Research*, Bunshah, R. F., Ed., John Wiley & Sons, 6 Part I, pp. 263–308.

Finegold, L. and M. A. Singer. 1986. The metastability of saturated phosphatidylcholines depends on the acyl chain length. *Biochim. Biophys. Acta,* 855:417-420.

Finegold, L. and M. A. Singer. 1991. Cholesterol-multilipid interactions in bilayers. *Chem. Phys. Lipids,* 58:169–173.

Finegold, L., W. A. Shaw, and M. A. Singer. 1990. Unusual phase properties of dilauroyl phosphatidylcholine (C12PC). *Chem. Phys. Lipids,* 53:177–184.

Hatta, I., S. Matuoka, M. A. Singer, and L. Finegold. A new liquid crystalline phase in phosphatidylcholine bilayers as studied by X-ray diffraction, submitted.

Hinz, H.-J. and Sturtevant, J. M. 1972. Calorimetric investigation of the influence of cholesterol on the transition properties of bilayers formed from synthetic L-α-lecithins in aqueous suspension. *J. Biol. Chem.,* 247:3697–3700.

Kan, C.-C. and R. Bittman. 1991. A DSC study of the effects of sterol side-chain structure on the phase transition of dipalmitoylphosphatidylcholine bilayers. *Biophys. J.,* 59:317a.

Mabrey, S., P. L. Mateo, and J. M. Sturtevant. 1978. High-sensitivity scanning calorimetric study of mixtures of cholesterol with dimyristoyl- and dipalmitoylphosphatidylcholines. *Biochemistry,* 17:2464–2468.

Mendelsohn, R., M. A. Davies, J. W. Brauner, H. F. Schuster, and R. A. Dluhy. 1989. Quantitative determination of conformational disorder in the acyl chains of phospholipid bilayers by infrared spectroscopy. *Biochemistry,* 28:8934–8939.

Mortensen, K., W. Pfeiffer, E. Sackmann, and W. Knoll. 1988. Structural properties of a phosphatidylcholine-cholesterol system as studied by small angle neutron scattering: ripple structure and phase diagram. *Biochim. Biophys. Acta,* 945:221–245.

Nagle, J. F. 1980. Theory of the main lipid bilayer phase transition. *Ann. Rev. Phys. Chem.,* 31:157–195.

Nagle, J. F. 1987. Definitive doubts. *Nature,* 329:682.

Schwarz, F. P. 1986. Biological standard reference materials for the calibration of differential scanning calorimeters: di-alkylphosphatidylcholine in water suspensions. *Thermochimica Acta,* 107:37–49.

Schwarz, F. P. 1991. Biological thermodynamic data for the calibration of differential scanning calorimeters: dynamic temperature data on the gel to liquid crystal phase transition of dialkylphosphatidylcholine in water suspensions. *Thermochimica Acta,* 177:285–303.

Scott, H. L. 1992. Lipid-cholesterol phase diagrams: theoretical and numerical aspects. In *Cholesterol in Membrane Models,* Finegold, L., Ed., CRC Press, Boca Raton, FL.

Singer, M. A. and L. Finegold. 1990a. Cholesterol interacts with all of the lipid in bilayer membranes: implications for models. *Biophys. J.,* 57:153–156.

Singer, M. A. and L. Finegold. 1990b. Interaction of cholesterol with saturated phospholipids: role of the C(17) side chain. *Chem. Phys. Lipids,* 56:217–222.

Singer, M. A., L. Finegold, P. Rochon, and T. J. Racey. 1990. The formation of multilamellar vesicles from saturated phosphatidylcholines and phosphatidylethanolamines: morphology and quasi-elastic light scattering measurements. *Chem. Phys. Lipids,* 54:131–146.

Tenchov, B. G., H. Yao, and I. Hatta. 1989. Time-resolved X-ray diffraction and calorimetric studies at low scan rates. I. Fully hydrated dipalmitoylphosphatidylcholine (DPPC) and DPPC/water/ethanol phases. *Biophys. J.,* 56:757–768.

Van Dooren, A. A. and B. W. Muller. 1981. Influence of experimental variables on curves in differential scanning calorimetry. *Thermochimica Acta,* 49:151–197.

Vist, M. R. and J. H. Davis. 1990. Phase equilibria of cholesterol/dipalmitoylphosphatidylcholine mixtures. *Biochemistry,* 29:451–464.

Wilkinson, D. A. and J. F. Nagle. 1981. Dilatometry and calorimetry of saturated phosphatidylethanolamine dispersions. *Biochemistry*, 20:187–192.

Yeagle, P. 1985. Cholesterol and the cell membrane. *Biochim. Biophys. Acta,* 822:267–287.

Yeagle, P. L. 1992. The biophysics and cell biology of cholesterol: an hypothesis for the essential role of cholesterol in mammalian cells. In *Cholesterol in Membrane Models,* Finegold, L., Ed., CRC Press, Boca Raton, FL.

Zuckermann, M. J., J. H. Ipsen, and O. G. Mouritsen. 1992. Theoretical studies of the phase behavior of lipid bilayers containing cholesterol. In *Cholesterol in Membrane Models,* Finegold, L., Ed., CRC Press, Boca Raton, FL.

Chapter 6

VISUALIZATION OF CHOLESTEROL DOMAINS IN MODEL MEMBRANES

Sek Wen Hui

TABLE OF CONTENTS

0-8493-4207-4/93/$0.00+$.50

ABSTRACT

The lateral distribution of cholesterol in membranes is determined by physical as well as by biological factors. The physical factors may be explained by the physical chemistry of model bilayers. Various models dealing with the scale of lateral phase separation of cholesterol in cholesterol/phospholipid mixtures are discussed. Phase-separated domains of cholesterol and phospholipids may be visualized by means of light and electron microscopy, utilizing several contrast-enhancing methods. Examples of observing cholesterol-bound solid domains in phosphatidylcholine monolayers by fluorescence microscopy, cholesterol-poor stripe domains by freeze fracture and diffraction contrast electron microscopy, are given. The complex nature of cholesterol distribution in biological membranes may preclude imaging by physical methods described here.

I. PHASE-SEPARATED DOMAINS IN MODEL MEMBRANES CONTAINING CHOLESTEROL

A. INTRODUCTION

It is well acknowledged that cholesterol does not distribute homogeneously in cell membranes, including the plasma membrane (Hui, 1988). The uneven or heterogeneous distribution of cholesterol, in both spatial and temporal senses, is meticulously maintained by the cell. Injury or death of the cell often results in the loss of the designated spatial distribution of lipids in its membranes, rendering the membrane to be no more than a lipid/protein mixture.

The spatial distribution of lipid components, including cholesterol, in model bilayers represents the inanimate assembly forms of lipids in biological membranes. The distribution of cholesterol in model lipid bilayers and monolayers is governed only by physical chemistry, such as the miscibility of its lipid components. However, even in this simplified sense, the microscopic spatial distribution of cholesterol in a mixed bilayer is by no means simple. One of the most studied mixture systems, the mixture of cholesterol with a saturated lecithin (dipalmitoylphosphatidylcholine [DPPC] or dimyristoylphosphatidylcholine [DMPC]), reveals an already complicated phase diagram. Within certain regions in the phase diagram, phase-separated domains coexist. The size and shapes of these domains have been calculated or deduced from calorimetric and spectroscopic experiments. In some cases these domains have been observed by microscopic methods.

I shall discuss in this chapter only the cases where these domains are observable microscopically. Since the sizes and shapes of these "macroscopic" domains, as observed by present-day microscopy methods, are determined by the molecular packing arrangements, we must first consider various molecular packing models. "Microscopic" heterogeneous distributions of cholesterol in the molecular scale, not observable by present-day microscopic methods, will not be considered.

B. MOLECULAR PACKING MODELS

The spatial distribution of cholesterol and other membrane lipids is governed by short- and long-range intermolecular forces. There is no evidence that cholesterol molecules form covalent bonds with themselves or with other molecules in membranes, although hydrogen bonding between cholesterol and the carbonyl oxygen of phospholipid molecules has been proposed (Presti et al., 1982; Huang, 1976). Many attempts have been made in the past to describe the distribution of cholesterol in phospholipid bilayers by model building. The free (ideal) mixing model assumes that the intermolecular forces between cholesterol/cholesterol, cholesterol/phospholipid, and phospholipid/phospholipid are of the same magnitude. Since the cross-section of both cholesterol and the acyl chains of phospholipids are known, Engleman and Rothman (1972) depicted that at a molar ratio of 1:2 (cholesterol/phospholipid), each cholesterol molecule is surrounded by seven acyl chains of phospholipid molecules. Above this ratio, cholesterol molecules may touch each other. Therefore, the macroscopic properties of the bilayer are expected to change abruptly at 33 mol% of cholesterol. (All subsequent percentages are mole percentages unless otherwise stated.) Based on monolayer measurements, Muller-Landau and Cadenhead (1979) pointed out that at 22% cholesterol, each cholesterol molecule is surrounded by two layers of acyl chains. In other words, each cholesterol molecule has an independent boundary ring of phospholipid molecules. Therefore, another discontinuity of bilayer properties should occur at this molecular ratio.

If the molecular affinity between molecules of the same species and between different species are not the same, phase separation will result. Martin and Yeagle (1978) expanded random models to include the formation of cholesterol dimers and predicted that corresponding changes of bilayer properties should occur at the cholesterol molar ratios of 31% and 47% for independent and shared rings of phospholipids around cholesterol dimers. Roger et al. (1979) presented two possible molecular packing models for the 1:1 molar mixture of cholesterol/phospholipid. In these models, cholesterol molecules are depicted to be aligned along a common axis. The parallel cholesterol-cholesterol alignment does not imply any complex between cholesterol and phospholipid molecules. Presti et al. (1982) proposed the existence of both 1:1 and 1:2 stoichiometric complexes between cholesterol and phospholipids. Their model also predicts discontinuities at 20 and 33 mol% of cholesterol due to the vanishing of the single and double acyl chain layers between rows of cholesterol/phospholipid complexes. The model also shows an alignment of cholesterol molecules.

All of the above models are based on the idea that large-scale separation between cholesterol-rich and cholesterol-poor areas is entropically unfavorable. In random mixtures, phospholipids may be treated as two subpopulations: one population consists of those with acyl chains adjacent to at least one cholesterol molecule, and one population of molecules with no chains making contact with cholesterol molecules. The first population, which are referred to as interphase phospholipids, whether complex forming with cholesterol or not. The interphase phospholipids are structurally different from the second population, which are referred to as bulk phospholipids. This is true if the interaction between cholesterol and methylene chains is confined to the first layer. Based on this definition, bulk phospholipids vanish at 20% of cholesterol regardless of models chosen. At 33 to 50%, increasing contacts between cholesterol molecules, i.e., cholesterol-rich domains, are formed. Thus 20%, 33%, and 50% are important turning points in cholesterol/phospholipid mixtures, regardless of models.

Based on a Monte Carlo computer simulation, Cruzerio-Hansson et al. (1989) predicted that the inclusion of up to 10% of cholesterol induces larger and more ramified domains at the phase transition of phospholipids. They also found that cholesterol molecules are preferentially located in the interfacial region during phase transition, as a consequence of cooperativity. By assigning an association constant between phospholipid and cholesterol molecules, according to measured calorimetric parameters, Friere and Snyder (1980) made a computer simulation of the distribution of these molecules. They derived the various domain sizes from the molar ratios and the mutual affinities of the molecules. For cholesterol and phosphatidylcholine the percolation of these domains links many smaller ones into a joined network at 22% cholesterol. This offers an alternative explanation for the functional discontinuity at this molecular ratio.

The question of miscibility should be focused on whether there is any large-scale cholesterol-rich area separated from the cholesterol-poor areas in a macroscopic sense, especially when cholesterol is more dilute; that is, do cholesterol molecules which are surrounded by, or forming complexes with, phospholipid molecules group together in large bands or patches or scatter randomly over the plane of the bilayer. With the proposed cholesterol alignment models, there is the possibility that the alignment is manifested as macroscopic cholesterol-rich stripe domains. If the aligned domains are retained in excess phospholipids, the latter will form ribbons between the aligned, cholesterol-rich domains. The geometry is then different from that of a randomly oriented percolation model.

C. SUPPORTING EVIDENCE FROM DIFFRACTION AND OTHER EXPERIMENTS

X-ray diffraction measurements provide supporting evidence on the degree of segregation of cholesterol-rich areas from cholesterol-poor areas. At 5% of cholesterol there is a dramatic increase of interlamellar spacing of gel-phase bilayers (Rand et al., 1980; Hui and He, 1983). This was interpreted as a dramatic increase of the amplitude of the expended ripple-phase range beginning at this molar ratio (Simon and McIntosh, 1991). It signals the limit of solubility of cholesterol in phospholipid, and the beginning of a two-phase state. If there is extensive segregation of cholesterol-rich and cholesterol-poor domains, coexisting diffraction signals representing two populations of phospholipids would be detectable simultaneously. Early X-ray diffraction results (Engleman and Rothman, 1972) show a sharp and broad diffraction line at 0.42 and 0.46 nm, respectively, coexisting up to 35% of cholesterol. The sharp, wide angle

FIGURE 1. Electron diffraction patterns of cholesterol/dimyristoylphosphatidylcholine mixtures at 4°C. (a) 2 mol% of cholesterol; (b) 20 mol% of cholesterol.

reflection at 0.42 nm represents the gel-phase PC. On the other hand, if the spatial separation of domains is of small scale, resembling an ideal mixture, then small, scattered, cholesterol-rich domains will expand at the expense of remaining pure PC domains, as more cholesterol is added. The sharp reflection of the gel-phase domain will become broadened as the "crystal" size of gel domains reduces. A more recent experiment (Hui and He, 1983) shows that the sharp reflection of the gel-phase phospholipid is gradually broadened as more cholesterol is added. This indicates that the gel-phase domains are becoming smaller even when the cholesterol content is much less than 20%. The difference in interlamellar spacing between gel-state and fluid-state multibilayers also vanishes at 20%. The disagreement probably arises from the "annealing" effect that allows the small domains to aggregate and merge over a period of time. Lipid paracrystals are known to form over weeks and months of storage (Hui and Strozewski, 1979). The time scale to reach equilibrium separation varies with conditions, and can be exceedingly long. Therefore, the observed domains may not represent equilibrium states. This may explain some disagreements among results derived from spectroscopy, diffraction, and imaging.

The question of aligned vs. random domains cannot be settled by X-ray diffraction or other techniques using bulk "powder" samples, since the area of cooperative alignment is expected to localize to micrometer size only. Electron beam is ideal for sampling such small areas. By using selected-area electron diffraction of free-standing bilayers maintained in an environmental chamber, Hui and He (1983) observed wide-angle electron diffraction patterns showing a marked azimuthal asymmetry (Figure 1), indicating a directional ordering of the domains within the area defined by the electron beam (about 5 μm in diameter).

The tilting of acyl chains in pure DMPC, which is seen as a deviation of hexagonal symmetry as the tilted lattice is cut by the Ewald "plane" (Hui, 1976; Hui, 1989), is removed by the addition of 2% or more cholesterol (Figure 1a). The hexagonal pattern is then transformed into two diffuse arcs as the phospholipid percentage is reduced further, indicating that the acyl chains are aligned so that ordering is higher in one direction (Figure 1b). The arcs broaden as the percentage of cholesterol becomes even higher, indicating that the stripes of phospholipid become thinner. At higher temperatures the bulk phospholipids undergo a phase transition to a fluid state that expresses a diffuse ring. At higher cholesterol percentages, cholesterol is free to form larger domains and express characteristic cholesterol diffraction spacings.

Other physicochemical studies also provide evidence for spatially separated domains in cholesterol-phospholipid mixtures. These include calorimetric and dilatometric studies of thermodynamic parameters. The results support that, below 5%, cholesterol is miscible with phospholipids. Five percent seems to be the limit of ideal solubility of cholesterol in phospholipids. There are some indications that two populations of phospholipids exist from 5% to 20% of cholesterol and that the excess phospholipids do form pure domains as predicted by the models. However, it is not possible to determine how large are the cholesterol-rich areas, nor can the question of aligned stripe domains vs. percolating domains be settled.

Recktenwald and McConnell (1981) showed that there is fluid/fluid immiscibility even at temperatures at which no gel-phase phospholipid exists. A study by fluorescence anisotropy (Lentz et al., 1980) confirmed phase boundaries at 20% and 33% of cholesterol at the gel-phase temperature of the phospholipid, and mixtures of cholesterol-rich and cholesterol-poor fluid phases at higher temperatures. Apparently, the phases of cholesterol/phospholipid mixture are more complicated than we once believed. The imaging of cholesterol-rich domains is only at the starting point of exploring this complicated system.

II. THE TASK OF IMAGING MICROSCOPIC CHOLESTEROL DOMAINS

With controllable, varied lateral pressure and the ease of lateral molecular translation, domains in monolayers on the air/water interface may grow to tens of micrometers across. These domains may be observable by light microscopes with the use of fluorescent probes that preferentially partition into fluid-phase domains (Peters and Beck, 1983; Losche et al., 1983; McConnell et al., 1984; Yu and Hui, 1992). The use of cholesterol-specific probes in monolayer studies has not been reported. The fluorescence of ergosterol (Rogers et al., 1979; Smutzer, 1986) may be a useful tool to image cholesterol-rich domains, although the fluorescence characteristics of this probe may be an obstacle.

X-ray diffraction results indicate that there are no large-scale phase separations in cholesterol-phospholipid bilayers. If the domains in question are smaller than the resolution of light microscopes, the observation of the geometry of phase-separated domains is best done by electron microscopy. However, a major problem is the lack of a reliable method to distinguish the cholesterol-rich domains from phospholipid domains. Since these domains are small (several nm to tens of nms), they may not be resolved by autoradiography. Most cholesterol-binding drugs, such as saponins (digitonin) and filipin, tend to sequester cholesterol from other parts of lipid membranes. Therefore, these drugs cannot be used as reliable labels for electron microscopy of lipid mixtures. A detailed discussion of the use of cholesterol markers for freeze fracture electron microscopy is given by Miller (1989).

Unlabeled cholesterol-rich domains do not appear different in morphology from most phospholipid bilayers in freeze fracture micrographs of lipid vesicles. Only when pure phospholipids domains are in the rippled (P_β or P_β') phases does the presumed smooth texture of cholesterol-rich domains stand out. This technique can only be applied to those binary mixtures containing cholesterol and one of the pure, saturated phospholipids, such as DPPC, DMPC, or DMPG, which form the P_β or P_β' phase. It has been suggested that cholesterol is sequestered along the ridges and valleys of the ripples, rather than being excluded from the rippled areas (Mortensen et al., 1988). Based on this interpretation, the phospholipids are presumed to intersperse between strips of cholesterol-rich areas, and there are no wide areas of separation of cholesterol-rich domains from pure phospholipid domains. The coherence length of phospholipid packing is therefore longer in one direction.

The diffraction pattern from pure phospholipids at the gel state and that from cholesterol-rich domains are different. Therefore, optical filtering or electron optical filtering may be applied to enhance the image contrast, utilizing the molecular packing difference as a marker. The diffraction contrast method was applied to visualize the domains of different phases in unsupported and hydrated bilayers of mixed cholesterol and phospholipid in an electron microscope (Hui and He, 1983; Hui and Parsons, 1975; Hui, 1981). The contrast is created by placing a filtering aperture in the diffraction plane of the objective lens in the electron microscope. Only the filtered beams corresponding to the molecular packing of selected domains are used to form the bright image, which stands out against dark areas corresponding to unselected beams (Hui, 1977; Hui, 1988; Hui, 1989).

III. MICROSCOPIC IMAGES OF CHOLESTEROL DOMAINS

A. FLUORESCENCE MICROSCOPY OF CHOLESTEROL DOMAINS IN MONOLAYERS

Using lipophilic fluorescent probes that favorably partition to the fluid domains of the air/water interfacial monolayer of phospholipids, solid domains of phospholipids may be distinguished as darker areas against the more brightly fluorescent fluid domains. With a small percentage of cholesterol added to the phospholipids, the line tension between solid and fluid domains is greatly reduced, resulting in the elongation of solid domains. The elongation is a result of the intradomain molecular alignment and dipole repulsion, which tend to favor thin domain strips. These forces act against the line tension, which tends to minimize the length of the domain boundary (Weis and McConnell, 1985; Keller et al., 1986; Gaub et al., 1986; McConnell and Moy, 1988). If an optically pure phospholipid is used, the strip-shaped domains express a spiral growth according to the chirality of the lipid. Otherwise, for a racemic lipid, the strips remain relatively straight (Weis and McConnell, 1985). Two examples of solid domains in monolayers of racemic DMPC and 2% cholesterol are given in Figure 2. The small amount of cholesterol is believed to be located primarily in the solid/fluid domains boundaries, resulting in altering the chemical potential difference between these two states of phospholipids (Weis and McConnell, 1985; Keller et al., 1986). With an increasingly higher percentage of cholesterol, the bulk phases are less distinguishable from the boundary layer. The long, thin, and sometimes spiraling solid domain geometry is replaced by more numerous and circular ones (Heckl and Mohwald, 1986; Heckl et al., 1988).

The domains of cholesterol/phospholipid monolayer is, in some sense, analogous to that in bilayers (Hui et al., 1975). However, similar domain geometry in bilayers has not been observed by fluorescence microscopy. It could be that the domains in bilayers are smaller due to various constraints of the bilayer configuration. Strip-shaped domains in cholesterol/phospholipid bilayers were indeed observed by electron microscopy, as described below. However, the conditions are not equivalent, and the analogy should be taken with caution.

B. ELECTRON MICROSCOPY OF CHOLESTEROL DOMAINS IN MONOLAYERS AND BILAYERS

1. Freeze Fracture Method

The freeze fracture technique was first applied by Kleeman and McConnell (1976) and later extended to quantitative measurement in a number of binary mixtures (Lentz et al., 1980; Copeland and McConnell, 1980). An example is given in Figure 3a–c. Under the condition that pure DMPC express the P_{β}' phase, as cholesterol is added to the mixture, smooth bands between ripple ridges increase proportionally to the amount of cholesterol added to the mixture until 20% cholesterol is reached and no more ripples are found. It is likely that these bands are the cholesterol-rich domains (Mortensen et al., 1988). The stripes of pure phospholipids are laterally isolated narrow domains, which remain until disappearing at 20% of cholesterol. The situation is not unlike that in the monolayer, which shows large scale solid stripes of solid DPPC being placed in between the more fluid cholesterol/PC domains (Figure 2). In the bilayer case the P_{β}'-phase DMPC bilayer domains are more rigid, and tend to remain in a straight direction normal to the chain tilting.

The freeze fracture morphology agrees with the diffraction findings that no large-scale patching of domains occurs and that the small domains are aligned unidirectionally, at least for P_{β}'-forming phospholipids. Yet even this picture is not always true. At 4°C, when pure DMPC is in the L_{β} or the L_c phase, the ripple width is no longer proportional to the cholesterol content (Figure 3d,e). Further-

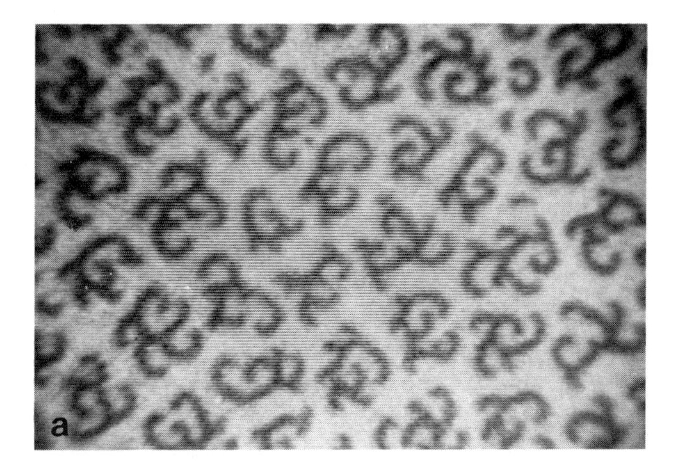

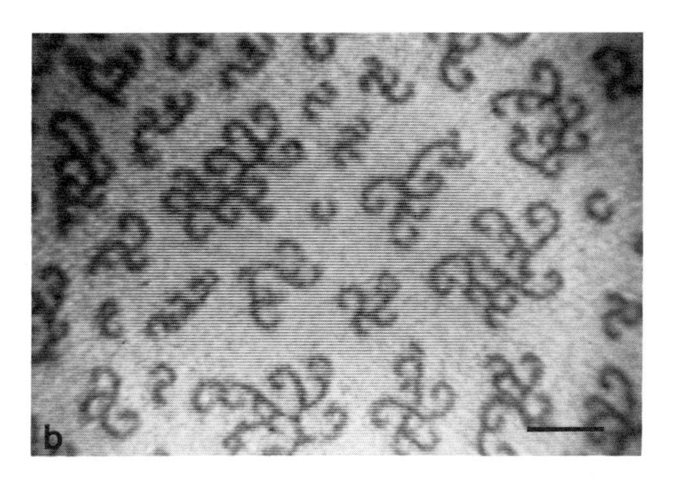

FIGURE 2. Fluorescence micrographs of 2 mol% cholesterol in a dimyristoylphosphatidylcholine monolayer labeled with 2 mol% of 2-(7-nitrobenz-2-oxa-1.3-diazol-4-yl)dodecanoly-3-dipalmitoyl-L-α-phosphatidylcholine. The monolayer is at the air/water interface at 21°C, at a surface pressure of (a) 14 dyne/cm, (b) 12. 5 dyne/cm. (Bar = 20 μm.)

more, the interpretation that stripes of pure phospholipid domains are gradually separating does not concur with the finding that the phospholipid diffraction line is gradually broadened due to diminishing domain size. Once separated, the coherency of each stripe is limited to the 15- to 25-nm width, and the shape-factor broadening should take effect immediately as the stripes begin to separate.

2. DIFFRACTION CONTRAST METHOD

In order to preserve the domain structure of the bilayer, the hydrated bilayer specimens must be protected by an environmental chamber in an electron microscope. Diffraction contrast electron micrographs were obtained from bilayers of various cholesterol/phospholipid ratios and at different temperatures.

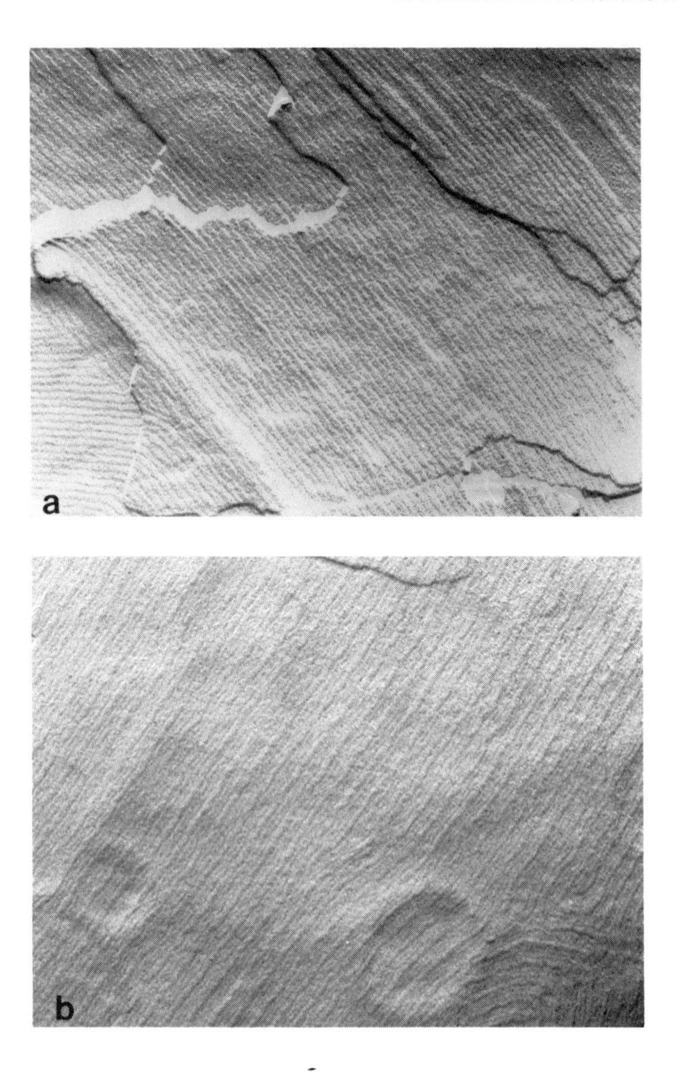

FIGURE 3. Freeze fracture electron micrographs of vesicles of cholesterol/dimyristoylphosphatidylcholine mixtures (a) 1 mol% of cholesterol at 21°C, (b) 5 mol% at 21°C, (c) 18 mol% at 21°C, (d) 1 mol% at 4°C, and (e) 5 mol% at 4°C. (Bar = 100 nm.)

The observed domains were indeed arranged as stripes (Figure 4), in contrast to the patch shapes observed from other mixed lipids (Hui, 1981). Usually the stripes were 20 to 30 nm in width, approximately equal to those observed by freeze fracture electron microscopy. They were somewhat aligned within local areas of several micrometers across. The geometry was similar to that observed in monolayers by fluorescence microscopy (Figure 2). However, the domains of the latter are much wider.

In general, according to microscopic observation, cholesterol does not seem to be ideally dissolved in phospholipid bilayers beyond 5 mole%. In a monolayer

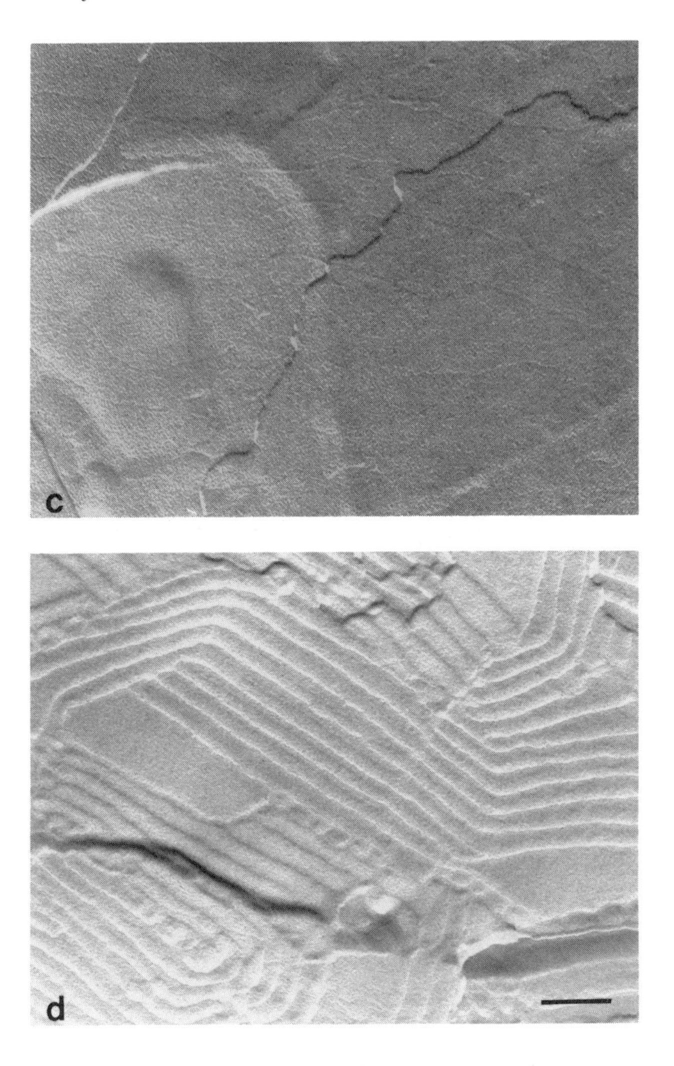

FIGURE 3. (continued).

or bilayer containing cholesterol and phospholipids, cholesterol-rich areas are formed, but these areas are initially very small and have a tendency to form aligned stripes with some phospholipids. At 20 mol% or higher cholesterol, phospholipids intersperse with cholesterol and lose their bulk properties unless there is careful annealing to encourage larger-scale separation, making use of the slight difference in association forces between like and unlike species.

IV. EPILOGUE

Although the evidence obtained using microscopy supports the stripe-domain model, all of these experiments were performed using saturated, equal-chain

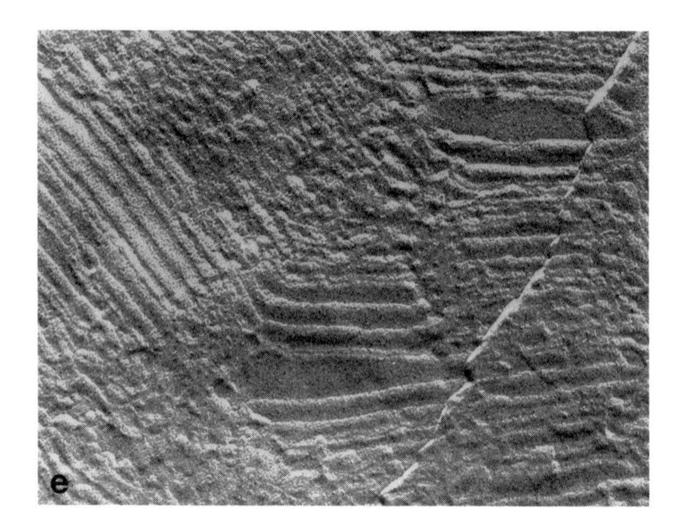

FIGURE 3. (continued).

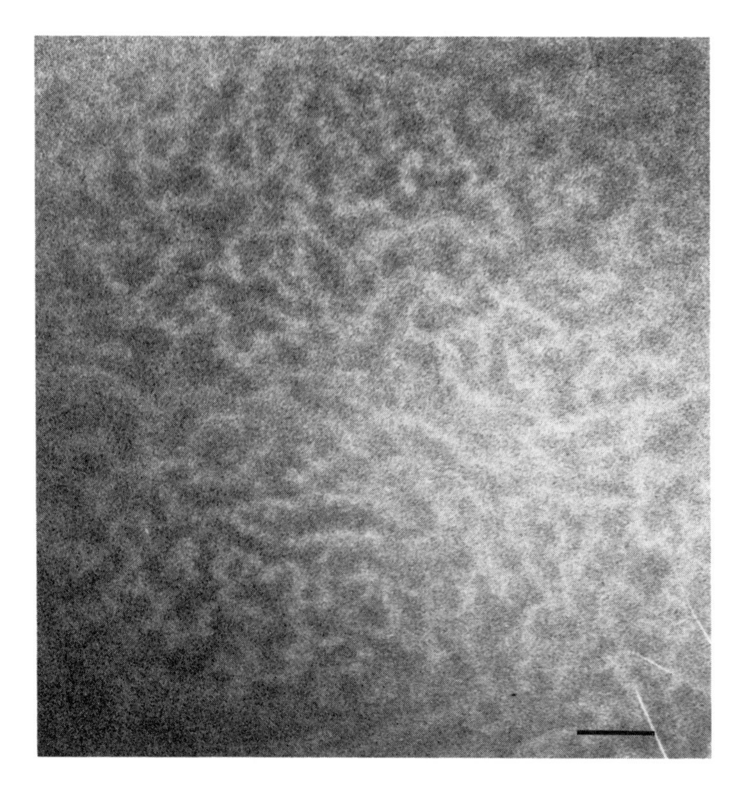

FIGURE 4. Diffraction contrast electron micrograph of a free-standing bilayer of choesterol/dipalmitoylphosphatidylcholine at 11°C. Light areas represent the gel-state domains selected by the diffraction contrast aperture. (Bar = 1 μm.)

phospholipids. In general, phospholipid/cholesterol domains may not be stripes, and the percolation model may still be a good approximation, although it has been disputed on the basis of ESR (Recktenwald and McConnell, 1981) and lateral diffusion (Rubenstein et al., 1979) results.

Recently cholesterol has been shown to have a destabilizing effect on lipid bilayers containing components that tend to form an inverted hexagonal phase (Tilock et al., 1982; Cheng et al., 1986; Cheetham et al., 1989). This effect is presumably due to the poor hydration of the cholesterol molecule (from one hydroxyl group) and its bulky hydrophobic segment. The destabilizing effect of cholesterol posts a limit on its content in membranes containing a large amount of nonbilayer-forming lipids (Hui, 1987; Hui and Sen, 1989). The local concentration of cholesterol may be associated with local instability in membranes. The interplay of cholesterol and membrane proteins in the stability of the bilayer could be a key factor in the predetermination of biomembrane composition (Hui and Sen, 1989).

So far we have considered only binary mixtures of cholesterol/phospholipids. Mixtures containing more than two components are expected to be even more complicated, and a pure physical chemical description may not be feasible. As indicated by electron diffraction results, the domains become smaller as the number of components becomes larger (Hui et al., 1980). The system may more closely resemble a random mixture. Any large-scale phase separation is likely due to biological factors rather than a spontaneous one predicted by the physical chemistry of lipids.

ACKNOWLEDGMENT

I wish to acknowledge the support of NIH grant GM-28120 for most of the work I reported here from my laboratory. The unpublished works I cited in this article (Figures 1–3) are the efforts of N.B. He, H. Yu, M.L. Hensen, and A. Corsi from my laboratory.

REFERENCES

Cheetham, J. J., E. Wachtel, D. Bach, and R. M. Epand. 1989. Role of stereochemistry of the hydroxyl group of cholesterol and the formation of nonbilayer structures inphosphatidylethanolamines. *Biochemistry*, 28:8928–8934.

Cheng, K. H., J. R. Lepock, S. W. Hui, and P. L. Yeagle. 1986. The role of cholesterol in the activity of reconstituted Ca-ATPase vesicles containing unsaturated phosphatidylethanolamine. *J. Biol. Chem.*, 261:5081–5087.

Copeland, B. R. and H. M. McConnell. 1980. The rippled structure in bilayer membranes of phosphatidylcholine and binary mixtures of phosphatidylcholine and cholesterol. *Biochim. Biophys. Acta*, 599:95–109.

Cruzeiro-Hansson L., J. H. Ipsen, and O. G. Mouritsen. 1989. Intrinsic molecules in lipid membranes change the lipid -domain interfacial area: cholesterol at domain interface. *Biochim, Biophys. Acta,* 979:166–176.

Engleman, D. M. and J. E. Rothman. 1972. Planar organization of lecithin-cholesterol bilayers. *Nature*, 247:3694–3697.

Gaub, H. E., V. T. Moy, and H. M. McConnell. 1986. Reversible formation of plastic two-dimensional lipid crystals. *J. Phys. Chem.,* 90:1721–1725.

Heckl, W. M. and H. Mohwald. 1986. A narrow window for observation of spiral lipid crystals. *Ber. Bunsenges Phys. Chem.,* 90:1159–1163.

Heckl, W. M., D. A. Cadenhead, and H. Mohwald. 1988. Cholesterol concentration dependence of quasi-crystalline domains in mixed monolayers of the cholesterol-dimyristoylphosphatidic acid system. *Langmuir*, 4:1352–1358.

Huang, C. 1976. Roles of carbonyl oxygen at the bilayer interface in phospholipid-sterol interaction. *Nature*, 259:242–245.

Hui, S. W. and D. F. Parsons. 1975. Direct observation of domains in wet lipid bilayers. *Science*, 190:383–384.

Hui, S. W., M. Cowden, D. Papahadjopoulos, and D. F. Parsons. 1975. Effects of temperature hydration and surface pressure on hydrated phospholipid single bilayers. *Biochim. Biophys. Acta*, 382:265–275.

Hui, S. W. 1976. The tilting of the hydrocarbon chains in a single bilayer of phospholipid. *Chem. Phys. Lipids,* 16:9–18.

Hui, S. W. 1977. Electron diffraction studies of membranes. *Biochim. Biophys Acta,* 472:345–371.

Hui, S. W. and C. M. Strozewski. 1979. Electron diffraction studies of human erythrocyte membrane and its lipid extracts. *Biochim. Biophys. Acta,* 555:417–425.

Hui, S. W., C. M. Stewart, M. P. Carpenter, and T. P. Stewart. 1980. Effects of cholesterol on the lipid organization in human erythrocyte membranes. *J. Cell Biol.*, 85:283–291.

Hui, S. W. 1981. Geometry of phase-separated domains in phospholipid bilayers by diffraction-contrast electron microscopy. *Biophys J.*, 34:383–385.

Hui, S. W. and N. B. He. 1983. Molecular organization in cholesterol-lecithin bilayers by X-ray and electron diffraction measurements. *Biochemistry*, 22:1159–1164.

Hui, S. W. 1987. Non-bilayer lipids: are they necessary in biomembranes. *Comments Mol. Cell. Biophys.,* 4:233–248.

Hui, S. W. 1988. In *The Biology of Cholesterol*, Yeagle, P. L., Ed., CRC Press, Boca Raton, FL, pp. 213–232.

Hui, S. W. 1989. Electron diffraction and diffraction contrast imaging of thin organic films. *J. Electr. Microsc. Tech.,* 11:286–297.

Hui, S. W. and A. Sen. 1989. Effects of lipid packing on polymorphic phase behavior and membrane properties. *Proc. Natl. Acad. Sci. U.S.A.*, 86:5825–5829.

Keller, D. K., H. M. McConnell, and V. T. Moy. 1986. Theory of superstructures in lipid monolayer phase transitions. *J. Phys. Chem.,* 90:2311–2315.

Kleemann, W. and H. M. McConnell. 1976. Interactions of proteins and cholesterol with lipids in bilayer membranes. *Biochim. Biophys. Acta*, 419:206–222.

Lentz, B. R., D. A. Barrow, and M. Hoechli. 1980. Cholesterol- phosphatidylcholine interactions in multilamellar vesicles. *Biochemistry*, 19:1943–1954.

Losche, M., E. Sackman, and H. Mowald. 1983. A fluorescence microscopic study concerning the phase diagram of phospholipids. *Ber. Bunsenges. Phys. Chem.,* 87:848–852.

Martin, R. B. and P. L. Yeagle. 1978. Models for lipid organization in cholesterol-phospholipid bilayers including cholesterol dimer formation. *Lipids,* 13:594–597.

McConnell, H. M., L. K. Tamm, and R. M. Weis. 1984. Periodic structures in lipid monolayer phase transitions. *Proc. Natl. Acad. Sci. U.S.A.,.* 81:3249–3253.

McConnell, H. M. and V. T. Moy. 1988. Shapes of finite two- dimensional lipid domains. *J. Phys. Chem.,* 92:4520–4525.

Miller, R. G. 1989. In *Freeze Fracture Studies of Membranes,* Hui, S. W., Ed., CRC Press, Boca Raton, FL, pp. 87–102.

Mortensen, K., W. Pfeiffer, E. Sackmann, and W. Knoll. 1988. Structural properties of phosphatidylcholine-cholesterol system as studied by small-angle neutron scattering: ripple structure and phase diagram. *Biochim. Biophys. Acta,* 945:221–245.

Muller-Landau, F. and D. A. Cadenhead. 1979. Molecular packing in steroid-lecithin monolayers. II. Mixed films of cholesterol with dipalmitoylphosphatidycholine and tetradecanoic acid. *Chem. Phys. Lipids,* 25:315–328.

Peters, R. and K. Beck. 1983. Translational diffusion in phospholipid monolayers measured by fluorescence microphotolysis. *Proc. Natl. Acad. Sci. U.S.A.,* 80:7183–7187.

Presti, F. T., R. J. Pace, and S. Chan. 1982. 1. Cholesterol- phospholipid interaction in membranes. 2. Stoichiometry and molecular packing of cholesterol-rich domains. *Biochemistry,* 21:3831–3835.

Rand, R. P., V. A. Parsegian, J. A. Henry, L. J. Lis, and M. McAlister. 1980. The effect of cholesterol on measured interaction and compressibility of dipalmitoylphosphatidylcholine bilayers. *Can. J. Biochem.,* 58:959–968.

Recktenwald, D. J. and H. M. McConnell. 1981. Phase equilibria in binary of phosphatidylcholine and cholesterol. *Biochemistry,* 20:4505–4510.

Rogers, J., A. G. Lee, and D. C. Wilton. 1979. The organization of cholesterol and ergosterol in lipid bilayers based on studies using non-perturbing fluorescent sterol probes. *Biochim. Biophys. Acta,* 552:23–37.

Rubenstein, J. L. R., B. A. Smith, and H. M. McConnell. 1979. Lateral diffusion in binary mixtures of cholesterol and phosphatidylcholines. *Proc. Nat. Acad. Sci. U.S.A.,* 76:15–18.

Simon, S. A. and T. J. McIntosh. 1991. Surface ripples cause the large fluid spaces between gel phase bilayers containing small amounts of cholesterol. *Biochim. Biophys. Acta,* 1064:69–74.

Smutzer, G., B. Crawford, and P. L. Yeagle. 1986. Cholesterol behavior in human serum low density lipoproteins: a fluorescent probe study. *Biochim. Biophys. Acta,* 862:361–366.

Snyder, B. and E. Freire. 1980. Compositional domain structure in phosphotidylcholine-cholesterol and sphingomyelin-cholesterol bilayers. *Proc. Nat. Acad. Sci. U.S.A.,* 77:4055–4059.

Tilock, C. P. S., M. B. Bally, S. B. Farren, and P. R. Cullis. 1982. Influence of cholesterol on the structural preferences of dioleoylphosphatidylethanolamine-dioleoylphosphatidylcholine systems: a phosphorus-31 and deuterium nuclear magnetic resonance study. *Biochemistry,* 21:4596–4601.

Weis, R. M. and H. M. McConnell. 1985. Cholesterol stabilizes the crystal-liquid interface in phospholipid monolayers. *J. Phys. Chem.,* 89:4453–4459.

Yu, H. and S. W. Hui. 1992. Merocyanine-540 as a probe to study the molecular packing of phosphatidylcholine monolayers: an epifluorescence microscopy and spectroscopy study. *Biochim. Biophys. Acta,* 1107:245–254.

Chapter 7

VIBRATIONAL SPECTROSCOPY OF CHOLESTEROL-LIPID INTERACTIONS

Timothy J. O'Leary

TABLE OF CONTENTS

0-8493-4207-4/93/$0.00+$.50

ABSTRACT

Cholesterol is a ubiquitous sterol molecule that constitutes a third or more of the cell plasma membranes of most eukaryotic (yeast or higher) organisms. Vibrational (infrared and Raman) spectroscopic methods provide powerful tools by which to assess the effects of cholesterol on both the conformation and environment of phospholipid headgroup, glycerol backbone, and acyl chain regions. Studies based on these techniques have shown that cholesterol does not abolish the gel-to-liquid crystalline transition of lipids, but rather broadens the temperature over which both gel and liquid crystalline phases coexist. Cholesterol does not, however, cause lipid molecules to form a phase intermediate between the gel and liquid crystalline phases. Instead, individual lipid molecules within cholesterol-containing bilayers are found within regions having structural characteristics similar to those of pure gel or liquid crystalline phases. Cholesterol tends to order the lipid bilayer near the headgroup region at all temperatures and tends to order the entire chain at high temperatures. At low temperatures, however, cholesterol disorders the acyl chain only near the methyl termini. Thus, cholesterol substantially modulates the effect of temperature changes on membrane structure. It therefore appears that one of the most important roles of cholesterol in the cell may be to make the membrane insensitive to changes in the cellular environment, enabling organisms to tolerate a much wider range of environments than would otherwise be possible.

I. INTRODUCTION

Cholesterol is a ubiquitous sterol molecule that constitutes a third or more of the cell plasma membranes of most eukaryotic (yeast or higher) organisms. Far from being the "bad" molecule portrayed by the advertisers of low-cholesterol food products, it seems that cholesterol is absolutely necessary for the functioning of all of our cells, as well as serving as a precursor for the synthesis of hormones and other biochemically important molecules. Although physicians and biochemists realize that cholesterol is essential to support life, the role it plays in cell membranes has yet to be completely understood.

To understand the role that cholesterol plays in the cell membranes and in model membranes, one must first understand some simple lipid biochemistry. Figure 1 illustrates a typical lipid molecule, a diacyl phosphatidylcholine. Although symmetric diacyl phosphatidylcholines are not a particularly large component of cell membranes, dipalmitoylphosphatidycholine (DPPC) constitutes most of the lipid lining the inside of the lung and is the lipid molecule that has been most extensively studied using thermodynamic and spectroscopic methods. This molecule has three distinct structural regions. At the polar end of the molecule is a region known as the *headgroup*. This region, which consists of a phosphate group bound (in this case) to a choline moiety, may easily be solvated by water and remains in contact with water when membranes are formed. The head group is attached to the *acyl chain* or *hydrocarbon* region via a glycerol backbone that constitutes the major component of the *interface region*.

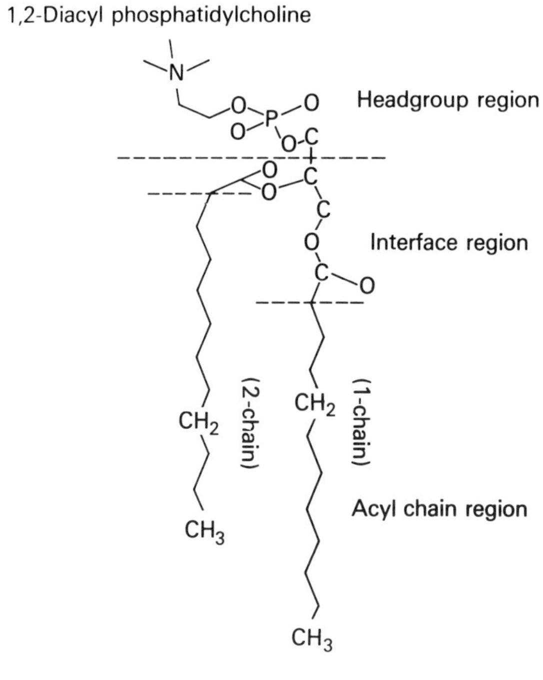

FIGURE 1. Structure of a diacylphosphatidylcholine molecule. The molecule is divided into distinct headgroup, interface (glycerol backbone), and acyl chain regions.

The hydrocarbon chains (which in this case are 16-CH_2-units long) attach to this glycerol backbone via an ester linkage, which is also considered part of the interface region. Thus, the acyl chain region is pure hydrocarbon and chemically resembles paraffin (candle wax).

Pure lipids, such as DPPC or dipalmitoylphosphatidylethanolamine (DPPE), may be readily dispersed in water or saline solutions, forming multilamellar lipid bilayer structures called liposomes. Lipids within the liposome are organized in such a way that their headgroups all come in contact with water, while their acyl chains only come in contact with other acyl chains (Figure 2). Thus, the interior of the lipid bilayer resembles pure paraffin in that, except for their attachments to the glycerol backbone, hydrocarbon chains come into contact with nothing except other hydrocarbon chains. Just as candle wax can exist as either a cool solid or a hot liquid, the lipid molecules within liposomes can exist in either a relatively solid "gel" phase or a relatively fluid "liquid crystalline" phase. In the gel phase the lipid hydrocarbon chains are mostly in an "all-*trans*" configuration, such as shown in Figure 1. In the liquid crystalline phase there is a considerable amount of *trans-gauche* isomerization around carbon-carbon bonds in the acyl chains, with the resulting formation of kinks and bends. In order to accommodate these changes, the lipid bilayer expands laterally, or perpendicularly to the hydrocarbon chain axis, and thins along this axis (Figure 2).

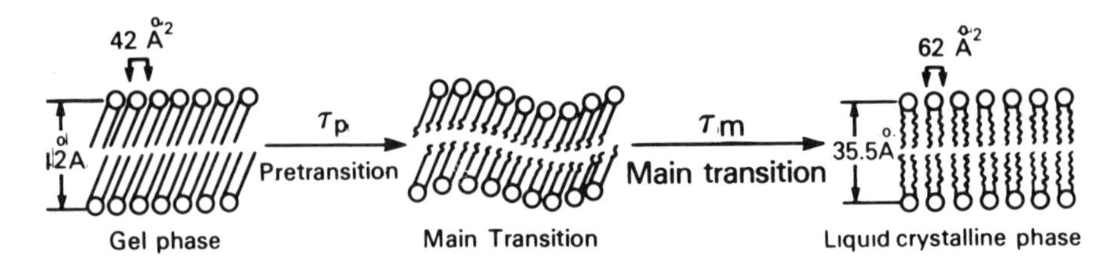

FIGURE 2. Structure of a membrane bilayer assembly. The headgroups are in contact with water outside the lipid bilayer, while the hydrocarbon tails of the lipids are in contact only with the hydrocarbon tails of other lipids. During the pretransition at approximately 35°C, the lipid undergoes a transition to a ripple phase, with only slight rotational disordering of the acyl chains. During the gel-to-liquid crystalline transition, formation of many *gauche* rotamers within the acyl chain is accompanied by lateral expansion and thinning of the lipid bilayer.

The transition between the gel and the liquid crystalline state typically occurs abruptly upon warming, at a temperature known as the gel-to-liquid crystalline phase transition temperature T_m. Since the formation of *gauche* rotamers in the hydrocarbon chains and the lateral expansion of the membrane are both cooperative, energy-requiring events, a sharp peak in the heat-capacity curve (Figure 3) is seen in differential scanning calorimetry experiments. Thermal measurements on intact cell membranes never reveal such sharp thermal transitions. Even in those rare organisms, such as mycoplasma, that exhibit any membrane phase transitions (most do not), the transitions are much broader than those seen in scanning calorimetry of pure lipid bilayers.

In this chapter we will provide evidence that one of the major contributions of cholesterol to the cell plasma membrane is to allow this membrane to exist in a state that is, in many ways, intermediate between the gel and liquid crystalline phases typically exhibited by pure lipids. Hence, the membrane is better able to maintain its structural stability than if it were a pure liquid crystal, yet the membrane allows freer diffusion of proteins and other membrane constituents than if it were a membrane in a pure gel state. Cholesterol achieves this effect both by disrupting the lateral packing of lipid molecules within the lipid bilayer and by creating voids within the acyl chain region of the bilayer, which, in turn, promote *trans-gauche* isomerization within the center of the bilayer, but not near the interface region. One consequence of this is to eliminate the gel-to-liquid crystalline phase transition seen by calorimetric methods in pure lipids. Among the findings of the infrared and Raman spectroscopic experiments is that this transition is not so much abolished as "spread out" over a wide temperature range. This, in turn, suggests that one biological effect of cholesterol is to minimize changes in the cell membrane that occur as a result of thermal fluctuations or other perturbations.

We now discuss vibrational spectroscopy using infrared and Raman methods. After a brief discussion of the physics underlying the use of these techniques, we discuss the spectroscopic features characterizing assemblies of pure phospholipid molecules, considering the interface and acyl chain regions separately.

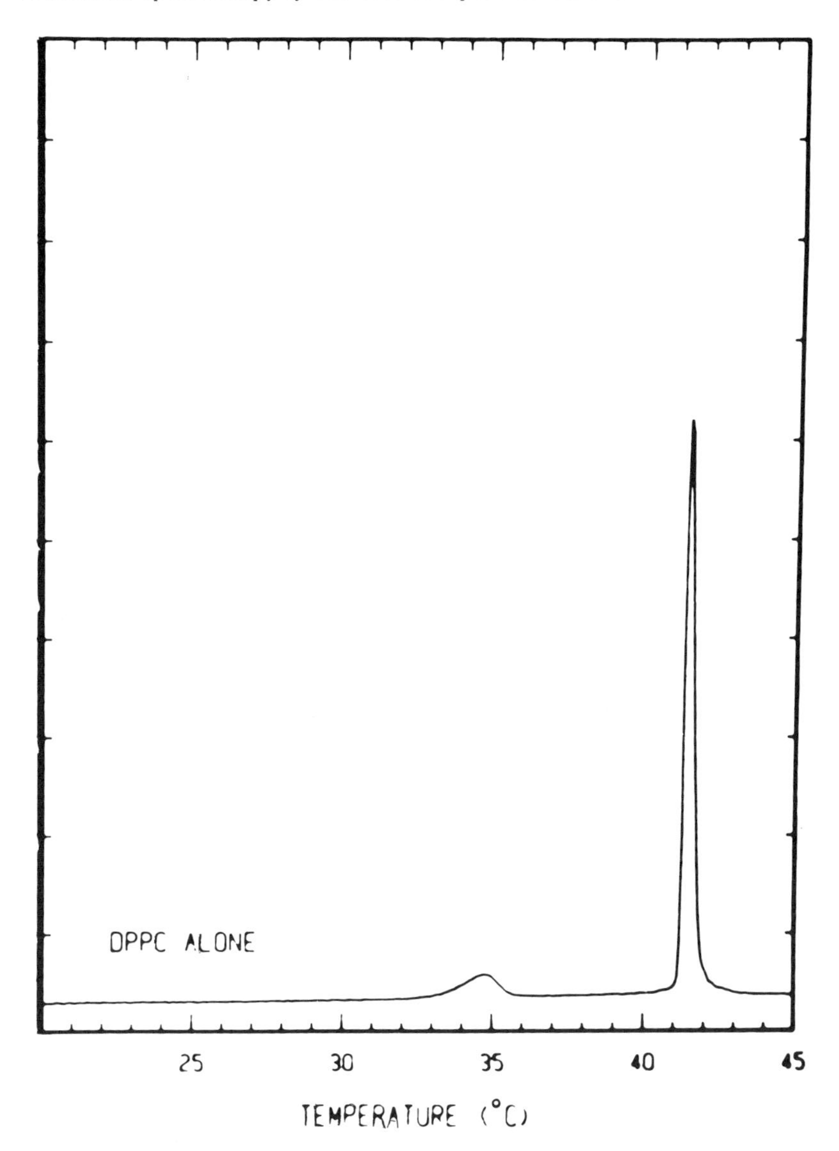

FIGURE 3. Thermogram, obtained by differential scanning calorimetry, demonstrating the phase transitions of dipalmitoylphosphatidylcholine (DPPC). The smaller peak at approximately 35°C corresponds to the lipid pretransition, while the larger peak at approximately 42°C corresponds to the gel-to-liquid crystalline (acyl chain melting) transition.

Then we discuss the effects of cholesterol upon both acyl chain melting and interface region structure. Finally, we propose a conceptual model for the effects of cholesterol on lipid assemblies, based upon the infrared and Raman spectroscopic results.

II. PRINCIPLES OF VIBRATIONAL SPECTROSCOPY

All molecules are continuously vibrating, occupying distinct energy levels that are reasonably well described by considering the molecule as a system of coupled harmonic oscillators. For a molecule consisting of N atoms, there are 3N-6 such vibrational modes (3N-5 for a linear molecule, such as CO_2). For example, water, with three atoms, has three distinct vibrational modes. Each of these vibrational modes may, in turn, exist in a variety of energy levels. At room temperature most molecules are found in the lowest energy level, or ground state. Molecules may be excited into higher vibrational levels by the absorption of radiation having an energy ($h\nu$) equal to the energy separating two energy states within one of these vibrational modes (assuming that certain physical requirements, called selection rules, dictated by quantum mechanics and symmetry considerations are met), where ν is the frequency of the radiation and h is Planck's constant. The most important of these rules is that the dipole moment of the molecule must change upon absorption of a photon (quantum of light). Specifically, the intensity of absorption is proportional to the square of the change in the dipole moment that occurs when a molecule vibrates. Thus, the infrared spectrum of a molecule reflects the frequencies of radiation that may cause a molecule to undergo a transition between two of its possible vibrational states. The spectrum thus obtained reflects both the chemical composition of the molecule and its molecular conformation. Typically, infrared absorption bands characteristic of particular chemical moieties, such as C=O or CH_2 groups are found within a relatively constant frequency range for all molecules (1670 to 1800 cm^{-1} for the C=O group, for example). The precise frequency of the absorption band depends not only on the chemical group, but also on its molecular environment. For example, the C=O group frequency is influenced significantly by the dielectric constant of the medium surrounding it (Mushayakarara and Levin, 1982). Similarly, the frequency of a CH_2 twisting mode in a hydrocarbon chain depends on the presence and location of *gauche* isomers in the hydrocarbon chain. Thus, infrared spectroscopy is a powerful tool by which to learn not only about the structure of a molecule, but also about its conformation and molecular environment.

Raman spectroscopy is similar to infrared spectroscopy in that it too depends upon changes in molecular vibrations and it too provides information not only about the chemical makeup of a molecule, but about its conformation and molecular environment. In Raman spectroscopy the sample is irradiated not with infrared photons, but with visible wavelength photons, typically from a laser source. In the Raman process, absorption of photons raises the molecule to an unstable intermediate quantum mechanical state called a "virtual level." As the molecule relaxes, the photons are emitted in all directions. Most of the emitted photons have the same frequency as the incident radiation (Raleigh scattering). A very small number of photons (10^{-8}) are shifted from the Raleigh line frequency by amounts that correspond to the frequencies of the molecular vibrational modes. The Raman spectrum is obtained by measuring the frequencies and intensities of these scattered photons. The intensity of this scattered light is proportional to the intensity of the incident radiation, to the frequency of the incident radiation raised to the fourth power, and to the square of the change in the polarizability of the molecule that occurs during a vibration.

While the Raman spectrum and the infrared spectrum are similar by virtue of the fact that they both reflect molecular vibrational energy states, they differ because the selection rules for infrared absorption and Raman scattering differ. For example, H_2 does not give an infrared spectrum because there is no change in the molecular dipole moment during vibration. It has a strong Raman spectrum, however, because there is a substantial change in the molecular polarizability. Similar differences are seen in the spectra of more complex molecules, resulting in complementary roles for infrared and Raman spectroscopy as tools in structural chemistry and biophysics.

An important feature of vibrational spectroscopic techniques is that they do not require one to introduce a probe molecule, such as a fluorophore, into the lipid bilayer to obtain a spectrum. As a result, one does not run the risk of perturbing the chemical structure around a probe molecule such that it ceases to reflect the structure of the lipid at large. In addition, infrared absorption and Raman scattering both occur very quickly in comparison with such motional events as *trans-gauche* isomerization or axial rotation of the lipid molecule. Thus, infrared and Raman spectra provide a "stopped action" snapshot that is not "blurred" by slow motions within the lipid bilayer.

The detailed mechanics of obtaining infrared and Raman spectra is beyond the scope of this chapter (Levin, 1984). Briefly, in a typical experiment, lipid and cholesterol are codissolved in chloroform in the appropriate proportions, dried under flowing nitrogen, and then dessicated under vacuum for 12 to 24 h. Water is added to the dried lipid to obtain a dispersion that is 50 to 90% water by weight, and the dispersion heated and cooled between room temperature and 60° about ten times. For infrared spectroscopy, 10 to 20 µl of this dispersion are placed in a transmission cell with an optical path length of 4 to 6 µm, or in an attenuated total reflection (ATR) cell. The windows or ATR crystal are made of calcium fluoride or zinc selenide, since water dissolves the KBr windows typically used for nonaqueous materials. A circulating-water cooling bath is used to regulate the temperature of the sample cell. Most modern spectrometers actually obtain the Fourier transform of the infrared spectrum, utilizing a blackbody light source, a moving-mirror optical path, and a liquid-nitrogen-cooled mercury-cadmium-telluride (MCT) detector. The system is interfaced to a small computer system, such as a laboratory workstation or a personal computer, which then computes the infrared spectrum. Once the sample has been placed in the cell, spectra may generally be obtained in 1 to 2 min per temperature. Suffice it to say that is much easier to obtain such spectra on lipid systems than to interpret the results.

III. SPECTROSCOPIC FEATURES OF PHOSPHOLIPID MOLECULES

Infrared and Raman spectra of lipid assemblies are quite complex (Figure 4). Understanding them is even more complex, in part because spectroscopists often refer to the features that appear in the spectrum in several different ways (as "features," "bands," or "modes," for example), and refer to these features by the approximate frequency at which it occurs. For example, the Raman scattering band resulting from the acyl chain methylene CH_2 symmetric stretching vibrational mode is often referred to as the "2850 cm^{-1} band," even though the band center may appear anywhere from about 2848 cm^{-1} to 2855 cm^{-1}. In the discussion that follows we will distinguish these approximate frequencies, which serve essentially to name a band, by placing them in italics. More precise frequencies, which serve to identify changes in band position, will remain in regular typeface.

To understand the spectral properties of these bilayers, it also helps to think of the lipid molecule as having three distinct regions: headgroup, interface, and acyl chain. Structural information about the head group region may be provided by phosphate PO_2^- stretching modes, appearing at *1250* cm^{-1} as well as by choline C–N stretching modes appearing at *717* cm^{-1}. Structural information about the interface region is best reflected in the C=O stretching modes found in the 1700- to 1800-cm^{-1} region. Information about acyl chain structure is reflected in the 1000- to 1200-cm^{-1} C–C stretching modes, by the approximately 1300-cm^{-1} CH_2 twisting modes, the 1400- to 1500-cm^{-1} CH_2 deformation modes, and by the 2800- to 3100-cm^{-1} C–H stretching modes (Figure 5). In order to better appreciate the kinds of information that can be obtained by analysis of the vibrational spectra of these molecules, it is useful to consider the headgroup, interface, and acyl chain regions separately. A number of excellent technical reviews of the infrared (Casal and Mantsch, 1984) and Raman (Levin, 1984) spectroscopy of lipid bilayers have been published that provide more detailed information on spectral analysis and interpretation. We briefly discuss the spectral features that characterize the acyl chain (III A) and interface (III B) regions.

A. THE ACYL CHAIN REGION

1. C–H Stretching Modes

The vibrational features observed between 2800 cm^{-1} and 3100 cm^{-1}, arising from hydrocarbon chain region C–H stretching modes, are among the most widely used and easily interpreted Raman spectroscopic features found in lipid assemblies. As seen in Figure 5 (upper panel), the three most intense features observed in unhydrated-crystalline or hydrated-gel states are the methylene C–H symmetric stretching, methylene C–H asymmetric stretching, and chain terminal methyl C–H symmetric stretching modes at *2850*, *2880*, and *2935* cm^{-1}, respectively. Upon transition to a liquid crystalline state (Figure 5, lower panel), the intensity of the *2880*-cm^{-1} band decreases relative to that of the *2850*-cm^{-1} band, while the intensity of the *2935*-cm^{-1} band increases. Simultaneously, there is an increase in the frequency of both the symmetric and asymmetric features.

The dramatic change in the relative intensities of the *2935*- and *2880*- cm^{-1} features allows one to readily monitor acyl chain melting by plotting the $I_{2935}/$

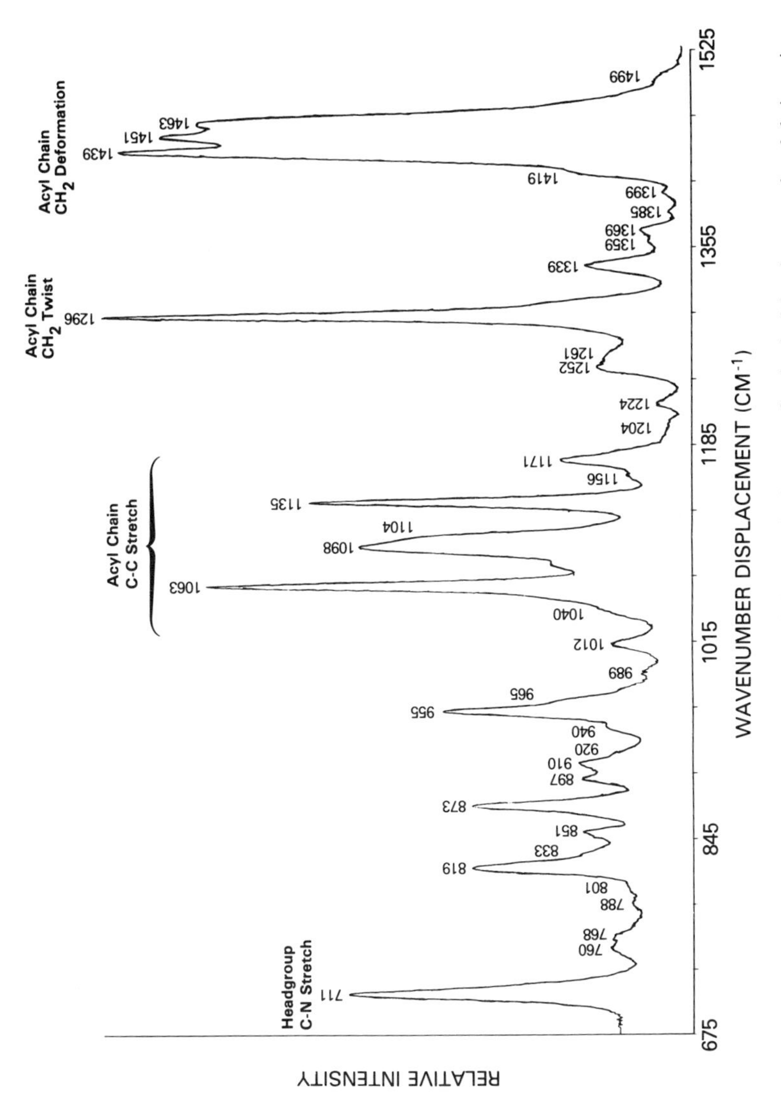

FIGURE 4. Raman spectrum of anhydrous DPPC. In addition to major features reflecting both headgroup and acyl chain regions, numerous smaller features reflecting headgroup, interface, and acyl chain structures are also present.

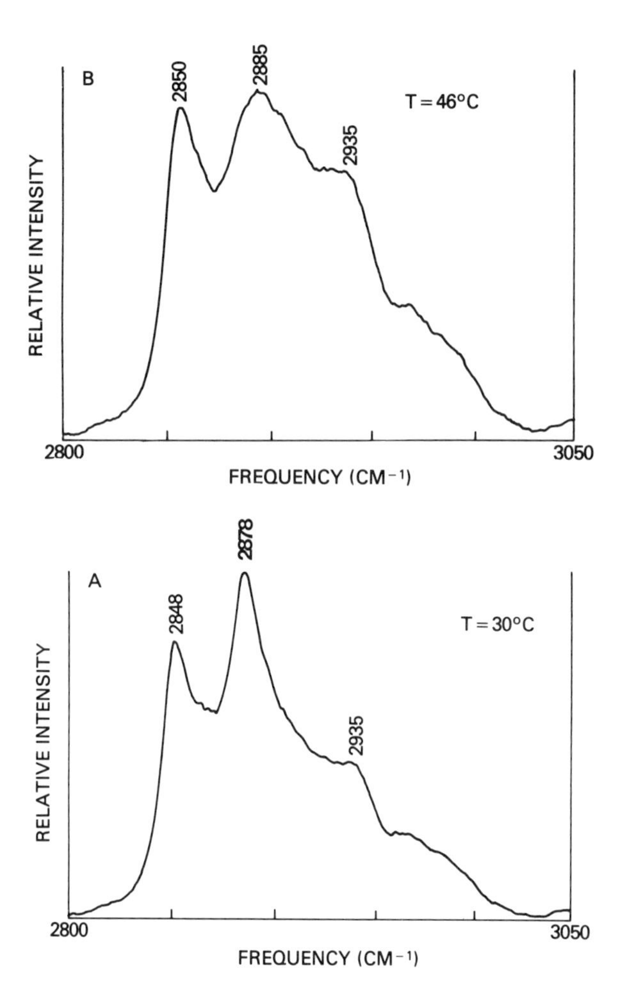

FIGURE 5. Raman spectra of the DPPC C–H stretching mode region in the gel (upper panel) and liquid crystalline (lower panel) phases. The changes in peak height ratios I_{2850}/I_{2880} and I_{2935}/I_{2880} are readily apparent and directly reflect acyl chain melting in the lipid assembly.

I_{2880} peak ratio (Figure 6). The change in this ratio at the gel-to-liquid crystalline phase transition is strongly correlated with the entropy of transition for symmetric chain saturated lipids (Huang, Lapides, and Levin, 1982). Since the entropy of the gel-to-liquid crystalline phase transition arises almost completely from the change in the number of bilayer *gauche* isomers, it is apparent that the I_{2935}/I_{2880} peak ratio is a sensitive indicator of lipid bilayer acyl chain disorder. Peak height ratios in this spectral region are also sensitive to chain packing and may be used to detect interdigitation in lipid bilayer assemblies (O'Leary and Levin, 1984).

Infrared spectra obtained in the CH stretching mode region (Figure 7) are also sensitive to hydrocarbon chain *trans-gauche* isomerization and may be used to monitor chain conformation, too. Unlike the Raman spectrum, peak-intensity ratios do not provide as much useful information as do the peak frequencies.

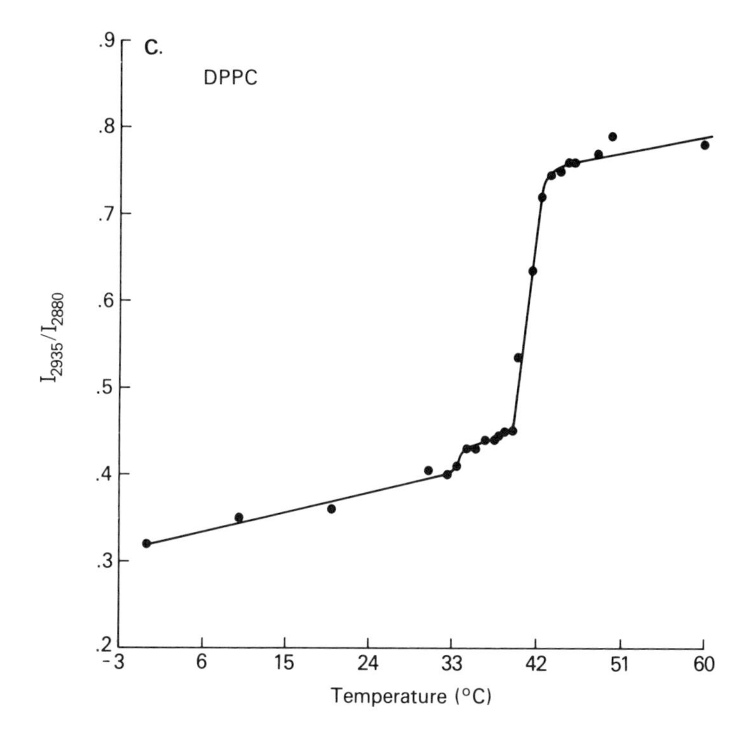

FIGURE 6. Melting profile of DPPC based upon the I_{2935}/I_{2880} peak height ratio. Low values of this ratio indicate a relatively ordered bilayer, such as that found in the gel phase, while high values indicate a disordered bilayer, such as that of the liquid crystalline phase. Both the pretransition at 35°C and the gel-to-liquid crystalline transition at 42°C are readily observed.

Either the symmetric or the asymmetric stretching-mode band may be used to monitor lipid melting, as the two respond similarly to chain isomerization. During the gel-to-liquid crystalline phase transition, the frequency maximum of the DPPC symmetric stretching-mode feature increases from approximately 2851 cm^{-1} to 2854 cm^{-1}. This is accompanied by significant broadening of the band (by approximately 2.5 cm^{-1} at three fourths of the peak maximum) (Casal and Mantsch, 1984). Smaller changes are seen at the pretransition, reflecting the smaller structural rearrangement occurring at this phase transition (O'Leary et al., 1986). Higher frequencies of this mode thus signify more disordered acyl chain structures; for example, hexagonal-phase phosphatidylethanolamines have a CH symmetric stretching peak centered at approximately 2854 cm^{-1}. Lower frequencies signify more ordered structures; dipalmiatoylphosphatidylethanolamine in the gel phase has a CH_2 symmetric stretching mode frequency of about 2849 cm^{-1}, indicating a more ordered conformation than seen in DPPC (Casal and Mantsch, 1984).

2. C–D Stretching Modes

Lipids that are selectively deuterated at a specific position along the acyl chain demonstrate symmetric and asymmetric C-D stretching modes at approximately

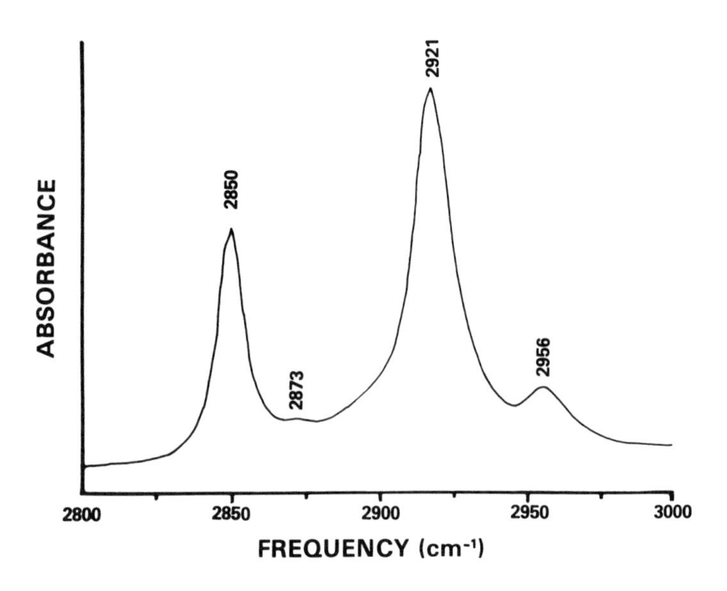

A

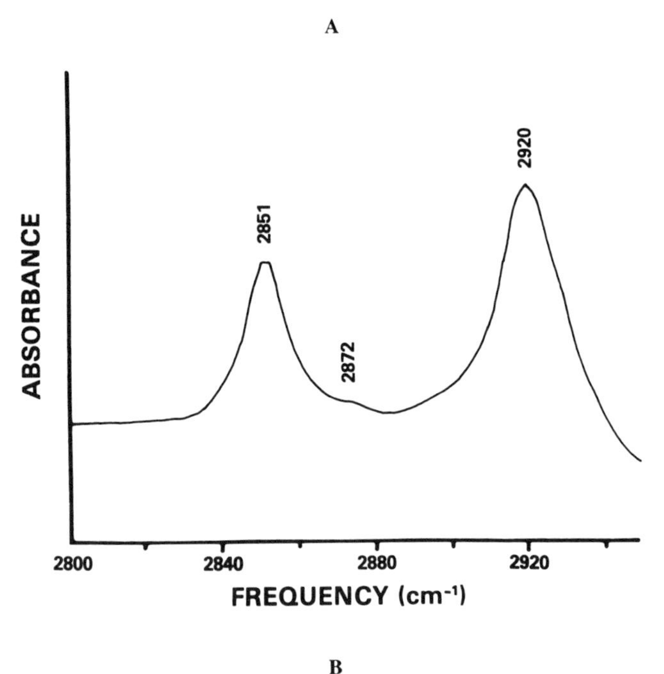

B

FIGURE 7. Fourier-transform infrared spectra of the DPPC C–H stretching mode region in the gel (upper panel) and liquid crystalline (lower panel) phases. A small change in the frequencies of both the symmetric (2850 cm^{-1}) and asymmetric (*2921* cm^{-1}) CH_2 stretching mode features is observed, which reflects acyl chain melting in the lipid assembly.

2100 and 2200 cm^{-1}, respectively. The precise frequencies and widths of these features reflect both the position along the acyl chain and the physical state of the lipid bilayer and may thus be used to give a detailed picture of acyl chain melting (O'Leary and Levin, 1986).

The frequency of a hydrocarbon chain CD_2 rocking mode depends upon the conformation of the chain in its immediate vicinity. In particular, CD_2 groups participating in a–CH_2*trans*–CD_2–*trans*–CH_2– sequence give rise to a band at *622* cm^{-1}, independent of the conformation of neighboring bonds. A band at *652* cm^{-1} results from chains in a –CH_2-*gauche*'–CH_2-*gauche*–CD_2–*trans*–CH_2–*trans*–CH_2– or –CH_2–*trans*–CH_2–*trans*–CD_2-*gauche*–CH_2–*trans*–CH_2– conformation. Finally, a band resulting from –CH_2–*trans*–CH_2–*trans*–CD_2–*gauche*–CH_2–*gauche*–CH_2– or –CH_2–*gauche*'–CH_2–*trans*–CD_2–*gauche*–CH_2–*gauche*–CH_2– sequences is found at *646* cm^{-1}. By carefully measuring the integrated band intensities of these three CD_2 rocking modes, one may calculate the relative numbers of CD_2 units participating in *gauche* and *trans* bonds. Thus, infrared spectroscopy is capable of giving a very detailed picture of acyl chain melting as a function of position along the acyl chain (Davies et al., 1990; Mendelsohn et al., 1989).

3. C–C Stretching Modes

Skeletal C–C stretching modes found at *1060, 1090*, and *1130* cm^{-1} provide a second set of spectroscopic features useful in monitoring acyl chain disorder (Figure 4). Although interpretation of the region is complicated by the presence of relatively weak phosphate and head group choline modes, it is clear that acyl chain melting is accompanied by decreases in the relative intensities of the *1060*- and *1130*-cm^{-1} features, which originate from all-trans chains, and an increase in the intensity of the *1090*-cm^{-1} feature, which arises from hydrocarbon chain segments containing *gauche* isomers. As a result, chain melting may be monitored using peak-height ratios, as for the C–H stretching modes. Typically, I_{1090}/I_{1130} has been employed, although other ratios, such as the ratio of the *1130*-cm^{-1} mode to that of the *717*-cm^{-1} choline C–N stretching feature, have also been used (Pink et al., 1981).

B. THE INTERFACE REGION

Although practical assessment of the interface region structure by vibrational spectroscopy depends only upon the interpretation of the spectral features arising from the *sn*–1 and *sn*–2 chain C=O groups, the interpretation of spectral changes in the 1700- to 1800-cm^{-1} C=O stretching mode region is difficult. The presence of two resolved, or partially resolved, peaks in this region, such as seen in dipalmitoylphosphatidylcholine, is indicative of differences in the packing of the two hydrocarbon chains. Presence of additional C=O stretching mode features may also reflect hydrogen bonding. The frequency and width of the carbonyl modes changes when the bilayer undergoes phase transitions. For example, peak-frequency changes of 1 to 2 cm^{-1} are seen at the subtransition, the pretransition, and the gel-to-liquid crystalline phase transition of dipalmitoylphosphatidylcholine. The changes are not easily interpretable, however, since they do not directly reflect either chain melting or packing, but rather their effects on the conformation of the interface region (Casal and Mantsch, 1984).

Some difficulties in interpreting this region may be overcome by the use of lipids that have ^{13}C incorporated in the *sn*–2 chain carbonyl moiety, since this shifts the frequency of the *sn*–2 carbonyl vibration to approximately *1685* cm^{-1}, well removed from that of the *sn*–1 chain at approximately 1737 cm^{-1}. This does not completely simplify interpretation of lipid C=O spectra, however, as changes in this region have been variously interpreted as resulting from differences in the dielectric constant of the carbonyl moiety's immediate environment, hydrogen bonding, or factor group splitting of equivalent carbonyls. Lack of agreement on the molecular changes that cause changes to occur in interface region spectral features necessitates great caution in interpreting changes resulting from the introduction of perturbants, such as cholesterol, into the bilayer. It appears to me that most of the published data on vibrational spectroscopic features in the carbonyl region may be crudely interpreted as resulting solely from changes in lipid packing, without reference to hydrogen bonding or, for that matter, to changes in glycerol backbone conformation. When the lipid bilayer takes on a planar geometry with lipid acyl chains perpendicular to the plane of the bilayer, as, for example, with dipalmitoylphosphatidylethanolamine (DPPE), only a single carbonyl region peak is observed. When the acyl chain axes are tilted with respect to the bilayer normal, additional carbonyl region features are observed (T.J. O'Leary and J.A. Centeno, unpublished data).

IV. EFFECTS OF CHOLESTEROL ON LIPID BILAYERS

A. ACYL CHAIN MELTING

Calorimetric studies of lipid-cholesterol interactions suggest that cholesterol nearly abolishes the lipid gel-to-liquid crystalline phase transition when the concentration is 30 to 50 mol% in the bilayer (Mabrey et al., 1978). Since almost no enthalpy of transition can be measured calorimetrically, one might be tempted to conclude that when cholesterol is present at high concentrations within a lipid bilayer, almost no change in bilayer structure occurs as the temperature changes. Infrared and Raman spectroscopic data from several investigators show that this is not the case. As long ago as 1971 Lippert and Peticolas (1971) demonstrated by Raman spectroscopy that the addition of 50 mol% cholesterol to DPPC multilayers significantly broadens, but does not abolish, the gel-to-liquid crystalline phase transition. Similar broadening of the transition, as observed by Raman spectroscopy, is seen in DPPC (Levin, 1984) and 1,2-di-O-hexadecyl-*sn*-glycero-3-phosphocholine (DHPC) (which is structurally identical to DPPC, except that the ester linkages between the glycerol backbone and the acyl chains are replaced by ether linkages) dispersions containing 18 to 25 mol% cholesterol (Levin et al., 1985). Below the pure liquid phase transition temperature, bilayers containing cholesterol tend to be somewhat less ordered than pure lipid bilayers, and above the phase transition temperature, those containing cholesterol are somewhat more ordered than those without. Nevertheless, it appears that at temperatures 30 to 40°C on either side of the phase transition temperature, the difference in the lipid order of bilayers that do or do not contain cholesterol is relatively small. In other words, as one heats the bilayer from about 40°C below the pure lipid phase transition temperature T_m to approximately 30°C above the pure bilayer Tm, the lipid molecules undergo a similar degree of disordering whether or not cholesterol is present. Infrared (Cortijo and Chapman, 1981; Umemura et al., 1980) and Raman (Asher and Levin, 1977; Levin, 1984) spectroscopic data comparing lipid bilayers containing 25 to 40 mol% cholesterol show similar results.

Although the differences in lipid order between cholesterol-containing and pure lipid at temperatures well away from the pure-lipid phase transition temperature is small, cholesterol tends to disorder the gel phase and order the liquid crystalline phase nearer to the phase transition temperature. The net result is to make the lipid bilayer as a whole less sensitive to temperature changes. At high cholesterol concentrations a change of a few degrees in temperature has little effect on lipid conformational order, whereas a similarly small temperature change can result in marked changes in lipid conformational order in the absence of cholesterol.

Whereas the effects of cholesterol on overall lipid conformational order can be readily appreciated on the basis of investigations on natural lipids, an appreciation for effects that cholesterol might have at different levels within the lipid bilayer requires the use of lipid molecules that have been selectively deuterated at specific sites along the lipid acyl chain. We (O'Leary and Levin, 1986) investigated the effects of high (50 mol%) cholesterol concentrations on dimyrostoylphosphatidylcholine (DMPC, a symmetric lipid with 14 CH_2 units on each acyl chain) bilayers selectively deuterated in the *sn*-2 chain at the 3, 6, and 10 positions using Raman spectroscopy of the *2100*-cm^{-1} CD_2 stretching modes (Figure 8).

To illustrate the interpretation of results on the selectively deuterated lipids, we consider the *sn*-2 chain selectively deuterated at the 6 position, in both the presence and absence of cholesterol. From Figure 8 we see that both the *2094*- and *2171*-cm^{-1} bands are broader for the more disordered liquid crystalline phase of the pure lipid (Figure 8B) than for the relatively ordered gel phase (Figure 8A). In the presence of cholesterol (Figures 8C and 8D) the bandwidths of these features are nearer to those of pure lipid in the gel phase than to those of pure lipid in the liquid crystalline phase. This indicates that the cholesterol is "ordering" the bilayer at the 6 position, both at a temperature in which the pure lipid would be in a liquid crystalline phase (Figure 8D) and at a temperature in which the lipid would normally exist in the gel phase (Figure 8C). From similar analysis of lipids deuterated at the 3 and 10 positions, we demonstrated that cholesterol tends to order the DMPC bilayer near the head group region at all temperatures and that it tends to order the entire chain at high temperatures. At low temperatures, however, cholesterol disorders the acyl chain only near the methyl termini.

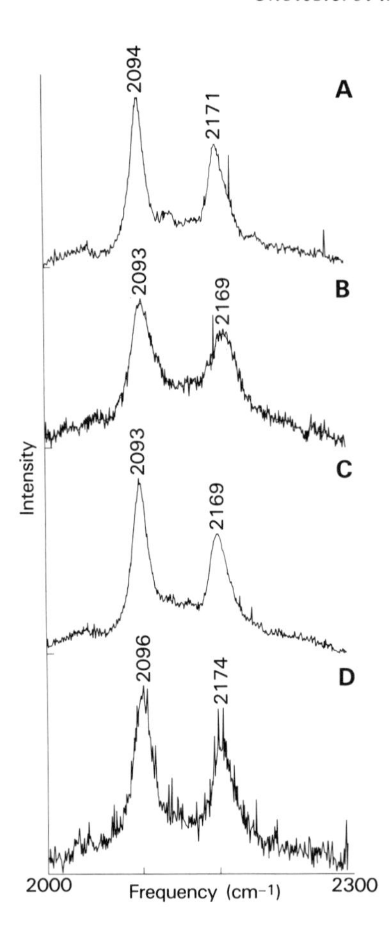

FIGURE 8. Raman spectra of hydrated 2-[6,6-D_2] dimyristoylphosphatidylcholine (DMPC) dispersions below the gel-to-liquid crystalline phase transition temperature (A) and above the gel-to-liquid crystalline phase transition temperature (B), contrasted with those of 2-[6,6-D_2] DMPC-cholesterol (1:1) mole ratio) dispersions at the same temperatures (C and D). (Reproduced from O'Leary and Levin, 1986. *Biochim. Biophys. Acta,* 854:321. With permission.)

These results from Raman spectroscopy have been corroborated for 33 mol% cholesterol-DPPC bilayer dispersions in an elegant infrared spectroscopic experiment carried out by Mendelsohn and co-workers (Davies et al., 1990). By measuring the integrated band intensities of the CD_2 rocking modes, they demonstrated that at high temperatures, cholesterol orders the bilayer at the 4, 6, and 10 positions. Unfortunately, they did not provide data comparing bilayers with and without cholesterol at temperatures below that of the pure-lipid phase transition.

Interestingly, the profound effects seen when cholesterol is added to hydrated-lipid bilayers are absent when cholesterol is added to anhydrous lipid. Bush, Adams, and Levin (1980) examined the effects of adding up to 30 mol%

cholesterol to anhydrous DPPC and found no disordering evidenced by changes in either C–C or C–H stretching region Raman spectra.

B. INTERFACE REGION

Several infrared and Raman spectroscopic investigations have been designed to answer two questions. (1) Does the hydroxyl group of cholesterol interact with either *sn*-1 or *sn*-2 chain carbonyl moieties via hydrogen bonding? (2) Does cholesterol perturb the structure of lipid molecules in the interface region? Bush, Adams, and Levin (1980), investigating the effects of adding 7, 14, 20, and 30 mol% cholesterol to anhydrous DPPC, found no perturbation of the *1720*- and *1738*-cm^{-1} C=O stretching features in the Raman spectrum. They concluded, therefore, that hydrogen bonding between the cholesterol 3β-OH and acyl carbonyl groups does not occur in the anhydrous bilayer. Bush, Levin, and Levin (1980) confirmed these results for cholesterol at concentrations ranging from 5 to 43 mol% using infrared spectroscopy. They further observed that addition of 50 mol% cholesterol to hydrated dispersions narrows the carbonyl band by about 25%, postulating that cholesterol reduces the conformational differences in the dielectric constant of the environment of the *sn*-1 and *sn*-2 chains. Again, they found no evidence of hydrogen bonding between the cholesterol 3β-OH and the lipid *sn*-2 carbonyl. Infrared spectroscopic studies have also shown that cholesterol does not hydrogen bond to acyl chain carbonyl groups in tripalmitoyl glyceride (Wong, Chagwedera, and Mantsch, 1989).

The conclusion that cholesterol does not hydrogen bond to the lipid carbonyl group has been disputed by Wong, Capes, and Mantsch (1989), who investigated the interactions of cholesterol with anhydrous L-α-1,2- dimyristoylphosphatidylcholine (DMPC), DPPC, and DHPC. Changes in the frequency of the carbonyl vibration were interpreted as demonstrating hydrogen bonding to both *sn*-1 and *sn*-2 carbonyl moieties. Since changes in carbonyl band frequencies may reflect not only hydrogen bonding and conformational alterations of the interface region, but also packing changes in the bilayer (such as transition from the P_{β}' to the L_{α} phase), this interpretation is controversial.

Green et al. (1987) also investigated the interactions of cholesterol with DPPC multilayers. They observed that in the presence of cholesterol, two distinct peaks (1738 cm^{-1} and 1734 cm^{-1}) may be attributed to the *sn*-1 carbonyl vibration, although only a single peak, at *1685* cm^{-1} may be attributed to the *sn*-2 carbonyl. The presence of two distinct *sn*-1 carbonyl features is observed over a wide temperature range, and the frequencies of the bands are typical of gel and liquid crystalline phase lipids, respectively (Figure 9). This data is interpreted to mean that both gel- and liquid-crystalline-phase lipids persist over a wide temperature range in the presence of high amounts of cholesterol.

C. HEADGROUP REGION

Bush, Adams, and Levin (1980) examined the effect of cholesterol on the *717* cm^{-1} C–N stretching modes of DPPC. Cholesterol at concentrations ranging from 7 to 30 mol% had no effect on this Raman band, indicating that head group conformation is unchanged in the presence of cholesterol.

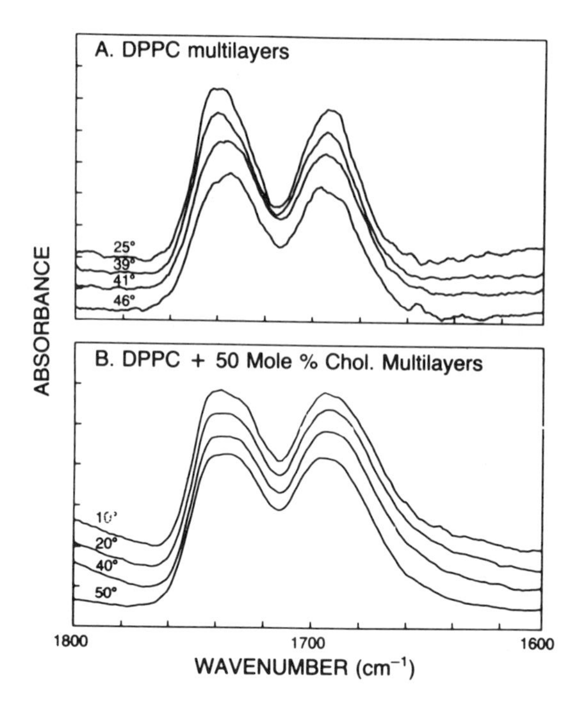

FIGURE 9. Spectra of hydrated (A) pure and (B) 50 mol% cholesterol 2-[1-^{13}C] DPPC bilayers. The spectra demonstrate that addition of cholesterol to the bilayer, or transformation to the liquid crystalline phase, causes significant broadening of both *sn*-1 and *sn*-2 chain features. The spectra of cholesterol containing bilayers resemble a combination of the gel and liquid crystalline phase spectra of the pure DPPC bilayer.

V. EFFECTS OF CHOLESTEROL ON COMPLEX MEMBRANE SYSTEMS

Yang et al. (1990) investigated the effects of cholesterol on the interaction of Ca^{2}-ATPase with 1-palmitoyl-2-oleoylphosphatidylethanolamine (POPE), using Fourier transform infrared (FTIR) spectroscopy and differential scanning calorimetry. By monitoring the frequency of lipid CH_2 symmetric stretching modes at *2850* cm^{-1}, they observed that the melting transition of the POPE component of a 103:12:1 (POPE:cholesterol:Ca^{2+}-ATPase) mixed bilayer occurred at lower temperatures compared to a protein-free binary mixture of POPE:cholesterol. On this basis they concluded that the enzyme preferentially sequesters 15 to 35 POPE molecules from the lipid mixture, thus excluding cholesterol from the immediate environment of the protein.

Rooney et al. (1984) used Fourier transform infrared spectroscopy and Raman spectroscopy to study the effects of cholesterol on human erythrocyte membrane structure. The erythrocyte membrane cholesterol content was modified by incubating the membranes with sonicated liposomes containing varying amounts of cholesterol. Using the lipid *2850*-cm^{-1} CH_2 symmetric stretching mode as a marker of acyl chain disorder, the authors concluded that cholesterol ordered erythrocyte membranes at temperatures between 5 and 35°C, with a relatively greater effect at higher temperatures. This result was not surprising, since the reported CH_2 stretching mode frequencies and C–C stretching-region spectra suggested that even at 5°C, the lipids were more nearly in a liquid crystalline than a gel state.

These authors have also examined the effect of cholesterol on the structure of proteins within the membrane. By comparing the intensity of a *940*-cm^{-1} α-helix peptide backbone C–C stretching Raman band with that of the *1004*-cm^{-1} phenylalanine ring vibration, they concluded that cholesterol-enriched erythrocyte membranes show a significant increase in protein helical structure compared to cholesterol-depleted membranes. A similar effect was found by Fong and McNamee (1987), who studied the effects of cholesterol on acetylcholine receptor secondary structure, using infrared spectroscopy. In this case, the *931*-cm^{-1} band resulting from α-helical structures and the *988*-cm^{-1} band resulting from β-sheet structures were compared with the *1029*-cm^{-1} phenylalanine band for determination of secondary structure. Cholesterol (mole fraction approximately 0.25) caused a small but definite (approximately 3%) increase in the α-helix content of the acetylcholine receptor.

VI. CONCLUSIONS

1. Cholesterol does not abolish the gel-to-liquid crystalline transition of lipids, but rather broadens the temperature over which both gel and liquid crystalline phases coexist. On average, lipid order at temperatures well above or below the pure-lipid phase transition temperature is little affected by the presence of cholesterol. Nearer the phase transition, however, cholesterol tends to disorder the gel phase and order the liquid crystalline phase.
2. Cholesterol does not cause lipid molecules to form a phase intermediate between the gel and liquid crystalline phases. Instead, individual lipid molecules within cholesterol-containing bilayers are found within regions having structural characteristics similar to those of pure gel or liquid crystalline phases.
3. Cholesterol tends to order the lipid bilayer near the headgroup region at all temperatures and tends to order the entire chain at high temperatures. At low temperatures, however, cholesterol disorders the acyl chain only near the methyl termini.
4. Although it is possible that the cholesterol 3β-OH hydrogen bonds to one of the lipid carbonyl moieties, there is no convincing vibrational spectroscopic evidence that it does so.
5. Cholesterol appears to stabilize α-helix structures within membrane proteins.

VII. SPECULATION

Why do eukaryotic cell membranes contain cholesterol? Although cholesterol may serve many roles in the cell membrane, it seems likely that one of the most important may be to make the membrane insensitive to changes in the cellular environment. Animals, particularly those with poorly developed temperature homeostasis, such as reptiles, undergo substantial changes in body (and hence membrane) temperature over relatively short time periods. Were the membrane to respond to these changes in the way pure lipids do, one might expect dramatic changes in membrane permeability and enzyme function, which might not be well tolerated. Cholesterol substantially modulates the effect of temperature changes on membrane structure, as well as the changes induced by such compounds as ethanol. This may enable organisms, including us, to tolerate a much wider range of environments than would otherwise be possible.

ACKNOWLEDGMENTS

The author thanks Philip Yeagle, Jeffrey Mason, Dianne O'Leary, and Len Finegold for their efforts in reading this manuscript and offering suggestions for its improvement.

REFERENCES

Asher, I. M. and I. W. Levin. 1977. Effects of temperature and molecular interactions on the vibrational infrared spectra of phospholipid vesicles. *Biochim. Biophys. Acta,* 468:63–72.

Bush, S. F., R. G. Adams, and I. W. Levin. 1980. Structural reorganizations in lipid bilayer systems: effect of hydration and sterol addition on Raman spectra of dipalmitoylphosphatidylcholine multilayers. *Biochemistry,* 19:4429–4436.

Bush, S. F., H. Levin, and I. W. Levin. 1980. Cholesterol-lipid interactions: an infrared and Raman spectroscopic study of the carbonyl stretching mode region of 1,2-dipalmitoylphosphatidylcholine bilayers. *Chem. Phys. Lipids,* 27:101–111.

Casal, H. L. and H. H. Mantsch 1984. Polymorphic phase behaviour of phospholipid membranes studied by infrared spectroscopy. *Biochim. Biophys. Acta,* 779:381–401.

Cortijo, M. and D. Chapman. 1981. A comparison of the interactions of cholesterol and gramicidin A with lipid bilayers using an infrared data station. *FEBS Lett.,* 131:245–248.

Davies, M. A., H. F. Schuster, J. W. Brauner, and R. Mendelsohn. 1990. Effects of cholesterol on conformational disorder in dipalmitoylphosphatidylcholine bilayers. A quantitative IR study of the depth dependence. *Biochemistry,* 29:4368–4373.

Fong, T. M. and M. G. McNamee. 1987. Stabilization of acetylcholine receptor secondary structure by cholesterol and negatively charged phospholipids in membranes. *Biochemistry,* 26:3871–3880.

Green, P. M., J. T. Mason, T. J. O'Leary, and I. W. Levin. 1987. Effects of hydration, cholesterol, amphotericin B and cyclosporine A on the lipid bilayer interface region: an infrared spectroscopic study using 2–1,^{13}C- dipalmitoylphosphatidylcholine. *J. Phys. Chem.,* 91:5099–5103.

Huang, C., J. R. Lapides, and I. W. Levin. 1982. Phase-transition behavior of saturated, symmetric chain phospholipid bilayer dispersions determined by Raman spectroscopy: correlation between spectral and thermodynamic parameters. *J. Am. Chem. Soc.,* 104:5926–5930.

Levin, I. W. 1984. Vibrational spectroscopy of membrane assemblies. In *Advances in Infrared and Raman Spectroscopy,* Clark, R. J. H. and R. E. Hester, Eds., Heyden, London, pp. 1–48.

Levin, I. W., E. Keihn, and W. C. Harris. 1985. A Raman spectroscopic study on the effect of cholesterol on lipid packing in diether phosphatidylcholine bilayer dispersions. *Biochim. Biophys. Acta,* 820:40–47.

Lippert, J. L. and W. L. Peticolas. 1971. Laser Raman investigation of the effect of cholesterol on conformational changes in dipalmitoyl lecithin multilayers. *Proc. Natl. Acad. Sci. U.S.A.,* 68:1572–1576.

Mabrey, S., P. L. Mateo, and J. M. Sturtevant. 1978. High-sensitivity scanning calorimetric study of mixtures of cholesterol with dimyristoyl- and dipalmitoylphosphatidylcholines. *Biochemistry,* 17:2464–2468.

Mendelsohn, R., M. A. Davies, J. W. Brauner, H. F. Schuster, and R. A. Dluhy. 1989. Quantitative determination of conformational disorder in the acyl chains of phospholipid bilayers by infrared spectroscopy. *Biochemistry,* 28:8934–8939.

Mushayakarara, E. and I. W. Levin. 1982. Determination of acyl chain conformation at the lipid interface region: Raman spectroscopic study of the carbonyl stretching mode region of dipalmitoyl phosphatidylcholine and structurally related molecules. *J. Phys. Chem.,* 86:2324–2327.

O'Leary, T. J. and I. W. Levin. 1984. Raman spectroscopic study of an interdigitated lipid bilayer: dipalmitoylphosphatidylcholine dispersed in glycerol. *Biochim. Biophys. Acta,* 776:185–189.

O'Leary, T. J. and I. W. Levin. 1986. Raman spectroscopy of selectively deuterated dimyristoylphosphatidylcholine: studies on dimyristoylphosphatidylcholine-cholesterol bilayers. *Biochim. Biophys. Acta,* 854:321–324.

O'Leary, T. J., P. D. Ross, and I. W. Levin. 1986. Effects of anesthetic alcohols on phosphatidylcholine bilayers: implications for the mechanism of the pretransition. *Biophys. J.,* 50:1053–1059.

Pink, D. A., T. J. Green, and D. Chapman. 1981. Raman scattering in bilayers of saturated phosphatidylcholines and cholesterol. Experiment and theory. *Biochemistry,* 20:6692–8.

Rooney, M. W., Y. Lange, and J. W. Kauffman. 1984. Acyl chain organization and protein secondary structure in cholesterol-modified erythrocyte membranes. *J. Biol. Chem.,* 259:8281–8285.

Umemura, J., D. G. Cameron, and H. H. Mantsch. 1980. A Fourier transform infrared spectroscopic study of the molecular interaction of cholesterol with 1,2-dipalmitoyl-sn-glycero–3-phosphocholine. *Biochim. Biophys. Acta,* 602:32–44.

Wong, P. T., S. E. Capes, and H. H. Mantsch. 1989. Hydrogen bonding between anhydrous cholesterol and phosphatidylcholines: an infrared spectroscopic study. *Biochim. Biophys. Acta,* 980:37–41.

Wong, P. T., T. E. Chagwedera, and H. H. Mantsch. 1989. Effect of cholesterol on structural and dynamic properties of tripalmitoyl glyceride. A high-pressure infrared spectroscopic study. *Biophys. J.,* 56:845–850.

Yang, J., G. L. Anderle, and R. Mendelsohn. 1990. Effects of cholesterol on the interaction of Ca2(+)-ATPase with 1-palmitoyl–2- oleoylphosphatidylethanolamine. An FTIR study. *Biochim. Biophys. Acta,* 1021:27–32.

Chapter 8

LIPID-CHOLESTEROL PHASE DIAGRAMS: THERORETICAL AND NUMERICAL ASPECTS

H. L. Scott

TABLE OF CONTENTS

0-8493-4207-4/93/$0.00+$.50

ABSTRACT

In the first part of this chapter, general aspects of modeling mixtures of complex molecular fluids are considered, with applicaton to lipid-cholesterol mixtures. By considering classical thermodynamic and Landau theory, it is shown that complex eutectic phase diagrams may be calculated with as few as three phenomenological parameters. A simple three-state statistical mechanical two-dimensional lattice model is then presented that can also exhibit complex phase diagrams and that can be interpreted as a model for lipid-cholesterol mixtures. Turning from the general discussion to a specific model, a Monte Carlo study of a three-dimensional lipid-cholesterol system, with atomic-level detail, is described. Using the results of the Monte Carlo simulations as a qualitative guide, a theoretical model for lipid-cholesterol mixtures is developed. The model exhibits all of the observed phases in these mixtures and unequivocally associates them with steric packing interactions. Finally, a brief general discussion of the interplay between large-scale computer simulation and analytical modeling is presented.

I. INTRODUCTION

The ubiquity of cholesterol in cell membranes has been documented elsewhere in this volume. Equally well documented in this volume and elsewhere (Presti, 1985) is the lack of a clear understanding of the role that this molecule plays in cell membrane structure and/or function. Because the cholesterol molecular structure is simple and the molecule itself is relatively small, it is unlikely that cholesterol can undergo major conformational changes, nor is cholesterol likely to be highly chemically active in most membrane environments. Therefore, the main influence it exerts in biomembranes should be a result of *physical* interactions with other membrane molecules. Since model lipid membranes exhibit sharp phase transitions (Nagle, 1980), a common way to study interactions between cholesterol and lipids is to monitor the phase properties of a lipid membrane as the concentration of cholesterol in the membrane is increased. The result of such an effort is a phase diagram (Vist and Davis, 1990), such as that described in the chapter by Davis in this book. This phase diagram is reproduced in Figure 1, as it will serve as a guide to the theoretical discussions in this chapter. The question to be addressed in this chapter is, What properties of cholesterol and phospholipids in model biomembranes are the cause of the observed phase behavior of these systems?

In order to begin to answer the above question, we must consider the nature of the phases found in lipid-cholesterol bilayer mixtures, the structures of individual lipid and cholesterol molecules, and the modes of interaction between them in a bilayer environment. Then models incorporating the structures and interactions can be constructed and analyzed by analytical (hopefully) and numerical means. If the model is sufficiently lucid it may then be possible to answer the question posed above by comparing the phase diagram predicted by

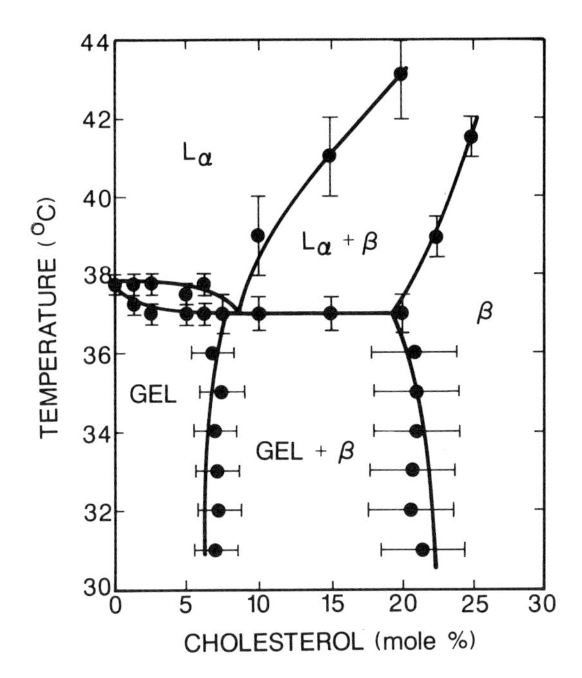

FIGURE 1. Experimental phase diagram for binary mixture of cholesterol and dipalmitoyl phosphatidylcholine determined by Vist and Davis and described in the chapter by Davis in this book. The notation used is L_α: fluid lipid-rich phase; L_β: gel lipid-rich phase; β: a cholesterol-rich phase in which lipid chains are isomerically ordered (in mainly all-*trans*) conformations, but are rotationally mobile. (Reproduced from Vist and Davis, 1990. *Biochemistry*, 29:451. With permission.)

the model with that measured experimentally and isolating those interactions in the model that play the key roles in the phase formation process.

Although simple by biological standards, phospholipids and cholesterol are physically complicated molecules whose interactions cannot be simply characterized. For this reason computer simulation plays a key role in identifying the components of these interactions that should be the most important in the modeling work. In this chapter we will first consider the general problem of constructing a theoretical model for a complex binary mixture. The aim of this part of the chapter will be to address a question posed to this author by the editor of this volume: namely, what are the minimal numbers of phenomenological parameters required to construct such a model? This question will be considered from the viewpoints of classical thermodynamics, Landau phenomenological theory, and statistical mechanics. Then a specific modeling effort involving both large-scale computer studies and analytical model calculations will be described. The chapter is organized as follows: first, the basic lipid and cholesterol structures are reviewed, and the experimental phase diagram is described briefly (see Chapter 4 for a full description). Then, the general thermodynamic considerations involving phase diagrams will be presented. This will be followed by a description of the numerical simulation procedure, which in turn will be

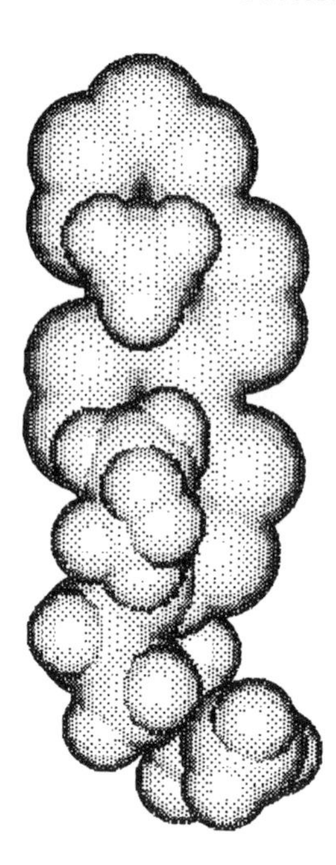

FIGURE 2. Space-filling model for cholesterol (minus the topmost hydrogen), showing the bulky and highly hydrophobic rings with methyl groups, and the flexible short chain-tail. The total length of the molecule is about 16 Å.

followed by a description of the analytical modeling procedure. The "Conclusions" section summarizes this chapter.

II. GENERAL

A. MOLECULAR STRUCTURES AND INTERACTIONS

Figure 2 shows a space-filling model of cholesterol. A typical phospholipid, dimyristoylphosphatidylcholine (DMPC) is shown in Figure 3. Figure 4a shows a cholesterol molecule with two neighboring hydrocarbon chains, and Figure 4b shows two cholesterol molecules with lipid chains in between them. Both the phospholipid and the cholesterol are amphiphilic molecules, having both hydrophilic and hydrophobic moieties. In a bilayer, both molecules should normally orient themselves with polar groups adjacent to the aqueous phase (which is the uppermost part of Figures 2 through 4). Hydrogen bonding is possible between the cholesterol hydroxyl and phospholipid glycerol moieties (Presti, 1985). However, in DMPC the polar group is far larger than the simple hydroxyl of cholesterol, and this, combined with the very hydrophobic sterol ring structure of cholesterol, ensures that cholesterol has its most significant interaction with the hydrocarbon chains of the phospholipids.

In a lipid bilayer environment the cholesterol-lipid nearest-neighbor configurations should typically resemble those depicted in Figure 4. Examination of these figures suggests that cholesterol pushes apart the lipid molecules, effec-

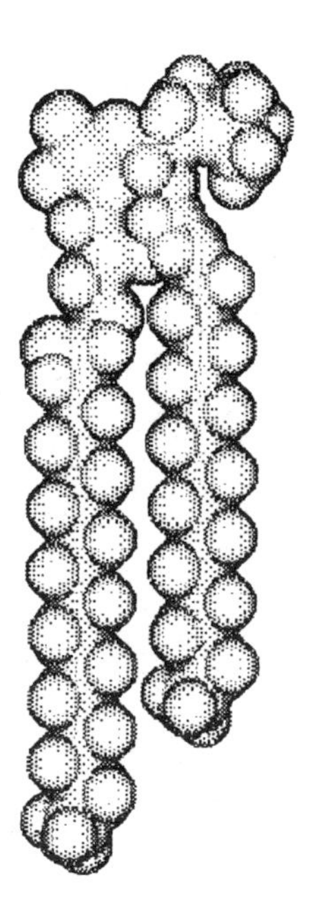

FIGURE 3. Space-filling model for a typical phospholipid, dipalmitoyl phosphatidylcholine, showing two hydrocarbon chains each 14 carbons in length, and a polar head group with phosphate and choline moieties. The total length of this molecule is about 27 Å.

tively reducing or eliminating some of the interactions between the lipid polar groups. Clearly, cholesterol changes the local environment for lipid hydrocarbon chains. In the absence of cholesterol each single hydrocarbon chain has, on average, six nearest neighbors of the identical structure and flexibility. A cholesterol molecule effectively provides a different packing arrangement for the six chains (from at least three lipid molecules) that are its near neighbors. In order to quantify the molecular packing, it is a good assumption to suppose that within the hydrophobic core of a lipid bilayer, the interactions are well described by a 6–12 type of potential between all of the atoms on all of the molecules. In this core, at fluid-to-gel state hydrocarbon densities typical of a bilayer, spacings between atoms are generally in the range of 4.9 to 6 Å (Tardieu et al., 1973). This is quite small compared to the length of the chains in the direction perpendicular to the bilayer plane, so that there is relatively little room for chain splay and tilting within the bilayer, without large-scale cooperative molecular rearrangement. Therefore, the short-range repulsive part of the 6–12 potential should have the greatest effect on the motional freedom of the lipid and cholesterol molecules. These same steric forces should, for the same reason, control the average shapes of the molecules. It is therefore natural to anticipate that introduction of cholesterol, with its rigid and bulky hydrophobic structure and small flexible chain "tail", into an otherwise single-component lipid bilayer will significantly alter the physical and thermal properties of the bilayer.

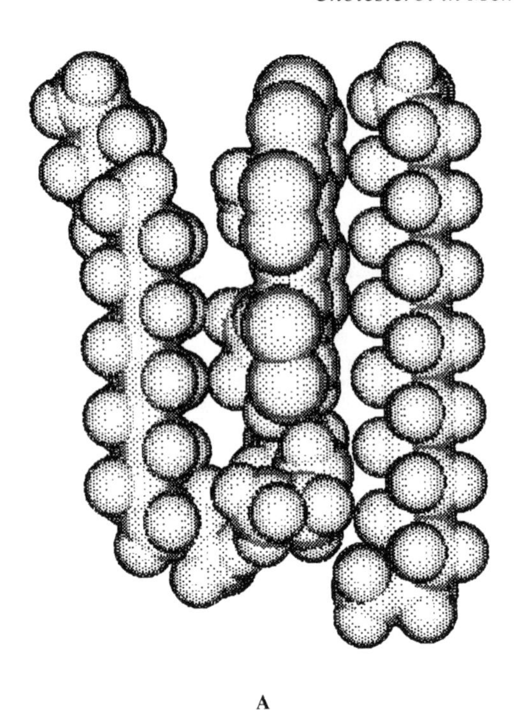

A

FIGURE 4. (A) A typical configuration of two C-14 hydrocarbon chains near a single cholesterol, illustrating the packing difficulties due to the shape of the cholesterol molecule. (B) A configuration of two cholesterol molecules with C-14 chains between them, illustrating steric hindrances forced upon lipid chains that are neighbors to two cholesterol molecules.

B. THERMODYNAMIC PRINCIPLES

In a single component lipid bilayer which contains no cholesterol the existence and detailed structure of several thermodynamic phases is very well documented (Nagle, 1980). These range from well ordered crystalline solid phases to corrugated "ripple" phases to disordered fluidlike phases. In a single component lipid bilayer the phase transition between disordered hydrocarbon chains (fluid) and ordered hydrocarbon chains (gel) phases is very striking and is easily seen by a wide variety of experimental methods. It is therefore natural to study lipid-cholesterol interactions by monitoring the lipid fluid-gel phase transition (called the "main" lipid phase transition) as the cholesterol concentration is varied.

In these terms, addition of even small (~5 mol%) amounts of cholesterol have a noticable effect. The main transition, as viewed in a thermal experiment such as differential scanning calorimetry, becomes broad and diffuse in temperature instead of sharp and narrow (Presti, 1985; Mabrey and Sturtevant, 1976; Estep et al., 1978; Demel et al., 1972; Jacobs and Oldfield, 1979). Above about 25 mol%, the excess enthalpy associated with the main transition vanishes. Spectroscopic measurements show that cholesterol affects the slower cooperative

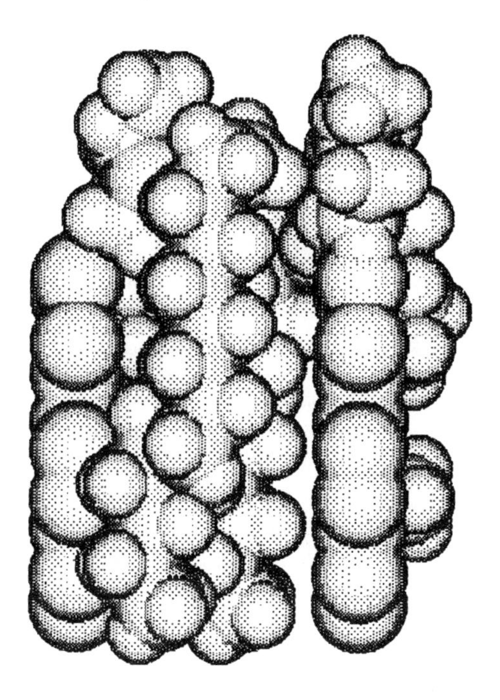

FIGURE 4B.

motions of lipid chains more than the faster single-chain and intrachain motions (Peng et al., 1989).

It is natural to attempt to synthesize results of many experiments into a phase diagram of the lipid-cholesterol system. This effort has been carried out by a number of investigators (Shimshick and McConnell, 1973; Recktenwald and McConnell, 1981; Lentz et al., 1980; Mortensen et al., 1988) and is reviewed in this volume in the chapter by Davis. In general, the phase diagrams deduced experimentally for the lipid-cholesterol bilayer system are nonideal and complex in nature, with regions of coexisting ordered phases and coexisting fluid phases bordered by a triple-point curve. The most recent phase diagram due to Vist and Davis (Vist and Davis, 1990) is illustrated in Figure 1 and shows several regions of two-phase coexistence, a line of triple points, and a four-phase coexistence point. The reader is referred to other chapters in this volume for a full discussion of experimental phase diagrams. Due to the complex nature of the phases involved in lipid-cholesterol systems, it is appropriate here to reconsider the general principles that underly the construction of phase diagrams for multicomponent systems.

The principal guideline for the construction of phase diagrams in any system is, of course, the Gibbs Phase rule (Callen, 1985)

$$f = r - M + 2 \tag{1}$$

where f is the number of degrees of freedom (i.e., the number of independent thermodynamic variables), r is the number of chemically distinct components, and M is the number of coexisting phases. For a given r and M, $f < 0$ means no $M-$ phase coexistence is possible, $f = 0$ means $M-$ phase coexistence may occur at a single point on the appropriate phase diagram, $f = 1$ means coexistence may occur along a curve, and so on. For lipid-cholesterol mixtures in bilayers *in excess water*, $r = 2$ for lipid and cholesterol components. This counting does not include the water component because, as long as the water content does not fall below the critical concentration for excess water, bilayer phase properties do not depend on this variable. For this system there are two thermodynamic degrees of freedom: the cholesterol composition and the temperature (the pressure on the system is due to the hydrophobic constraining effect of the water and, in excess water, is fixed). Therefore, $f = 2$, 1, or 0, depending on whether neither, one, or both of the thermodynamic variables are held fixed. In a lipid-cholesterol mixture it is therefore possible to have four coexisting phases at a point in x-T space ($M = 4$, $f = 0$), three coexisting phases along a contour ($M = 3$, $f = 1$), or two coexisting phases in a region of the x-T plane ($M = 2$, $f = 2$). As can be seen in Figure 1, all of these situations occur in lipid-cholesterol mixtures.

With a view towards theoretical modeling of lipid-cholesterol systems, we will now consider how the above constraints generally affect theoretical approaches to the problem. The goal in the following sections is to illustrate two phenomenological approaches to the problem of describing a complex phase diagram with a theoretical model, and to attempt to find a "minimal model" (i.e., a model with the fewest arbitrary parameters) that will do the trick.

III. PHENOMENOLOGICAL MODELS

A. THERMODYNAMIC MODELS

It is possible to construct phase diagrams for *ideal* systems from thermodynamic reasoning alone. However, in order to construct phase diagrams for *nonideal* systems the thermodynamic arguments must be supplemented with phenomenological expressions for the nonideal aspects of the problem, since these contributions are too complex and poorly understood to be treated exactly. This type of effort was carried out some time ago in the context of lipid bilayers, by Lee (Lee, 1977).

To illustrate the approach consider first an ideal mixture of two species, A and B, each of which may exist in two phases, 1 and 2. Let x_A^1, x_A^2, x_B^1, and x_B^2 denote the molar concentrations of A and B in the two phases. For thermal equilibrium at a temperature T (at fixed pressure), the chemical potentials μ of the phases must be equal. Assuming complete miscibility in both phases of A and B, we have

$$\mu_A^1 = \mu_A^2$$
$$\mu_B^1 = \mu_B^2 \tag{2}$$

For ideal materials (Callen, 1985):

$$\mu_A^1 = \mu_A^{1\,0} + RT\ln x_A^1$$
$$\mu_A^2 = \mu_A^{2\,0} + RT\ln x_A^2 \tag{3}$$

with similar expressions for component B. The quantities $\mu_A^{1\,0}$ and $\mu_B^{1\,0}$ are the chemical potentials of the pure phases of component A at temperature T. Phase equilibrium requires that

$$\ln\frac{x_A^1}{x_A^2} = \frac{\mu_A^{2\,0} - \mu_A^{1\,0}}{RT}$$
$$\ln\frac{x_B^1}{x_B^2} = \frac{\mu_B^{2\,0} - \mu_B^{1\,0}}{RT} \tag{4}$$

Together with the constraints

$$x_A^1 + x_B^1 = 1$$
$$x_A^2 + x_B^2 = 1 \tag{5}$$

it is possible to solve for the components of A and B in the two phases and to construct the phase diagram. This system has no arbitrary phenomenological parameters, as it is possible to determine the chemical potential differences for the pure systems from experimental data (Lee, 1977):

$$\ln\frac{x_A^1}{x_A^2} = \frac{\Delta H_A}{R}\left(\frac{1}{T} - \frac{1}{T_A}\right) \tag{6}$$

In this expression ΔH_A and T_A are the enthalpy change and the transition temperature for pure component A. A similar expression may be written for component B and, using the constraint equations for the concentrations, the concentrations of the coexisting phases may be determined. The mathematical structure of the above expression is such that lens-shaped coexistence regions will always be predicted (Lee, 1977). This a straightforward consequence of the ideal mixing assumption in both phases.

In order for complex phase diagrams to be derived from thermodynamic reasoning, it is necessary to include approximately the effects of nonideal interactions. A simple way to accomplish this is to suppose total solid-phase immiscibility while still maintaining ideal mixing in the fluid phase. Then one has an ideal fluid mixture of A and B in which the A molecules are in equilibrium with a solid A phase, and the B molecules with a solid B phase. It is only necessary to consider concentrations of the fluid components. It is easy to show (Lee, 1977) that one gets a simple eutectic phase diagram from this procedure, with a triple-point line that extends across the entire $T-x_A$ diagram through a eutectic point where the two fluidus curves meet. Below this line, which defines the eutectic temperature, there are only solid A and B phases that do not mix.

In real eutectic systems (the lipid-cholesterol phase diagram resembles a complex eutectic), there is nonideal mixing in all phases. At the thermodynamic level one constructs phase diagrams for such cases by modifying Equation 3 to include activity coefficients γ_A, γ_B for the two species so that

$$\begin{aligned} \mu_A^1 &= \mu_A^{1\,0} + RT\ln x_A^1 \gamma_A^1 \\ \mu_B^1 &= \mu_B^{1\,0} + RT\ln x_B^1 \gamma_B^1 \end{aligned} \tag{7}$$

with similar expressions for phase 2. In terms of the activity coefficients, the nonideal component of the free energy is

$$G_{nonideal} = RT\left(x_A^1 \ln\gamma_A^1 + x_A^2 \ln\gamma_A^2 + x_B^1 \ln\gamma_B^1 + x_B^2 \ln\gamma_B^2\right) \tag{8}$$

Since direct information about the behavior of $G_{nonideal}$ cannot be obtained from thermodynamic reasoning alone a phenomenological expression is required. Following (Lee, 1977) we set

$$\begin{aligned} RT\ln\gamma_A^1 &= \rho_0^1\left(x_A^1\right)^2 \\ RT\ln\gamma_B^1 &= \rho_0^1\left(x_B^1\right)^2 \\ RT\ln\gamma_A^2 &= \rho_0^2\left(x_A^2\right)^2 \\ RT\ln\gamma_B^2 &= \rho_0^2\left(x_B^2\right)^2 \end{aligned} \tag{9}$$

where the ρ parameters represent energy differences between AB pairs and AA and BB pairs in the two phases. Equating the chemical potentials and solving for the concentrations of the coexisting phases at fixed temperature yields phase diagrams that are, in general, lens-shaped, but may be double-lobed (for the case where multiple solutions of the equations exist at the given temperature). The shapes of the calculated phase diagrams depend strongly on the values of ρ_0^1 and ρ_0^2.

If these two parameters are non-negative an additional possibility is the existence of a misciblity gap in the solid phase (Lee, 1977). This gap signifies coexisting solid phases, and its signature is a concavity-violating lobe in the solid-phase free energy vs. concentration curve at the given temperature. As in

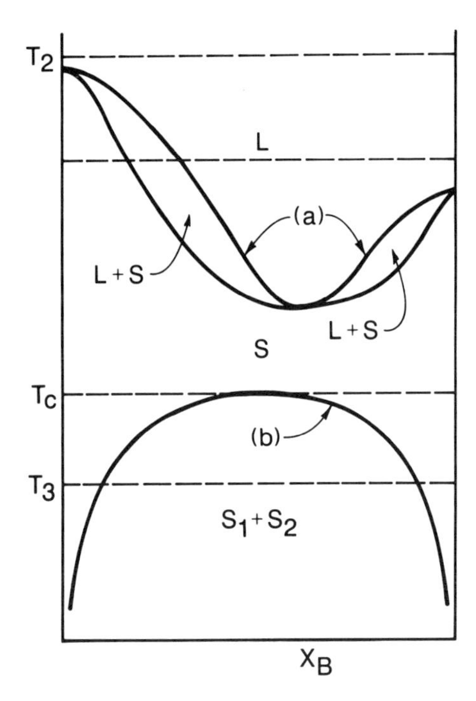

FIGURE 5. An illustration of the process of construction of a phase diagram using thermodynamic arguments. (Reproduced from Lee, 1977. *Biochim. Biophys. Acta,* 472:285. With permission.) (a) (top figure) A double lens-shaped fluidus-solidus curve that outlines the coexisting phase region for fluid and solid phases. (b) (top figure) The solvus curve for the system, obtained from as double tangent construction on the solid-phase free energy. This curve outlines the coexistence region between two solid phases. (c) (bottom figure) The phase diagram resulting from the combination of (a) and (b). Stable coexistence regions are outlined in a solid border. The combination produces a line of triple points (three-phase coexistence) and a eutectic point E (four-phase coexistence).

the simple van der Waals gas, the unstable lobe is bypassed with a double tangent construction from which the concentrations of the coexisting solid phases are identified. The locus of such concentrations for various temperatures is the solvus curve. If the solvus curve intersects the fluidus-solidus coexistence region (calculated as described in the previous paragraph), one obtains a symmetric eutectic phase diagram. These possibilities are schematically illustrated in Figure 5.

Since the experimental lipid-cholesterol phase diagram is a nonsymmetric eutectic, a "minimal model" based solely upon thermodynamic reasoning would require at least one additional phenomenological parameter to account for the asymetry. The answer, at the thermodynamic level, to the question posed at the beginning of this section is therefore "three." At *least* three free phenomenological parameters (e.g., ρ_0^1, ρ_0^2, and the asymmetry parameter) are required to devise a reasonable thermodynamic model for the lipid-cholesterol phase diagram.

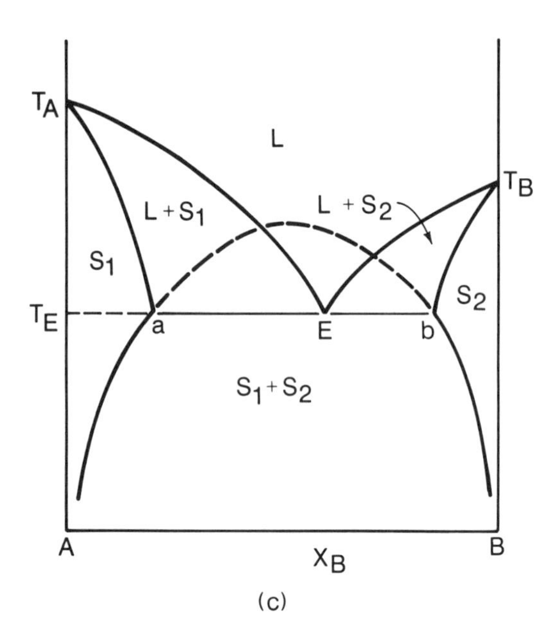

FIGURE 5. (continued).

Additional parameters in excess of these three reduce the thermodynamic modeling to an exercise in curve-fitting, given the accuracy of present experimental data.

B. LANDAU THEORY

A phenomenological alternative to the purely thermodynamic model is the Landau theory (Landau, 1937). This simple theory was originally proposed in 1937 to show how phase transitions and critical points may arise from a postulated form for the free energy functional of the system. Since the theory begins with an expression for the free energy of the system in question, there is no direct connection to intermolecular forces, and so this type of modeling, like that of the previous section, is purely phenomenological. It is useful, however, in that one can determine the *mathematical* properties of the free energy that lead directly to phase transitions in the system. This can serve as a guide for subsequent statistical mechanical modeling in which *physical* properties of the specific system are utilized. The application of the Landau theory to binary and ternary mixtures was carried out by Griffiths (Griffiths, 1974), although earlier free energy modeling (not strictly based upon the Landau theory) was done by Meijering (Meijering, 1963). It is instructive to review this work here in order to see just how little phenomenology is required to obtain phase diagrams for systems such as lipid-cholesterol mixtures. We follow the description given by Knobler and Scott (Knobler and Scott, 1984).

For a simple symmetrical system the basic assumption of the Landau theory is that, near a phase transition point, the free energy may be written in terms of an *order parameter*, η. The order parameter is a thermodynamic variable associated with the symmetry of the system, which is nonzero in the *ordered* phase and vanishes in the *disordered* phase. An example of an order parameter is the zero field magnetization of a ferromagnet or, in the case of pure lipids, the fraction of chains with all *trans* C–C bonds (although the latter is not a symmetrical system). In the Landau theory the free energy is expressed as a Taylor series in η,

$$G = A + B\eta^2 + C\eta^4 + D\eta^6 + \ldots \tag{10}$$

where the coefficients A, B, C, and D are functions of the remaining thermodynamic variables of the system, such as temperature and pressure. Inclusion of only even powers of η in the expansion insures that $G(-\eta) = G(\eta)$ for a symmetrical system.

In equilibrium the value of the order parameter η is that which minimizes G at the given temperature and pressure. It is easy to see that minimization with respect to η yields the roots

$$\begin{aligned} \eta &= 0 \\ \eta^2 &= [-C + (C^2 - 3BD)^{1/2}]/3D \end{aligned} \tag{11}$$

In general, the root for which G is an absolute minimum describes the equilibrium state of the system, and if two or more roots yield the same absolute minimum, then one has coexisting phases in the model. For the present case, when $C = -2\sqrt{BD}$ all three roots give $G = A$, so that the model has a three-phase coexistence point. In order to get a three-phase coexistence point, terms up to sixth order in the free energy expansion are required. The Landau approach is extended to mixtures by allowing the coefficients A, B, C, and D to be functions of the concentration x of one of the components as well as the temperature:

$$G = A(x,T) + B(x,T)\eta^2 + C(x,T)\eta^4 + D(x,T)\eta^6 + \ldots \tag{12}$$

By proper choice of the functions $A(x,T)$, $B(x,T)$, $C(x,T)$, $D(x,T)$ it is possible to construct phase diagrams for any symmetric two-component system. Inclusion of odd powers of η in the free energy expansion leads to many more phase diagrams (Griffiths, 1974).

An attempt to remove some of this arbitrariness from the Landau approach in the case of lipid mixtures was made by Priest (Priest, 1980). Priest began by considering a single component lipid bilayer. In this theory the order parameter is S, the total fraction of *gauche* bonds per chain, so that S vanishes in the *gel* phase, and $S \approx 0.3$ to 0.4 in the fluid phase. Using arguments from polymer statistics and some semi-empirical energetic reasoning, a nonsymmetric cubic Helmholtz free energy polynomial for a single chain system is constructed which has the form:

$$F = AS - \frac{1}{2}BS^2 + \frac{1}{3}CS^3 \qquad (13)$$

where A, B, and C are predetermined from the energetic and statistical assumptions and so are not entirely arbitrary. Rather, they are expressed in terms of three other parameters that are directly related to the intermolecular interaction energy. By appeal to experimental data for polyetheylene and to the vanishing of the chain melting enthalpy at a chain length of about 11 carbons, the number of free parameters in a single-component version of the theory is reduced to one. An appropriate choice for this single parameter yields transition temperatures and enthalpies for pure lipid systems with chain lengths between 12–22 carbons, which agree very well with experiment.

Priest (Priest, 1980) has extended the model to binary mixtures of lipids differing in chain length by adding entropy and energy of mixing terms to the free energy

$$G(S,X) = F(S) + \frac{RT}{2}\left[X\ln X + (1-X)\ln(1-X) + X\Delta U + \Delta L\frac{\sigma}{RT}\left(\frac{1}{2} - S\right)X(1-X)\right] \qquad (14)$$

where ΔL is the chain length difference, X is the concentration of the long chain component, and ΔU is the chemical potential difference for the two types of lipid. Phase diagrams are calculated by minimizing $G(S, X)$ with respect to S and X. At a given temperature between the phase transitions of the two pure components, there exists a value of ΔU for which the free energies of fluid and gel phases are equal. The concentrations X_l and X_g at which this occurs define the phase boundaries. The parameter σ is freely adjustable. By an appropriate choice of σ the calculated phase diagrams agree well with experiment for lipid mixtures for which $\Delta L = 2, 4$, and 6. In the case $\Delta L = 6$ Priest finds that G may have three equal minima and a eutectic type phase diagram. Figure 6 shows the resulting phase diagram for $\Delta L = 6$.

The remarkable thing about the phase diagram in Figure 6 is that it reveals phase sequences that are strikingly similar to those observed in lipid-cholesterol mixtures. Namely, the phase diagram shows a region of fluid-gel coexistence, a region of gel-gel coexistence, and a small distinct region of gel-gel coexistence that differs from the larger gel-gel coexistence area. If the *long*-chain component is identified with cholesterol in the β phase, Figure 6 bears a surprisingly strong resemblance to the lipid-cholesterol phase diagram of Vist and Davis (Vist and Davis, 1990) in Figure 1. If the cholesterol β phase is thought of as an ordered phase that is stable at temperatures above the lipid chain melting temperature, then cholesterol should indeed be considered the "long-chain component" in the Priest model. It seems clear that minor manipulation of the parameters would allow this model to be even better adapted to the lipid-cholesterol problem with

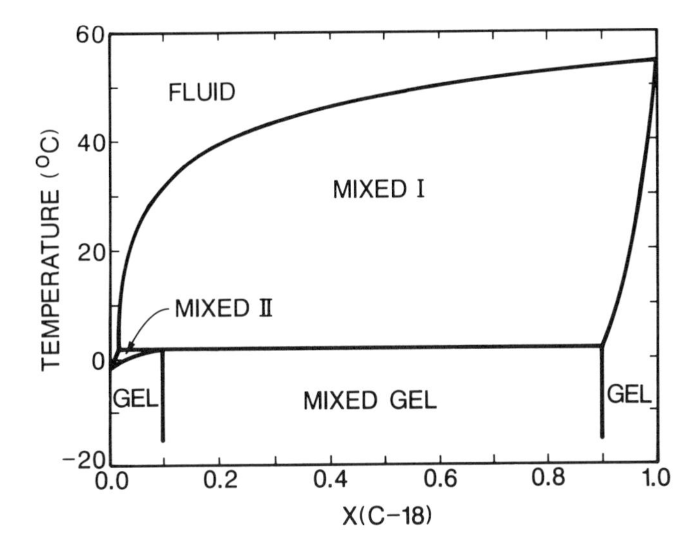

FIGURE 6. Eutectic type phase diagram calculated by Priest (1980) for a binary mixture of lipids with 12-carbon chains and 18-carbon chains, using the Landau phenomenological theory. (Reproduced from Priest, 1980. *Mol. Cryst. Liq. Cryst.*, 60:167. With permission.)

no additional parameters. Thus, the phase diagram for a lipid-cholesterol mixture can be fitted within the Landau theory by appeal to independent experimental data and only *two* totally free parameters.

IV. MICROSCOPIC MODELS

The thermodynamic and Landau modeling approaches described in the preceding sections do not allow any direct insight into the mechanisms by which intermolecular interactions in a system lead to complex phase diagrams. For this, microscopic models treated by statistical mechanics are required. In this section we first describe a "minimal" statistical mechanical model that is able to predict an eutectic type phase diagram. Then in the subsequent subsections, we describe a large-scale numerical simulation of lipid-cholesterol interactions in bilayers and show how this work led to the construction of an analytical model that yields a reasonable lipid-cholesterol phase diagram.

A. A SIMPLE STATISTICAL MECHANICAL MODEL

For a statistical mechanical model to provide a description of a system that may have three phases in coexistence, a minimum of three degrees of freedom are clearly required. While there are many ways to define such a model, the most straightforward is probably a three-state lattice gas model (Furman et al., 1977). To envision a two-dimensional version of this model, consider an area A that is partitioned into cells of equal volume. Further, suppose that there are three types of molecules in the system, and each cell holds exactly one molecule at all times. If instead of three distinct molecules, we identify the three species by solid lipid, fluid lipid, and cholesterol, we have a general model for lipid-cholesterol mixtures.

For the general case we define p_i^α to be a state variable equal to 1 if cell i is occupied by a molecule of species α and zero otherwise. The Hamiltonian for the model is (Furman et al., 1977):

$$H = \sum_{i,j} E_{\alpha\beta} p_i^\alpha p_j^\beta \tag{15}$$

where the sum runs over nearest neighbor pairs only. If μ_α is the chemical potential of a type α molecule, then in the mean field approximation the free energy per molecule of the system is (Furman et al., 1977)

$$F / N = kT \sum_{\alpha} A_\alpha \ln A_\alpha + \sum_{\alpha,\beta} E_{\alpha\beta} A_\alpha A_\beta - \sum_{\alpha} \mu_\alpha A_\alpha \tag{16}$$

where A_α is the ensemble average of the state variable p^α (which does not depend on the lattice site for a translationally invariant system), and k is Boltzmann's constant. The free energy functional must be minimized over the mean field variables A_α in order to determine the thermodynamic equilibrium state. If multiple solutions are found to the minimization equations, then the solution set for which F is an absolute minimum represents the thermodynamic equilibrium state.

This procedure allows, as in phenomenological models, for multiple-phase equilibrium when multiple solutions of the minimization equations have identical free energies. The phase diagram for this model has been explored thoroughly by Furman, Dattagupta, and Griffiths (Furman et al., 1977), and (in the mean field approximation) can be mapped isomorphically onto a thermodynamic model proposed many years earlier by Meijering (Meijering, 1963; Knobler and Scott, 1984). For phenomenological modeling purposes, there are only adjustable parameters for either model. For the lattice gas model the parameters are related to the differences in the energies $E_{\alpha\beta}$ by (Furman et al., 1977):

$$a = \frac{1}{2}(E_{22} + E_{33} - 2E_{23}) \tag{17}$$

$$b = \frac{1}{2}(E_{22} + E_{11} - 2E_{21})$$

$$c = \frac{1}{2}(E_{11} + E_{33} - 2E_{13})$$

Variation of the parameters and the temperature leads to a wide variety of ternary phase diagrams (Meijering, 1963; Knobler and Scott, 1984). It can be shown that the mean field free energy above is also equivalent to the Landau free energy under the symmetry condition $b = c$ (Knobler and Scott, 1984). Also, the statistical mechanical formulation of the model in Equation 16 can be equivalently recast into a spin-one Ising model or into a lattice gas model for a binary mixture. For the lipid-cholesterol application, one would assign the three states, as described in the first paragraph of this section. The significant point is that a three-state statistical mechanical model solved in the mean field approximation is fully capable of predicting a large variety of ternary system phase diagrams. The three intrinsic adjustable parameters are sufficient to fit most, if not all, binary and ternary mixture phase diagrams reasonably well. Additional parameters in a multistate statistical mechanical model will, of course, make the fitting procedure easier and perhaps closer to experimental data, but will not likely add any new insights.

A problem with a general statistical mechanical model like the above is that the state variables p_i^α are really too simple to contain many details of atomic or molecular structure. For example, $p_j^\alpha = 1$ could describe a magnetic state, a molecular orientation, or a molecular species at the site j. This ambiguity, coupled with the mean field nature of the analysis, tends to obscure the true microscopic input in the model. Because of the complex structure and interactions of phospholipids and cholesterol, it is quite difficult to progress beyond the above type of modeling effort. However, progress may be made by using high-speed computers to study the detailed interactions between molecules in a lipid-cholesterol mixture. Then it may be possible to discern which aspects of these interactions are most important in the phase transitions. Ideally, it would then be possible to construct analytical models that include only the *precalculated* interactions and that, with proper analysis, could predict accurate phase behavior with no additional adjustable parameters. In the next section such an effort is described.

B. MONTE CARLO STUDIES OF LIPID-CHOLESTEROL INTRERACTIONS

In order to gain insights at the atomic or molecular level, as suggested above, we began a program of Monte Carlo simulation of lipid chains constrained to bilayer geometries. The objective of these studies was (and still is) to explore the *configuration space* of the molecules in a lipid bilayer. The effort began in 1977 with a simulation study of chains of hard spheres with one end attached to an interface (Scott, 1977). The original work considered a system of six chains in a monolayer geometry, each having twelve tetrahedrally bonded hard spheres. The method was first applied to lipid-cholesterol interactions several years later (Scott and Kalaskar, 1989). Current work simulates up to 200 chains, 100 in each half of a bilayer, with 12–18 carbons per chain interacting via Lennard-Jones 6–12 potentials.

Since the Monte Carlo algorithm used for lipid chain simulation has been described in detail elsewhere (Scott, 1990), we will only briefly review the methodology and concentrate on the results obtained. The Monte Carlo simulations of cholesterol-lipid interactions (Scott, 1991) involved the following model system. An array of 100 hydrocarbon chains was generated on the computer by positioning the topmost carbon of each chain in a plane and generating the remaining carbons in each chain, using rotation operators. Then between 1 and

13 chains were replaced by cholesterol molecules. The positions of the cholesterol molecules were chosen so that several different lipid-cholesterol microenvironments could be sampled in different simulations. The Monte Carlo method was used for this project in order to sample a large region of the phase space of the system in a reasonable amount of computer time. The following are the salient points of the model and the method used in the Monte Carlo calculations:

- In each Monte Carlo move, a single chain was translated in the plane of the monolayer by a small amount. At the same time the chain was rotated about its long axis by a small amount, and one or two bonds were picked at random for *gauche* rotations.
- The rotational-isomeric model was used for the *gauche* rotations, that is, the only allowed rotation angles were 0 (±120)°. This model has the advantage that only the energetically most-favored angles are allowed, so that configuration sampling is more energetically efficient. The disadvantages are that intermediate angles may be more important in densely packed systems, and MC rejection rates are high (because a change in the isomeric state by ±120° may represent a large displacement of the end segments of the chain).
- The end chains on the cholesterols also were allowed to rotate, but the cholesterols were not translated in the monolayer plane.
- The chains and the cholesterol molecule interacted via optimized 6–12 potential functions between all nonhydrogen atoms within a cutoff distance of 15 Å.
- The standard Metropolis Monte Carlo algorithm (Binder, 1977) was used. Simulations were run for systems of 1, 3, 7, and 13 cholesterol molecules in a layer. This allowed for sampling of lipid chain conformations for isolated chains, for chains that were near neighbors to a single cholesterol, and for chains that were near neighbors to two or more cholesterol molecules.
- Averages of the order parameters for each bond were calculated, as well as fluctuations in the averages. These statistics, along with graphical displays of snapshots of generated configurations, were the major results of the simulations.

The order parameters S for each bond n on a chain are defined by the average

$$S_n = \frac{1}{2}\left\langle 3\cos^2\theta - 1\right\rangle \tag{18}$$

where θ is the angular deviation of bond n from its orientation with respect to the bilayer normal when the entire chain is in the all-*trans* conformation.

The order parameter profiles (plots of S_n vs. n) calculated for chains that were not neighbors of a cholesterol ("bulk chains") match experimental data obtained for pure-lipid bilayers by deuterium NMR (after correction by a scale factor for chain tilting not included in the Monte Carlo algorithm) (Seelig, 1983). The average order parameters for chains that were near neighbors to a single cholesterol did not differ significantly from those of the bulk chains. However, analysis of the fluctuations in these averages revealed that the rotational freedom of chains is reduced by the presence of the cholesterol. Since the simulations all begin with an array of all-*trans* chains, chains near a cholesterol initially have enough free volume available to undergo isomeric conformational changes along with the bulk chains, but at some point in the computer runs, due to the space-filling effects of the cholesterol, the Monte Carlo rejection rates for the cholesterol neighbors increases to over 90%, meaning that these chains are more restricted in their motions than the bulk chains (rejection rates 75 to 85%). This restriction is more striking for the case where a chain is a near neighbor to two or more cholesterol molecules. Then the order parameter profile changes to resemble a more-ordered chain, and near total immobility sets in for the upper segments of these chains. The basic qualitative conclusion to be drawn from the computer calculations is that cholesterol reduces the ability of any neighboring lipid hydrocarbon chains to undergo *trans-gauche* isomerism. At higher concentrations, cholesterol also forces neighboring chains into conformations with increased fractions of *trans* bonds.

Given the limited number of simulation models and cholesterol concentrations sampled, it is not easy to obtain a quantitative measure of the effect of cholesterol on lipid chain conformational space. This space is very large, even in the rotational-isomeric model, and therefore the goal of obtaining a completely precalculated interaction potential for lipids and cholesterol from the simulations cannot be reached at this time. For this reason it is necessary to rely upon qualitative insights gained from the simulations, in devising a theoretical model for lipid-cholesterol mixtures. In the next subsection such a model is described.

C. A THEORETICAL MODEL FOR LIPID-CHOLESTEROL MIXTURES

The results of the Monte Carlo simulations described above were of such a nature that they could be readily incorporated into a generalization of a theoretical model for lipid bilayer phase transitions, developed earlier by Scott and co-workers (Cheng and Scott, 1979; Scott, 1981). This model begins by considering the two monolayers of a lipid bilayer to be independent, so that the theoretical model need only consider one monolayer of the bilayer. Then the monolayer is reduced to a true two-dimensional system by projecting the lipid molecules onto a plane parallel to the plane of the original bilayer. This reduces the monolayer to a two-dimensional system in which the molecules are flat "shadows" of the original lipids. This system is a two-dimensional *fluid* of irregularly shaped objects that interact through short-range steric and longer-range attractive forces. Other theoretical approaches (Zuckermann et al., this volume) involve two-dimensional *lattices* of point molecules, which interact via mean fields.

In a true projection the shadows of all the lipid molecules in different conformational states will differ in shape because of the varying lateral extension of the chains due to *gauche* rotations. Given the very large number of possible conformations, such a system is far too complex to study analytically, even in two dimensions. However, to a good approximation, large subgroups of different lipid conformational states cast shadows that are of very nearly the same shape. This makes it possible to consider statistically a very large number of lipid conformational states within a much smaller set of shadows. In practice the procedure used was to suppose that the shapes of all possible shadows were circularly capped rods of fixed width, but whose length varied according to the conformation of the lipid molecule casting the shadow. Because this representation of a lipid monolayer is a mapping of a large number of molecular states onto a few shadow states, a degeneracy was calculated (from elementary combinatorial arguments) for each shadow shape included in the model. In this manner a quasi-three-dimensional lipid monolayer is modeled as a two-dimensional sea of hard rods of various shapes, with internal degeneracies associated with each possible shape. Figure 7 shows a typical set of rods in a two-dimensional plane.

Unfortunately, this system is still too complex to be studied directly using the methods of statistical mechanics. In order to approximately calculate the thermodynamic properties of the two-dimensional system, one must determine the energy and entropy of the system at each temperature. If the set $\{\alpha_i\}$ denotes the fraction of the hard rods in the *i*th state, then the energy of the system may be expressed in terms of these variables if the internal energy of each rod can be determined. This energy is simply calculated from the average number of *gauche* bonds associated with the rod state by the mapping. It is more difficult to calculate the entropy of packing, and no exact method exists. Therefore, the approximate method of scaled particle theory (Reiss et al., 1959) was used. In this approximation the free energy of the system has the form:

$$G \setminus RT = \sum_k \alpha_k \ln \frac{\alpha_k}{\omega_k} + \ln\left(\frac{\rho}{1-\rho\sum_i a_i A_i}\right) + \frac{1}{2}\left(\frac{\rho}{1-\rho\sum_i \alpha_i A_i}\right) \times$$
$$\sum_{j,l} \alpha_j \alpha_l \left(\pi \frac{d_j d_l}{2} + l_l d_j + lj_d l + l_j l_l \,|\sin\theta_{j,l}|\right) + \tag{19}$$
$$\sum_k \alpha_k \varepsilon_k - \frac{C}{RT(A-38)^{1.5}} + \frac{N_A \Pi A}{RT}$$

where α_k is defined above as the fractional occupation of the *k*th hard rod shadow state, ω_k and ε_k are the degeneracy and internal energy of that state, d_k and l_l are the diameter (width) and length of the circular-capped hard rod of state *k*, respectively, and A_k is the area of that rod state. *C* is a phenomenological attractive interaction constant, *A* is the average area per molecule of the full system, N_A is Avogadro's number, and Π is a lateral pressure applied to the system, originating from hydrophobic interactions that cause the bilayer to form

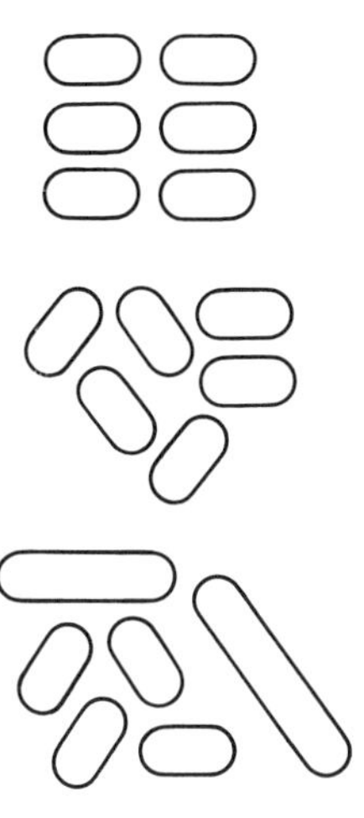

FIGURE 7. Schematic drawing of several hard-rod "shadow" states for the two-dimensional projection of a monolayer of a lipid bilayer, which forms the basis for a theoretical model for lipid phase transitions and lipid-cholesterol mixtures. (Top) Rods in shortest-length state and all in the same orientatonal state. (Middle) Rods in shortest-length state, but in different orientational states. (Bottom) Rods in different length and orientational states. (Reproduced from Scott, 1977. *Biochim. Biophys. Acta,* 469:264. With permission.)

initially. The physical state predicted by this model at fixed temperature T and pressure Π is found by minimizing G over the occupation variables α_k.

The full details of the minimization procedure are described in earlier references (Scott, 1991; Cheng and Scott, 1979; Scott, 1981) and will not be repeated here. It will suffice to note that the interaction parameter C could be adjusted so that the model produced a consistent picture of the phase transition in a pure lipid bilayer, and in bilayers containing binary mixtures of similar (chain length difference ~ 4 carbons) lipids. Figure 7 shows schematic pictures of the hard-rod orientations and length states possible in the various phases present in the model. These phases form a fully orientationally and conformationally ordered state (Figure 7a) to a conformationally ordered but orientationally disordered state (Figure 7b) to a conformationally and orientationally disordered state (Figure 7c).

This model offers a very straightforward vehicle for incorporation of the effects of cholesterol on the lipid chains, as observed in the MC simulations. Namely, the cholesterol reduces the conformational freedom of lipid chains and therefore the degeneracies $\{\omega_k\}$. The amount by which degeneracies are reduced should monotonically increase with cholesterol concentration. A simple way to quantify this effect into the model is to multiply each ω_k by a linear or quadratic function that decreases with the cholesterol concentration. A quadratic polynomial was chosen to represent the effects of isolated cholesterol molecules (the linear term in x_c, the cholesterol concentration) and neighbor pairs of cholesterol molecules (the quadratic term in x_c:

$$f(x_c) = 1 - ax_c - bx_c^2 \tag{20}$$

For shadow states with small degeneracies [less than 300 (Scott, 1991)] the quadratic term was omitted, since the effect should be less drastic on these states. Finally, the cholesterol molecule itself was projected as a shadow in the model plane, with a greater diameter and smaller rod length than the lipid molecules. This is a consequence of the methyls and the rings on the cholesterol, which give it a rather rounded shadow.

Figure 8 shows the calculated phase diagram for the model. The three free parameters that could be adjusted are the coefficients a and b in the function $f(x_c)$ and the length-to-width ratio of the cholesterol shadow in the model. The results turned out to be relatively insensitive to the parameters a and b as long as they were positive and of order 1. The results were very sensitive to the dimensions assigned to the cholesterol shadow, probably because of the overemphasis of the hard-core interactions within the scaled particle theory. The interesting result is that for a fairly wide-range of values for the cholesterol shadow length and width, the model contains all three of the phases seen experimentally (Vist and Davis, 1990). Namely, there is a fluid lipid phase (L_α) (depicted schematically in Figure 7c), an ordered lipid phase (L_β) (depicted schematically in Figure 7a), and a phase in which lipid chains are in all-*trans* conformations, but in which there is orientational disorder of the shadows (the β phase) (depicted schematically in Figure 7b). Furthermore, unlike thermodynamic models, Landau theories, or mean field lattice models, this model is sufficiently simple that the calculated thermodynamic phase properties are directly related to the microscopic interactions in the model. Of course, there are sacrifices in such a simple model. The steric interactions are certainly exaggerated, and this leads to complete solid-phase immiscibility that is not observed experimentally. Also, the two- and three-phase coexistence regions do not occur at the same cholesterol concentrations as in the experimental data. However, the qualitative resemblance of the calculated phase diagram to the experimental phase diagram in a simple and transparent model lends support to the notion that the underlying mechanism for lipid-cholesterol phase properties is indeed the short-range repulsive interactions and the resulting restrictive effect of cholesterol on the lipid chain conformations.

V. CONCLUSIONS

In this chapter we have attempted to address the question, How does one best calculate a phase diagram theoretically for a binary mixture of lipid and cholesterol in a bilayer with the minimum number of adjustable parameters and the maximum theoretical clarity? By consideration of thermodynamic, Landau theory, and mean-field thermodynamic model approaches, we have shown that at this level of approximate modeling, one can obtain interesting and qualitatively accurate phase diagrams with only three free parameters.

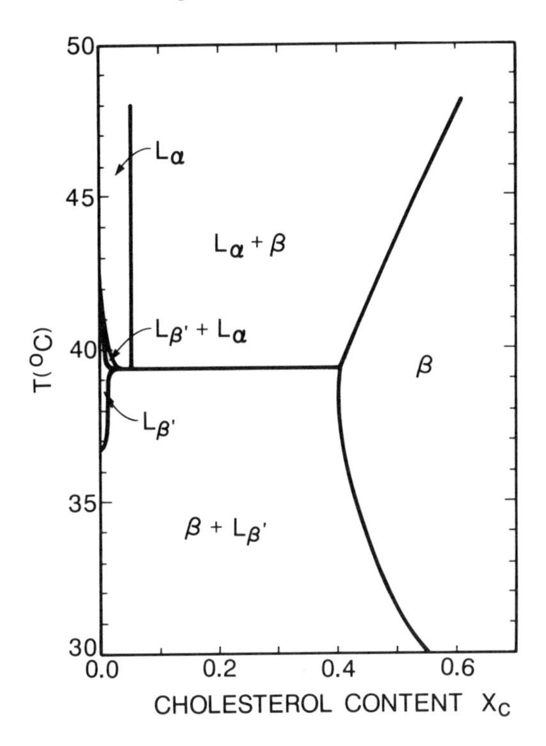

FIGURE 8. Phase diagram for a lipid-cholesterol mixture calculated for the hard-rod shadow model using scaled particle theory. (Reproduced from Scott, 1977. *Biochim. Biophys. Acta*, 469:264. With permission.)

We have also attempted to address the question of the role that large-scale numerical simulations can play in theoretical modeling. This role was illustrated by our own work in which the Monte Carlo method was employed to produce a microscopic picture of the manner in which lipid chains and cholesterol interact in a monolayer geometry. Then a theoretical model was employed in which it was possible to directly incorporate the restrictions on the lipid chains observed in the simulations. The qualitative agreement between the calculated phase diagram and the experimental phase diagram, coupled with the simplicity of the model, provide a plausible picture of the manner by which microscopic molecular interactions lead to macroscopic phase behavior.

The basic conclusion that we may draw from the material presented in this chapter is that purely phenomenological models for lipid cholesterol mixtures with more than three or four free parameters are basically curve-fitting exercises. Monte Carlo or molecular dynamics simulations can be critically important in the construction of theoretical models, but *only if the models simulated are accurate in atomic detail.* Then insights into the true microscopic interactions may be gained. Since these interactions are far too complex to describe by simple two-state variables, such as used in Section A, the numerical modeling may point to the *components* of the interactions that are most important for phase transitions in the mixtures. This may eventually allow for the construction of a statistical mechanical model that is sufficiently clear and free of arbitrary parameters and that contains predictive power to guide future experimental work.

ACKNOWLEDGMENT

We thank Dr. W. S. McCullough, Department of Physics, Oklahoma State University, for a critical reading of the manuscript.

The numerial work described herein utilized the Cray XMP and YMP systems at the National Center for Supercomputing Applications at the University of Illinois.

REFERENCES

Binder, K., Ed. 1977. *Monte Carlo Methods in Statistical Physics*, 2nd ed., Springer-Verlag, Berlin.

Callen, H. B. 1985. *Thermodynamics and an Introduction to Thermostatistics*, John Wiley & Sons, New York.

Cheng, W. H. and H.L. Scott. 1979. A theoretical model for lipid mixtures: phase transitions and phase diagrams. *Biophys. J.*, 28:117–132.

Davis, J. H. 1992. The molecular dynamics, orientational order, and thermodynamic phase equilbria of cholesterol/phosphatidylcholine mixtures: ^{2}H nuclear magnetic resonance. In *Cholesterol in Membrane Models*, Finegold, L., Ed., CRC Press, Boca Raton, FL.

Demel, R. A., K. R. Bruckdorfer, and L. L. M. van Deenen. 1972. Structural requirements of sterols for the interaction with lecithin at the air-water interface. *Biochim. Biophys. Acta*, 255:311–320.

Estep, T. N., D. B. Mountcastle, R. B. Biltonen, and T. E. Thompson. 1978. Studies on the anomalous thermotropic behavior of aqueous dispersions of dipalmitoylphosphatidylcholine-cholesterol mixtures. *Biochemistry*, 17:1984–1989.

Furman, D., S. Dattagupta, and R. B. Griffiths. 1977. Global phase diagram for a three-component model. *Phys. Rev.*, B15:441–464.

Griffiths, R. B. 1974. Thermodynamic model for tricritical points in ternary and quaternary mixtures. *J. Chem. Phys.*, 60:195–206.

Jacobs, R. and E. Oldfield. 1979. Deuterium nuclear magnetic resonance investigation of dimyristoyllecithin- and dipalmitoyllecithin-cholesterol mixtures. *Biochemistry*, 18:3280–3285.

Knobler, C. M. and R. L. Scott. 1984. Multicritical points in fluid mixtures: experimental studies. In *Phase Transitions and Critical Phenomena*, vol. 9, Domb, C. and J. Lebowitz, Eds., Academic Press, London, pp. 164–231.

Landau, L. D. 1937. Physik. Z. Sowjetunion, 11:26–46. Reprinted in *Collected Papers of L.D. Landau*, ter Haar, D., Ed., Pergammon Press, London, 1965. See also Lifshitz, E. M. and L. P. Pitaevski. *Statistical Physics*, 3rd Ed., part 1, Pergamon Press, London, pp. 451–459.

Lee, A. G. 1977. Lipid phase transitions and phase diagrams. II. Mixtures involving lipids. *Biochim. Biophys. Acta*, 472:285–344.

Lentz, B. R., D. A. Barrow, and M. Hoechli. 1980. Cholesterol-phosphatidylcholine interactions in multilamellar vesicles. *Biochemistry*, 19:1943–1954.

Mabrey, S. and J. M. Sturtevant. 1976. Investigation of phase transitions of lipids and lipid mixtures by high sensitivity differential scanning calorimetry. *Proc. Nat. Acad. Sci. U.S.A.*, 73:3862–3866.

Meijerling, J. L. 1963. Miscibility gaps in ferromagnetic alloy systems. *Philips Res. Rep.*, 18:318–330.

Mortensen, K., W. Pfeiffer, E. Sackmann, and W. Knoll. 1988. Structural properties of a phosphatidylcholine-cholesterol system as studied by small angle neutron scattering: ripple structure and phase diagram. *Biochim. Biophys. Acta*, 945:221–245.

Nagle, J. F. 1980. Theory of the main lipid bilayer phase transition. *Ann. Rev. Phys. Chem.*, 31:157–195.

Peng, Z. Y., N. Tjandra, V. Simplaceanu, and C. Ho. 1989. Slow motions in oriented phospholipid bilayers and effects of cholesterol or gramicidin. *Biophys. J.*, 56:877-885.

Presti, F. T. 1985. The role of cholesterol in membrane fluidity. In *Membrane Fluidity in Biology*, vol. 4; *Cellular Aspects*, Aloia, R. C. and J. M. Boggs, Eds., Academic Press, New York, pp. 97–146.

Priest, R. G. 1980. Landau phenomenological theory of one and two component phospholipid bilayers. *Mol. Cryst. Liq. Cryst.*, 60:167–184.

Recktenwald, D. J. and H. M. McConnell. 1981. Phase equilibria in binary mixtures of phosphatidylcholine and cholesterol. *Biochemistry*, 20:4505–4510.

Reiss, H., J.L. Lebowitz, and H.L. Frisch. 1959. Statistical mechanics of rigid spheres. *J. Chem. Phys.*, 31:369–380.

Scott, H. L. 1977. Monte Carlo studies of the hydrocarbon region of lipid bilayers. *Biochim. Biophys. Acta*, 469:264–271.

Scott, H. L. 1981. Lecithin bilayers: a model which describes the main and lower phase transitions. *Biochim. Biophys. Acta*, 643:161–167.

Scott, H. L. and S. Kalaskar. 1989. Lipid chains and cholesterol in model membranes: a Monte Carlo study. *Biochemistry*, 28:3687–3692.

Scott, H. L. 1990. Computer aided methods for the study of lipid chain packing in model membranes and micelles. In *Molecular Description of Biological Membrane Components by Computer Aided Conformational Analysis*, Brasseur, R., Ed., CRC Press, Boca Raton, FL, pp. 123–150.

Scott, H. L. 1991. Lipid-cholesterol interactions: Monte Carlo simulations and theory. *Biophys. J.*, 59:445–455.

Seelig, J. 1983. The description of membrane lipid conformation, order, and dynamics by ^{2}H NMR. *Biochim. Biophys. Acta*, 737:117–143.

Shimsick, E. J. and H. M. McConnell. 1973. Lateral phase separations in binary mixtures of cholesterol and phospholipids. *Biochem. Biophys. Res. Commun.*, 53:446–451.

Tardieu, A., V. Luzzati, and F. C. Reman. 1973. Structure and polymorphism of the hydrocarbon chains of lipids: a study of lecithin-water phases. *J. Mol. Biol.*, 75:711–733.

Vist, M. R. and J. H. Davis. 1990. Phase equilibria of cholesterol/dipalmitoylphosphatidylcholine mixtures: ^{2}H nuclear magnetic resonance and differential scanning calorimetry. *Biochemistry*, 29:451–464.

Zuckerman, M. J., J. H. Ipsen, and O. G. Mouritsen. 1992. Theoretical studies of the phase behavior of lipid bilayers containing cholesterol. In *Cholesterol in Membrane Models*, Finegold, L., Ed., CRC Press, Boca Raton, FL.

Chapter 9

THEORETICAL STUDIES OF THE PHASE BEHAVIOR OF LIPID BILAYERS CONTAINING CHOLESTEROL

M. J. Zuckermann, J. H. Ipsen, and O. G. Mouritsen

TABLE OF CONTENTS

0-8493-4207-4/93/$0.00+$.50

I. INTRODUCTION

A. THE FLUID LIPID-BILAYER COMPONENT OF CELL MEMBRANES

The plasma membranes of cells are molecular composite materials of a very stratified nature (Bloom et al., 1991). The core of the membrane is the fluid lipid bilayer, which is a key structural motif of the membrane. Towards the cell interior, the lipid bilayer is attached to a cytoskeleton that is a very extended polymeric filament that is important for maintaining the shape of the cell. On the exterior side the membrane is covered by a glycocalyx that is made up by the carbohydrate moities of membrane proteins and lipids. The glycocalyx is thought to be involved in cell-cell recognition and receptor functions. A major part of the biological activity, on the molecular level, in all cellular organisms is associated with membranes, in particular with the fluid-lipid bilayer component. The lipid bilayer acts as a permeability barrier, which ensures that the cell can distinguish between outside and inside and sustain the chemical and electrical gradients necessary for physiological function; the bilayer is also an important regulator of the functions of proteins, enzymes, and receptors attached to the membrane.

In order to support physiological functions, it is mandatory that the lipid bilayer be in a "fluid" state, that is, a state that assures high lateral mobility of the molecules embedded in the membrane. Hence, the bilayer needs to be in a liquid thermodynamic phase. However, the bilayer needs, at the same time, to provide a permeability barrier and to have a large mechanical coherence. It is at this point that the physical chemistry of the lipid bilayer becomes important. Over a timespan of more than 10^9 years, nature has evolved a type of lipid bilayer with a physical chemistry that leads to materials characteristics that in every respect outperform every manmade material (Bloom et al., 1991). In the lipid bilayers of eucaryotic plasma cell membranes, *cholesterol* in animals and *ergosterol* in plants are a universally present lipid species. The presence of sterols has implied a major evolutionary advantage for the development of eucaryotes by the unique capacity of sterol of controlling membrane permeability, lateral bilayer organization, membrane fluidity, and mechanical strength.

Despite its great importance for membrane properties and despite the fact that cholesterol-containing membranes have been an extensively studied topic, the understanding of the physical effects of cholesterol on lipid bilayers has been a longstanding problem in the physical chemistry of model membranes (Bloom et al., 1991; Demel and De Kruyff, 1976; Yeagle, 1985; Presti, 1985). In recent years some substantial progress has, however, been made both on the experimental and theoretical side. It is some of this progress that forms the motivation of the present chapter.

B. OUTLINE OF THIS CHAPTER

In this chapter the reader is given a description of recent theoretical studies for the phase behavior of lipid bilayers containing cholesterol, using two-dimensional lattice models. The theoretical studies are described in close connection with pertinent experimental data. The chapter begins with a general overview of phase transitions in lipid systems. It continues with a detailed formal description of some statistical mechanical lattice models used to describe the phase behavior of pure lipid systems, and of how these models can be extented to include the effects of cholesterol. It is shown that the extended model provides a physical description and interpretation of the experimental phase diagram of DPPC-cholesterol bilayers and predicts the correct experimental behavior for the specific heat and the hydrophobic thickness of these systems, without the neccessity of invoking complexing. Computer simulations for lipid-cholesterol bilayers, using a reduced version of the model, are discussed, and it is shown that this model can be used to predict the effects of cholesterol on the dynamic heterogeneity of lipid bilayers with low concentrations of cholesterol near the main phase transition. Furthermore, the computer simulations permit a study of thermal fluctuations near the main phase transition and of how these fluctuations influence the specific heat, the lateral compressibility, and the ionic permeability of lipid bilayers containing cholesterol.

II. RELEVANT EXPERIMENTAL AND THEORETICAL STUDIES

A. EXPERIMENTAL STUDIES

A large number of experimental investigations have been devoted to membrane systems containing cholesterol, and the results of these studies have been used to answer three fundamental questions: (1) What is the effect of cholesterol on the main phase transition of pure lipid bilayers? The main phase transition, which takes place at a temperature T_m, is the transition that takes the bilayer from a crystalline solid (gel) phase with conformationally ordered acyl chains to a liquid (liquid-crystalline) phase with conformationally disordered acyl chains (Chapman et al., 1968). (2) What is the effect of cholesterol on the rotational and translational motions of the lipid molecules in the liquid-crystalline state? (3) What is the effect of cholesterol on the conformation of the acyl chains of the lipid molecules? The experimental techniques used to answer these questions include, among others, calorimetry (Ladbrooke et al., 1968; Mabrey et al., 1978; Estep et al., 1978; Imaizumi and Hatta, 1984; Melchior et al., 1980), magnetic resonance spectroscopy (Recktenwald and McConnell, 1981; Kusumi et al., 1986; Presti and Chan, 1982; Presti et al., 1982; Stockton and Smith, 1979; Dufoure et al., 1984; Vist and Davis, 1990; Davis, 1988), fluorescence depolarization (Alecio et al., 1982; Smutzer and Yeagle, 1985; Van Ginkel et al., 1986), infrared and Raman spectroscopy (Lippert and Peticolas, 1971; Pink et al., 1981; Levin et al., 1985; O'Leary and Levin, 1986; Rooney et al., 1986), neutron- and X-ray scattering (Knoll et al., 1985; McIntosh, 1978; Hui and He, 1983), electron microscopy (Knoll et al., 1985; Copeland and McConnell, 1980; Hicks et al., 1987; Lentz et al., 1980), and micromechanics (Evans and Needham, 1986).

From these experimental studies a general consensus has been reached concerning the answers to these fundamental questions. Addition of cholesterol leads to a more conformationally ordered state of the acyl chains of the lipid molecules (Recktenwald and McConnell, 1981; Kusumi et al., 1986; Presti and Chan, 1982; Presti et al., 1982; Stockton and Smith, 1979; Dufoure et al., 1984; Vist and Davis, 1990; Davis, 1988). Within the liquid-crystalline phase, cholesterol has only a small effect on the molecular translational and rotational motions (Lindblom et al., 1981; Rubenstein et al., 1979; Kawato et al., 1978). At low cholesterol concentrations a slight broadening of the main phase transition occurs, whereas at higher concentrations the liquid phase of the pure bilayer is dramatically stabilized. These results were only obtained with considerable effort, and the determination of the lipid-cholesterol phase diagram in excess water turned out to be a particularly elusive problem. Recently Vist and Davis (Vist and Davis, 1990; Davis, 1988) presented a careful study of DPPC-cholesterol multibilayer systems, using both deuterium NMR and DSC experiments. The system was therefore characterized on both the molecular and the thermodynamic level. The study was performed in the temperature range of 20 to 60°C and from 0 to 30 mol% cholesterol. Analysis of the data from both of these experiments, in addition to results from ESR measurements (Recktenwald and McConnell, 1981), leads to the phase diagram of Figure 1. Sections of the phase diagram are in agreement with data obtained by several other techniques (Copeland and McConnell, 1980; Evans and Needham, 1986), as also shown in Figure 1.

In Figure 1 and in the text below we use a notation for the phases that describes both the two-dimensional translational order and the chain-conformational order occurring in the phases: the gel phase is referred to as the solid-ordered or ***so*** phase, where *s*, for solid, refers to the crystalline order, and **o**, for ordered, to the average acyl-chain conformation. Similarly, the liquid-crystalline phase at low cholesterol concentration is denoted liquid-disordered or *l***d** whereas at high cholesterol concentration there is a liquid-ordered or *l***o** phase. The nomenclature is motivated by the fact that several independent experimental studies have demonstrated the simultaneous occurrence of a liquid-crystalline phase and high acyl-chain order. Micromechanical studies (Evans and Needham, 1986) show that the bilayer at high cholesterol concentrations behaves as a liquid with no surface shear rigidity, but a greatly reduced membrane-area compressibility. The NMR measurements show that the deuterium order parameters approach the value 0.5, which is typical of an all-*trans* rotating chain (Vist and Davis, 1990; Davis, 1988). The NMR spectra also show that diffusion is fast enough to average dipolar couplings just below the main transition temperature (Ulmius et al., 1975). A fast translational diffusion in the *l***o** phase is demonstrated by photobleaching fluorescence studies (Rubenstein et al., 1979).

The phase diagram of Figure 1 shows several remarkable features that indicate that cholesterol interacts with lipid molecules in an unusual way. Normally a solid is not as good a solvent as a liquid, and the addition of an impurity results in a sizable freezing-point depression. The two-dimensional

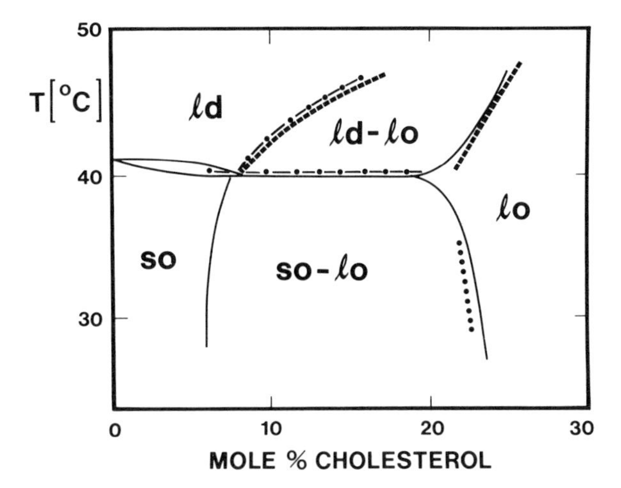

FIGURE 1. Experimental phase diagram for the DPPC-cholesterol system as determined by NMR spectroscopy and differential scanning calorimetry (————) (Vist and Davis, 1990; Davis, 1988), EPR spectroscopy (– – –) (Recktenwald and McConnell, 1981), freeze-fracture (····) (Copeland and McConnell, 1980), and micromechanics (·—·—·—) (Needham, D. and E. Evans, unpublished data). Note that the NMR and calorimetry studies were carried out on perdeuterated DPPC for which T_m is about 38°C. The experimental data have been scaled accordingly to facilitate the comparison within a single figure. The notation for the different phases is explained in the text. The various phases are denoted by *so* (solid-ordered), *l***d** (liquid-disordered), and *l***o** (liquid-ordered).

solid and fluid phases do not initially respond in this way to cholesterol, which, according to the phase diagram, is almost as soluble in the solid-ordered as the liquid-disordered phase at low concentrations. However, the situation completely changes for high cholesterol concentrations, where the liquid phase of the bilayer is stable down to temperatures substantially lower than T_m. Therefore, in the high concentration regime, it is clear that cholesterol strongly favors the liquid phase over the solid phase, in contrast to the low-concentration behavior. There is a third feature of the phase diagram that deserves attention. In a concentration range of about 10 to 20% cholesterol at temperatures above T_m, there are indications of a phase separation where two liquid phases of the bilayer coexist. In Section IV, the cholesterol-rich liquid phase will be described in terms of a liquid-ordered phase.

These three remarkable features of the phase diagram of Figure 1 indicate that there are very specific lipid-cholesterol interactions. Another experimental result that gives the same indication is the occurrence of thermal anomalies in the specific heat of DPPC-cholesterol bilayers, as seen by differential scanning calorimetry (Mabrey et al., 1978; Estep et al., 1978; Imaizumi and Hatta, 1984). The anomalous specific heat is composed of a sharp component with an associated entropy that decreases with increasing cholesterol concentration, and a broad component that is negligible at low cholesterol concentrations. The sharp component exhibits a maximum at a specific cholesterol concentration and vanishes again at higher concentrations.

B. THEORETICAL STUDIES

We now examine the status of theoretical work for lipid-cholesterol bilayer systems, using the classification given in the introduction. In this context it should be stated that lipid molecules considered in this article are mostly phospholipids, i.e., they are amphiphiles that usually consist of two hydrophobic acyl (hydrocarbon) chains ether-linked by a glycerol backbone to a PC polar head that is quite bulky. The acyl chains are flexible in the sense that they can support induced *gauche* bonds. In compressed monolayers or bilayers, the number of *gauche* bonds and the resulting chain conformations are limited by steric effects, since these systems are relatively close packed. In contrast the polar head of cholesterol is a hydroxyl group, and its hydrophobic tail is predominantly rigid, with five closed rings at the top and a butyl group at the end. In addition, the cholesterol molecule has a larger cross-sectional area than a rigid acyl chain, and therefore has difficulty in fitting into a solid phase of the lipids. Finally, the hydroxyl group of a cholesterol molecule in a bilayer is known to be on the level of the carbonyl group of the lipid molecule just below the glycerol backbone. The other end of the cholesterol is taken to lie at the level of the twelfth carbon atom if the acyl chain is saturated. It is therefore reasonable to suppose that the cholesterol has a rigidifying effect on neighboring acyl chains in the liquid-crystalline phase, and this conclusion is borne out by the NMR data of Vist and Davis (1990). In the rest of this section we will describe how the above characteristics of cholesterol are incorporated into theoretical models.

Several theoreical models (Martin and Yeagle, 1978; Slater and Caillé, 1981; Snyder and Freire, 1980; O'Leary, 1983; Jackson, 1976; Owicki and McConnell, 1979; Jähnig, 1981; Engelman and Rothman, 1972; Owicki and McConnell, 1980; Pink and MacDonald, 1988) have been proposed for the phase behavior of lipid-cholesterol bilayers other than those described in this section. These include the models based on complex formation, and the reader is referred to the original articles. Some more detailed molecular interaction models are also discussed in Chapter 8 (Finegold, 1992) in this volume. The models described in this section are partially based on the work of Marcelja (1976), who used a nonspecific interaction between lipid and cholesterol molecules in conjunction with his mean-field treatment of the main phase transition. He found that the order parameter only varied strongly over a distance of three to four lipid layers around the cholesterol molecule in the region of the main phase transition.

Pink and Carroll (1978) were the first authors to propose a two-state lattice model [based on an idea due to Doniach (1978)] for the acyl chains in lipid-cholesterol bilayers. The two conformational states plus their interactions describe the thermodynamics of pure lipid bilayers, and the cholesterol molecules are assumed to interact differently with neighboring lipid molecules in different conformational states. A fit in the mean-field approximation to the latent heat of DMPC-cholesterol bilayers resulted in the prediction of a critical point at a concentration of 24 mol% cholesterol. The interaction parameters obtained from the best fit indicated that a cholesterol molecule prefers rigid lipid acyl chains as neighbors. Pink et al. (1981) used an extension of their ten-state conformational lattice model to understand their own Raman data. The ten-state conformational model, which is briefly described in Section III below, comprises nine conformational states of an acyl chain that are compatible with the gel or *so* phase, and one "excited" state that is compatible with the liquid-crystalline or *l***d** phase. This model was successfully used to analyze both thermodynamic properties and Raman data for pure DPPC bilayers and thermodynamic properties for other pure PC bilayers. The extension to cholesterol involved the reclassification of the nine gel-like states into two groups: one group of five "ground" states, which includes the all-*trans* state, and a second group of four "intermediate" states. Acyl chains in the ground or excited states were assumed to interact very weakly with neighboring cholesterol molecules, whereas acyl chains in the intermediate states were assumed to interact very strongly with neighboring cholesterol molecules, for steric reasons. Values for the lipid-cholesterol interaction parameters were obtained by fitting to the Raman data via mean-field theory and were used to predict the phase diagram and the specific heat for such systems. The phase diagram has the following features: (1) a small horizontal phase separation region between *s***o** and *l***d** phases for low cholestrol concentrations at T_m for the pure system. (2) A wide miscibility gap for $T < T_m$ with *s***o**-*l***d** phase separation at higher temperatures, a phase separation between a lipid-rich *s***o** phase, and a cholesterol-rich phase in which the acyl chains are mostly in the "intermediate" conformational states. A calculation of the specific heat (D. A. Pink, A. Georgallas, M. J. Zuckermann, and M. O. Steinitz, unpublished calculations) for DPPC-cholesterol bilayers, using the parameters found in the fit to the Raman data, yielded a single component of the specific heat, with the broad component absent. The fitted parameters were also used by Tessier-Lavigne et al. (1982) to calculate the order parameter in the neighborhood of cholesterol and protein molecules at low concentrations close to T_m. It was found that cholesterol molecules cause their neighboring acyl chains to become more statically ordered. However, the effect is maximal for second-nearest neighbors and is only present up to third-nearest neighbors in agreement with the work of Marčalja (1976). Finally, Mouritsen et al. (1983) reported the results of numerical Monte Carlo simulations dynamics on a two-state conformational model of the type proposed by Pink and Carroll (1978) described above. They were able to fit to the latent heat of transition of DPPC-cholesterol bilayers as a function of cholesterol concentration. However, only a single component was obtained for the specific heat, and the authors concluded that intermediate configurations needed to be introduced into the formalism in order to account for the experimental data. This will be discussed further in Section IV.

III. THE PINK-POTTS MODEL FOR PHASE TRANSITIONS IN LIPID BILAYERS

This section is based on Pink et al. (1981), Zuckermann and Mouritsen (1987), Mouritsen and Zuckermann (1987), and Ipsen et al. (1989) and is organized as follows. First we give a detailed description of the mathematical model for lipid monolayer and bilayer phase transitions in terms of two Hamiltonians: one for the conformational states of the acyl chains and the other for the Q-state Potts model, which mimics the positional degrees of freedom. We then give a detailed description of the mean-field approximation and present the mean-field phase diagram for this model appropriate for pure DPPC bilayers.

A. FORMALISM

The mathematical model treated in this section is based on the ten-state conformational model of Pink (Pink et al., 1981) which describes chain melting in lipid monolayers and bilayers, combined with a multistate Potts model that is used to treat the positional degrees of freedom of lipid monolayers and bilayers in an approximate manner. The ten-state model has been described in great detail in previous publications (Pink et al., 1981; Caillé et al., 1980) and requires, at most, a brief introduction. This model is a two-dimensional lattice model that accounts for the intrachain conformational energy, the rotational isomerism of the chains, and the van der Waals interactions between chains. The ten chain conformations allowed for in the model include an all-*trans* ground state, eight gel-like excited states, and a highly excited state of low conformational order characteristic of the fluid phase. Each conformational state, n, is characterized by a cross-sectional area per molecule, A_n, an internal conformational energy, E_n, which is related to the number of *gauche* defects, and the degeneracy, D_n, which can be regarded as a density of states. The ten-state model on its own describes a phase transition from a conformationally ordered to a conformationally disordered phase, which is driven by the difference between the high conformational entropy of the excited state and the considerably lower conformational entropies of the gel-like states.

The multistate Potts model used here to mimic the positional degrees of freedom is a modification of the standard high-Q-state Potts model that has had great success in the description of grain growth in polycrystalline aggregates. The value of the Potts index is typically chosen to be 30, which represents the large-Q limit. The standard Potts model accounts for the energy of the grain boundaries of a metastable distribution of crystalline domains, each of which is characterized by a Potts state. In the modified Potts model, each lattice site represents an acyl chain of the lipid molecule and carries two independent sets of states, i.e., Q Potts states and ten configurational states. The Potts variables again describe the orientation of crystalline domains with which the chain on the lattice site is associated. A domain boundary energy is modeled by allowing neighboring acyl chains to interact with an energy $J_P > 0$ if they are in different Potts states and if each chain is in one of the first nine gel-like states. Otherwise the interaction is zero. This is reasonable, since the tenth conformational state is repesentative of the liquid crystalline phase that is fluid and that therefore cannot have a granular nature. The modified Potts model therefore replaces, in a very approximate manner, the translational degrees of freedom of the solid phase. It cannot account for more involved effects such as, for example, bond orientational order in hexatic phases.

The total Hamiltonian of the bilayer can now be written as follows:

$$H_{TOT} = H_{CONF} + H_P \tag{1}$$

H_{CONF} is the Hamiltonian of the ten-state Pink model for the conformational states of the chains and can be written in the following form:

$$H_{CONF} = H_{C0} + H_{CC} + \Pi \sum_i \sum_{n=1}^{10} A_n L_{in} \tag{2}$$

In Equation 2, Π represents the internal pressure exerted on a lipid bilayer due to interfacial effects (hydration, polar head interactions, etc.). i is a site index for the lattice, and L_{in} is an occupation variable that is unity when the ith chain is in the nth conformational state, and zero otherwise. H_{C0} is the Hamiltonian for the intrachain energies, E_n, and degeneracies, D_n, and is given by:

$$H_{C0} = \sum_i \sum_{n=1}^{10} E_n L_{in} \tag{3}$$

where $E_n = E_n - k_B T \ln D_n$. H_{CC} is the Hamiltonian for the effective van der Waals interaction between neighboring chains and is written as follows:

$$H_{CC} = -J_0 \sum_{<i,j>} \sum_{m,n=1}^{10} I_m I_n L_{im} L_{jn} \tag{4}$$

where i and j are nearest neighbor indicies. J_0 is the coupling constant of the van der Waals interaction between acyl chains in the all-*trans* state and I_m depends on A_m. A detailed expression for I_n can be found in Pink et al. (1981) and Caillé et al. (1980). The values of all parameters pertaining to H_{CONF}, except J_0, are those that were previously determined from applications to the chain melting transition of lipid bilayers.

H_P is the Hamiltonian for the modified Q-state Potts model and is written as follows:

$$H_P = J_P \sum_{<i,j>} \sum_{m,n=1}^{9} \sum_{p,q=1}^{Q} (1 - \delta_{pq}) L_{ip} L_{jq} L_{im} L_{jn} \tag{5}$$

J_P is the coupling constant of the Potts interaction and L_{ip} is an occupation variable that is unity when the chain on the ith lattice site is associated with the pth Potts state.

B. MEAN-FIELD RESULTS

The Hamiltonian of Equation 1 was used in conjuction with the mean-field approximation to obtain the phase diagram in terms of temperature and Potts interaction, J_P. The resulting phase diagram, in terms of temperature and coupling ratio, J_P/J_0, appropriate for DPPC bilayers is almost identical to the mean-field approximation phase diagrams for the two-state model presented in our previous work (Zuckermann and Mouritsen, 1987). The values of the parameters of the ten-state model used in the calculation correspond to saturated acyl chains with 16 carbons (see Pink et al., 1981). It is shown in the article by Zuckermann and Mouritsen (1987) that the system exhibits a triple point for a particular value $J_P = J_P^*$. For $J_P > J_P^*$, there are two phases separated by one phase boundary. For bilayers the phase at high temperatures is characterized by conformationally disordered chains, i.e., acyl chains in the tenth conformational state and Potts disorder. We identify this phase as the disordered (*l***d**). The phase at low temperatures is characterized by conformationally ordered chains, i.e., acyl chains in one of the first nine conformational states, and Potts order. We identify this phase as the solid-ordered (*s***o**) phase. For $J_P < J_P^*$, a second phase boundary appears in addition to the first one and results in the presence of an intermediate phase that is characterized by conformational ordered chains and Potts disorder and that we will refer to as the liquid-ordered (*l***o**) phase. Though this phase does not appear to occur in pure lipid bilayers, we show in Section IV that it is very important in our interpretation of the effect of cholesterol on lipid bilayers.

IV. A LATTICE MODEL FOR LIPID-CHOLESTEROL BILAYERS*

A. FORMALISM

The advantage of the model descibed in Section III is that it gives a description of lipid bilayers in terms of two degrees of freedom, i.e., one for the chain conformations in terms of rotational isomerism, and the other for the positional order in terms of the Potts variables. This description is particularily important for bilayers containing intrinsic impurities, since a suitably extended model allows the impurity to affect both degrees of freedom differently. Cholesterol, in particular, can be regarded as a "crystal breaker", from the point of view of the positional degrees of freedom, and a "rigidifier" of the acyl chains relative to the conformational degrees of freedom. The "crystal breaker" effect is included into the model as follows: whenever a lipid acyl chain in a gel-like state has cholesterol as a nearest neighbor, the Potts interaction between the acyl chain and the cholesterol molecule is taken to be much smaller than the Potts interaction between acyl chains. Furthermore, the Potts interactions between the acyl chain and other gel-like acyl chains in its immediate neighborhood are weakened by the presence of cholesterol. The effect of these "crystal breaking" properties of cholesterol is to reduce or destroy the effective value of the overall Potts interaction and therefore to promote the stability of the intermediate *l***o** phase at higher cholesterol concentrations.

In the case when cholesterol destroys the Potts interaction between the gel-state acyl chains and itself, the effect of both "crystal breaking" and rigidifying

* Ipsen et al., 1989; Ipsen et al., 1987; Ipsen et al., 1990a; Ipsen et al., 1990b.

of acyl chains by the cholesterol molecules can be included into the Hamiltonian of Equation 1 by the addition of the following three terms:

$$H_{cp} = \Pi \sum_i A_c L_{ic} \tag{6}$$

$$H_{cc} = -\frac{J_0}{2} \sum_{<i,j>} \sum_{m,n=1}^{10} I_{cc} L_{ic} L_{jc} \tag{7}$$

$$H_{cC} = -J_0 \sum_{<i,j>} \sum_{m=1}^{10} I_c I_m \left(L_{ic} L_{jm} + L_{im} L_{jc} \right) \tag{8}$$

Here H_{cp} describes the interfacial pressure-area contribution due to cholesterol, H_{cc} is the Hamiltonian representing the interaction between two neighboring cholesterol molecules, and I_{cc} is the strength of this interaction relative to J_0. H_{cC} is the Hamiltonian representing the interaction between the cholesterol and lipid chains. I_c is the (constant) shape factor assigned to the cholesterol molecule. It is related to the van der Waals interaction between the hydrophobic part of the cholesterol molecule and the corresponding part of a lipid chain or another cholesterol molecule. The cholesterol cross-sectional area (Engelman and Rothman, 1972) is $A_c = 32$ Å^2. The interactions of Equations 6–8 are not specific for cholesterol, but may equally well represent the interaction with other small, stiff membrane-bound amphiphilic molecules that are hydrophobically smooth. We assume for convenience that $I_{cc} = I_c^2$. The introduction of cholesterol therefore involves only a single new parameter, I_c. L_{ic} is the site occupation variable for a cholesterol molecule at lattice site *i*, and the concentration of cholesterol is given by

$$x = \frac{< L_{ic} >}{(2 - < L_{ic} >)} \tag{9}$$

The mean-field approximation is then used in conjunction with the modified Hamiltonian to obtain a form for the free energy that now is a function of T, Π, and x. The phase diagram is obtained by minimizing a suitable form for the free energy, as in Ipsen et al., 1987, to which we refer the reader for mathematical derivations and details of the parameters used.

B. MEAN-FIELD RESULTS

Figure 2 shows the phase diagram for lipid-cholesterol bilayers, which was calculated from the mean-field theory in conjunction with the Hamiltonian formalism given by Equations 1–9 above (Ipsen et al., 1987). The conformational model for the acyl chains was therefore the ten-state model due to Pink et al. (1981), and the parameters for the lipid-cholesterol interactions were chosen so as to give a phase diagram that was qualitatively as close as possible to the experimental phase diagram of Figure 1. The diagram of Figure 2 shows a narrow

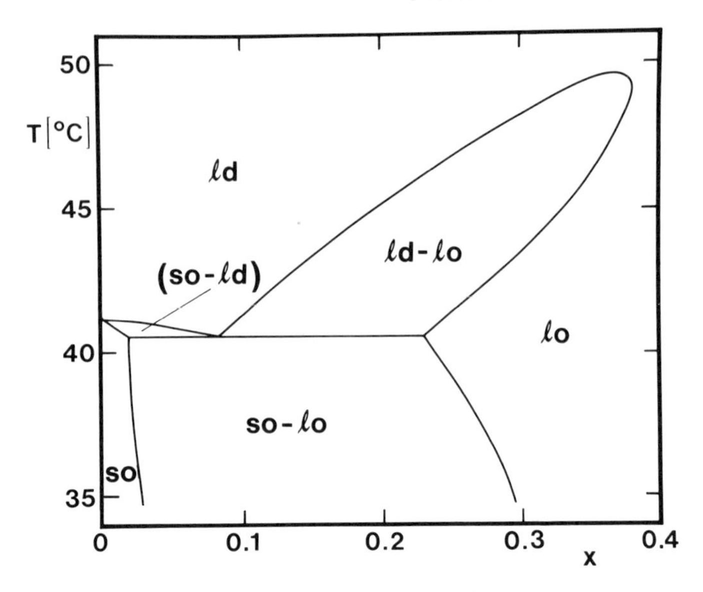

FIGURE 2. Theoretical phase diagram for DPPC-cholesterol bilayers. The various phases are denoted by *s***o** (solid-ordered), *l***d** (liquid-disordered), and *l***o** (liquid-ordered).

coexistence region (*s***o**-*l***d**) at low cholesterol concentration, a three-phase line at intermediate cholesterol concentration, and the occurence of a new phase, *l***o**, at high cholesterol concentration. A distinct feature of the phase behavior is the occurrence of an eutectic point. Finally, the (*l***o**-*l***d**) coexistence region at high temperatures terminates in an upper critical point. Qualitatively similar phase diagrams were obtained by Ipsen et al. (1987) from both a microscopic two-state conformational model and a phenomenological thermodynamic model. From Figures 1 and 2 it can be deduced that the unusual behavior of cholesterol in relation to lipid bilayers can be understood in the following terms: (1) cholesterol dissolves most easily in liquid phases, and (2) cholesterol prefers neighboring acyl chains to be orientationally ordered, i.e., with conformations characteristic of the solid phase. Cholesterol resolves this dilemma by preferring both lipid phases equally up to a certain cholesterol concentration. This leads to almost horizontal phase lines for low x and very little phase separation, cf. Figure 1. Above about 10 mol% the cholesterol-induced acyl-chain ordering in the liquid phase and the cholesterol-induced crystal breaking in the solid phase reach a point where the conformational order is similar to that of the gel phase. Cholesterol then prefers a liquid-ordered (*l***o**) phase that is characterized by positional disorder and conformational order. This leads to massive phase separation, cf. Figure 2. At this point the conformational and crystalline degrees of freedom decouple, giving way to the unusual phase behavior of lipid-cholesterol mixtures.

The specific heat can be derived from the free energy of the microscopic interaction model of Equations 1–9 using standard thermodynamic relations (Ipsen et al., 1990a). The numerical differentiation of the free energy was

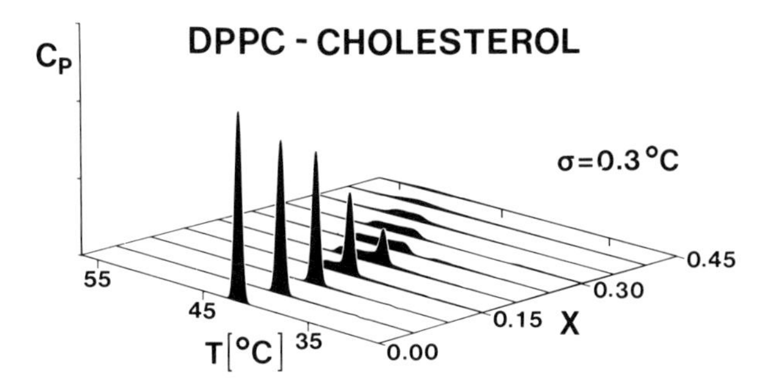

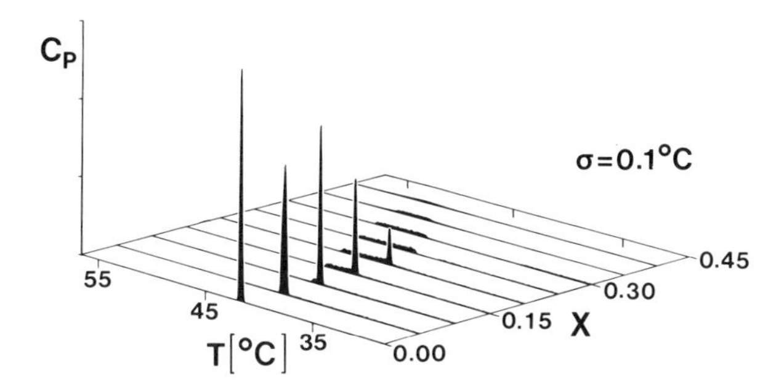

FIGURE 3. Theoretical specific heat, $C_P(T,x)$, vs. temperature and composition for DPPC-cholesterol bilayers. Results are shown for two different widths, σ, of the resolution function: (a) $\sigma = 0.3$°C and (b) $\sigma = 0.1$°C. C_P is in arbitrary units, and the data in case (a) have been expanded vertically by a factor of 2.5. For the sake of clarity the area between the curve and the baseline has been shaded.

facilitated by means of convoluting the free energy function with a "temperature" Gaussian of width σ, which represents the instrumental resolution. The delta-function singularity in the specific heat predicted for pure-lipid bilayers at T_m is therefore broadened into a Gaussian by this smoothing procedure.

The structure of the phase diagram of Figure 2 is clearly reflected in the calculated specific heat scans that are shown in Figures 3 and 4. Figure 3 shows the overall behavior for two different widths of the resolution function, whereas Figure 4 gives a more detailed picture of the behavior at low concentrations and in the neighborhood of the eutectic point. The more narrow the resolution function is, the more details become available for the specific heat. From these figures we make the following observations about the specific heat anomalies as the cholesterol content is increased:

- The sharp peak at zero cholesterol concentration with the initial width of the resolution function decreases slightly in intensity as the cholesterol concentration, x, is increased, and the peak position moves towards

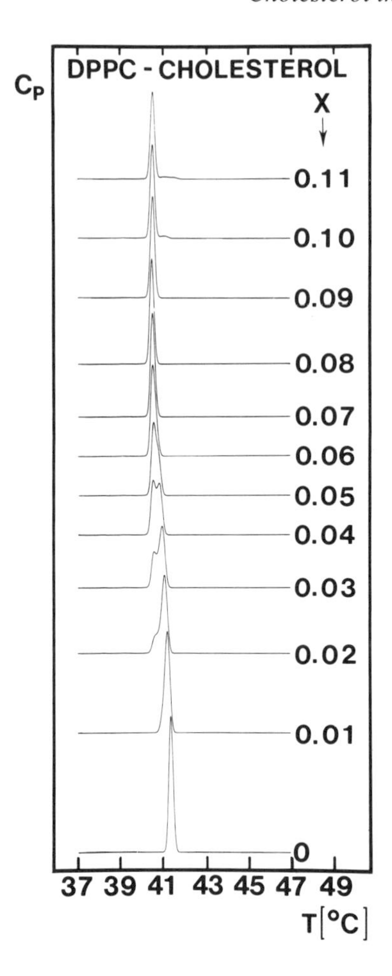

FIGURE 4. Theoretical specific heat, $C_p(T,x)$, vs. temperature for different compositions of DPPC-cholesterol bilayers at low concentrations and around the eutectic point, cf. Figure 2. The width of the resolution function used is $\sigma = 0.1$°C. C_P is in arbitrary units.

somewhat lower temperatures as the three-phase line is approached. The peak position closely follows the *l***d**-(*s***o**-*l***d**) phase boundary. There is a small amount of asymmetry towards the *s***o**-(*s***o**-*l***d**) boundary.

- As x is increased towards the eutectic point, the peak intensity first decreases and then increases again. For high resolution the specific heat can be resolved into two distinct peaks following the two phase boundaries. These two peaks have a maximum splitting at the low-x terminus of the three-phase line. The splitting decreases towards the eutectic point. For a lower resolution the specific heat peak will appear as a broadened peak in this concentration range, and the width of the peak first increases and then decreases again as the eutectic point is approached.
- At the eutectic point, the peak passes through a maximum, and it attains the width of the Gaussian resolution function. The intensity of the peak is

down by a factor that is much greater than $(1 - x)$ of the intensity at $x = 0$ due to the unusual properties of the *l***o** phase.

- Above the eutectic point a second broad component of the specific heat can be discerned. For poor resolution (see, e.g., Figure 3a) this new feature reduces to an upper shoulder of the main peak. When x is increased, the sharp specific heat component steadily decreases in intensity without appreciably changing in width and with a position that follows the three-phase line. At $x = x^*$, which marks the high-x terminus of the three-phase line, the sharp component has vanished. At the same time, the broad component first increases in width and then decreases again while its intensity steadily decreases. The broad component matches the extent of the (*l***d**-*l***o**) coexistence region. Beyond the high-x terminus of this coexistence region, the broad component is almost gone. However, there is still a measureably finite specific heat immediately above the terminus due to critical fluctuations. The center of this contribution moves towards higher temperatures as x is increased.
- The *l***o**-(*s***o**-*l***o**) and *s***o**-(*s***o**-*l***o**) phase lines give only very weak signals in the specific heat, since these lines do not involve acyl-chain disordering entropy.

We have extracted the heat content of each of the two specific heat contributions separately. Obviously, there is no unique way of performing such a separation, since the two components overlap. However, in view of the experimental interest in such a separation (Estep et al., 1978; see also Chapter 5 by Finegold and Singer in the present volume), we have carried out such a separation using the following rationale: the total heat content in the range from 32 to 52°C is obtained by integration. The heat content in the sharp component is approximated by the integral of a Gaussian fitted to the sharp component. The heat content in the broad component is then taken to be the difference between the total heat content in the specified range and that of the sharp component. The results obtained by this procedure are given in Figure 5, which gives the enthalpy of the two components relative to the transition enthalpy of the pure system. It can be seen that the enthalpy of the sharp component decreases linearly and vanishes at a cholesterol concentration corresponding to the terminus, x^*, of the three-phase line, cf. Figure 2. In contrast, the broad component has an increasing enthalpy content up to about x^* and then decreases again and becomes vanishingly small above $x \approx 0.5$. Figure 5 also shows the total enthalpy of the mixture, as obtained by integrating over the specified range from 32 to 52°C. The theoretical data in Figure 5 are very similar to the experimental data (Estep et al., 1978).

The cross-sectional area per molecule, A, the hydrophobic thickness, d, and the average acyl chain order parameter, S, were calculated in the mean-field approximation as functions of temperature, T, and cholesterol concentration, x, by Ipsen et al. (1990b). The variation of A with T and x is shown Figure 6. This figure shows that A has two distinct features: a sharp change at the three-phase line, which decreases towards zero as the teminus of the three-phase line at $x =$

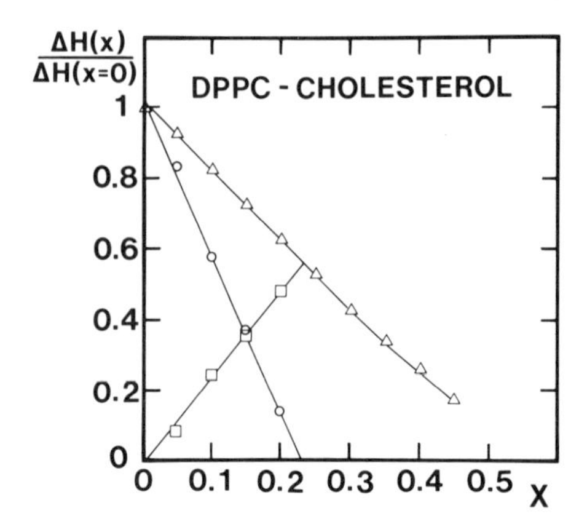

FIGURE 5. Relative enthalpy content $\Delta H(x)/\Delta H(x = 0)$ vs. composition of DPPC-cholesterol bilayers for the sharp (o) and the broad (❑) components of the specific heat, cf. Figure 3, shown together with the total enthalpy (Δ) in the range from 32 to 52°C.

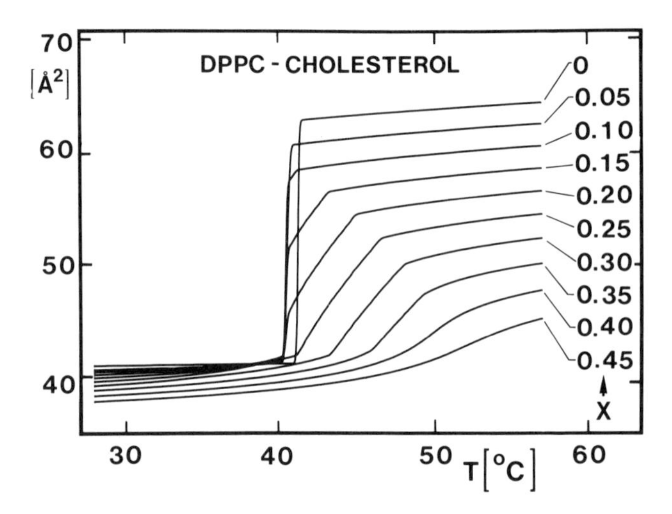

FIGURE 6. Average cross-sectional area, A, per molecule of a DPPC bilayer containing cholesterol. Theoretical results for A are plotted as a function of temperature for different cholesterol concentrations, x.

0.23 is approached, and a flattened portion corresponding to the (*l***o**-*l***d**) coexistence region. The data of Figure 6 clearly demonstrate the strong condensation effect of cholesterol above the main phase transition. Furthermore, the expansion effect of cholesterol can be clearly seen in Figure 6 at low cholestrol concentrations.

The expansion and condensation effects of cholesterol are also reflected in the data for the hydrophobic thickness or average acyl chain length shown in Figure 7. This figure shows that cholesterol thins the bilayer below the main phase transition and thickens it above the main phase transition. The data plotted for d

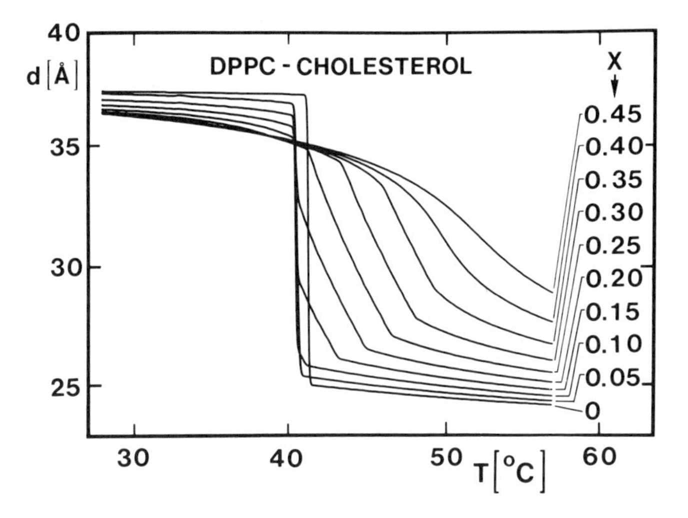

FIGURE 7. Hydrophobic thickness, *d*, of a DPPC bilayer containing cholesterol. Theoretical results for *d* are shown as a function of temperature for different cholesterol concentrations, *x*.

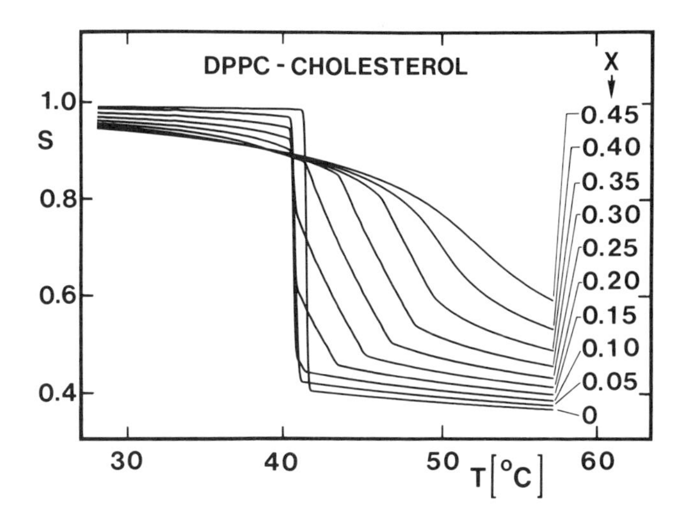

FIGURE 8. Average acyl-chain order parameter, *S*, for a DPPC bilayer containing cholesterol. Theoretical results for *S* are plotted as a function of temperature for different cholesterol concentrations, *x*.

in the two-phase region is the averge hydrophobic thickness. These effects are also reflected in the average acyl-chain order parameter shown in Figure 8. This figure shows that cholesterol disorders the lipid chains below the main phase transition and orders them above the main phase transition. It is important to note that these effects on *S*, as obtained from the calculations, are highly nontrivial results, since the lipid-cholesterol interactions used in our model are nonspecific, being van der Waals interactions.

Ipsen et al. (1990b) have shown that the hydrophobic thickness can be related to the NMR order parameter of the CH bond, which be obtained from the "first moment" of the ^{2}H NMR "half-spectrum" of phospholipid molecules. In a recent publication, Nezil and Bloom (1991) used the effect of cholesterol on the hydrophobic thickness of the membrane, as predicted in Figure 6, to manipulate the thickness of deuterated POPC bilayers containing polypeptides. They examined two types of bilayers: a thin bilayer consisting of pure DOPC, and a thick bilayer composed of DOPC plus 30% cholesterol. They showed that this amount of cholesterol thickens the bilayer from 27 to 32 Å. Using this observation Nezil and Bloom were able to test the effect of hydrophobic thickness on the phase behavior of lipid-polypeptide bilayers, as predicted by the mattress model of Mouritsen and Bloom (1989).

V. NUMERICAL SIMULATIONS OF THE EFFECT OF CHOLESTEROL ON LIPID BILAYERS

In this section we go beyond the mean-field description of Section IVB by presenting results from computer-simulation calculations on a model of lipid-cholesterol intercations. These simulations* take proper account of the thermal density fluctuations suppressed by the mean-field theory. Mouritsen et al. (1983) performed computer simulations of lipid-cholesterol mixtures using a two-state model with exclusively acyl chain conformational degrees of freedom. Only interactions between lipid chains in the lower state and cholesterol were considered. It was found from the simulations that such a simple model provides an incorrect description of the phase transition. In an attempt to model more accurately some aspects of the cooperative phenomena in membranes containing cholesterol, without having to adopt the full ten-state model with crystalline variables described in the previous section, Cruzeiro-Hansson et al. (1989) studied a simpler model that is designed to give the correct phase behavior in the low concentration regime where the peculiar properties of the (*l***o**) phase have not yet manifested themselves, i.e., for concentrations $x \leq 10$ mol%.

The model studied by Cruzeiro-Hansson et al. (1989) is given by

$$H_{TOT} = H_{C0} + H_{cp} + H_{CC} + H_{cC} + H_{CC} + H_{cc} \qquad (10)$$

Equation 10 is the same as the Hamiltonian used in the previous section,

* Computer-simulation methods applied to the study of cooperative phenomena in lipid membranes are reviewed in the article by Mouritsen (1990). Monte Carlo simulation techniques have now become standard tools in the approximate solution of the statistical mechanical problem involved in deriving the thermodynamic properties of microscopic, molecular interaction models. There exists no commercially available software for such calculations, and workers in the field most often prefer to develop and write their own computer programs. In principle, these programs are quite simple, and the art of the simulation discipline is, as in an experiment, related to the proper processing and interpretation of the data. The simulations described in the present chapter can conveniently be carried out on modern fast workstations, e.g., of the type Apollo 10000, Ardent Titan, or the HP 700 series, and the cpu time required is typically of the order of some hundreds of hours.

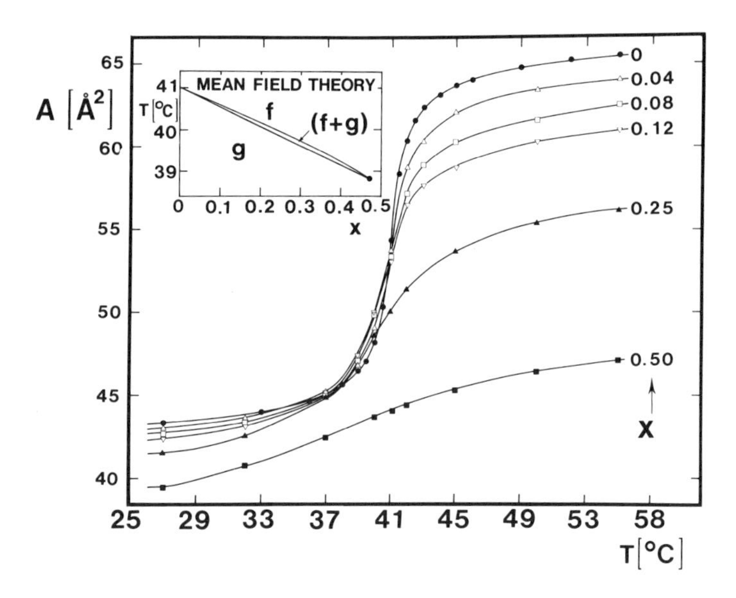

FIGURE 9. Cross-sectional area, $A(T)$, per molecule of the mixture for a DPPC lipid bilayer containing cholesterol in different concentrations, x, as obtained from Monte Carlo simulations on the ten-state Pink model for DPPC bilayers. The insert shows the corresponding phase diagram calculated in the mean-field approximation; g and f denote gel and fluid lipid phases, respectively.

Equations 1–8, with the Potts interaction, Equation 5, removed. I_c was fixed by Cruzeiro-Hansson et al. (1989) by the requirement that the phase diagram of the model resemble that of DPPC-cholesterol mixtures, cf. Figures 1 and 2, for x up to about 10 mol%, i.e., with a very narrow coexistence region and a modest freezing-point depression. A simple mean-field calculation suggests $I_c = 0.45$ to be a suitable choice and leads to the phase diagram shown as an insert in Figure 9. The coexistence region between gel and fluid bilayer phases terminates, according to this model, in a critical end point at $x \cong 0.48$. Despite the fact that the model of Cruzeiro-Hansson et al. should not be considered a proper model of the phase equilibria in the DPPC-cholesterol system above $x \cong 0.1$, results will be presented below for concentrations beyond this limit. The reason for this is twofold: firstly, the model describes general aspects of lipid membranes with intrinsic molecules, and secondly, the model may in fact, also at high concentrations, account for the dominant effect of cholesterol on the general phase behavior, since these effects are closely coupled to the three-phase line that the simple phase diagram of Figure 9 follows rather closely.

A. MEMBRANE AREA, CHAIN STATISTICS, AND LATERAL COMPRESSIBILITY

The Monte Carlo data for the average cross-sectional area per molecule of the DPPC-cholesterol mixture, $A(T)$, are shown in Figure 9. This set of data is very similar to the experimental results for DMPC obtained by Needham et al. (1988) from micromechanical measurements on giant single vesicles. Despite the

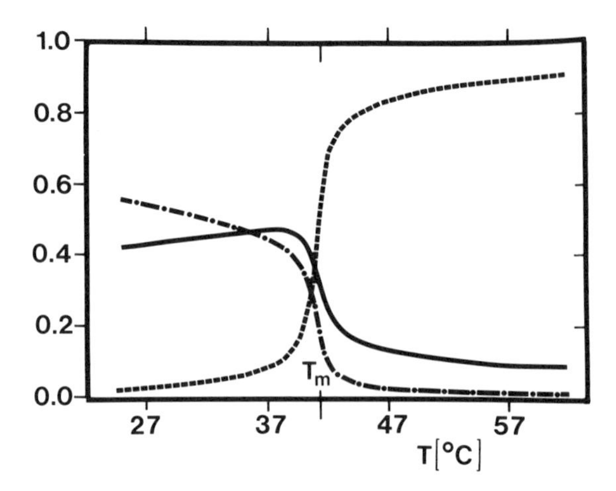

FIGURE 10. Temperature dependence of chain-state occupation variables as obtained from Monte Carlo simulations on the ten-state Pink model of a DPPC bilayer with cholesterol at a concentration of $x = 0.12$. (-----): all-*trans* state $< L_{1i} >$; (- - -): fluid state $< L_{10i} >$; and (———) intermediate states $\sum_{\alpha=2}^{9} < L_{\alpha i} >$.

difference in lipid type and despite the fact that the model is not designed to be valid at high concentrations, there is an amazing similarity between the two data sets. Figure 9 shows that cholesterol leads to a broadening of the transition and that there is a rather complicated relationship between $A(T)$ and x below the transition.

The changes seen in the average cross-sectional area per molecule in Figure 9 reflect the influence of cholesterol on the acyl-chain conformational order that is observed experimentally via the NMR order parameter (Vist and Davis, 1990). This is also clearly demonstrated within the model by the effect on the chain-state occupation variables, as seen in Figure 10 for $x = 0.12$. By comparison with the corresponding results for the pure-lipid bilayer, it is seen that the intermediate chain states are promoted in the transition region by cholesterol. Furthermore, the all-*trans* state is suppressed below the transition, and the fluid state is suppressed above the transition. These findings are in accord with the conventional interpretation of spectroscopic data for lipid-cholesterol membranes (Presti, 1985).

Turning then to the effects of cholesterol on the lateral isothermal compressibility, $\chi(T)$, the Monte Carlo data in Figure 11 demonstrate that the pronounced peak at the transition temperature changes in a very characteristic way: the peak height is reduced as x is increased, and at the same time the intensity of $\chi(T)$ is increased away from the transition. Hence, the lateral density fluctuations are suppressed at the transition point, but enhanced away from the transition. This is a rather remarkable effect that is supported by similar results for the specific heat. Since there are no signs of a phase separation within the resolution limits of the computer simulation, (Cruzeiro-Hansson et al., 1989) the peak position of

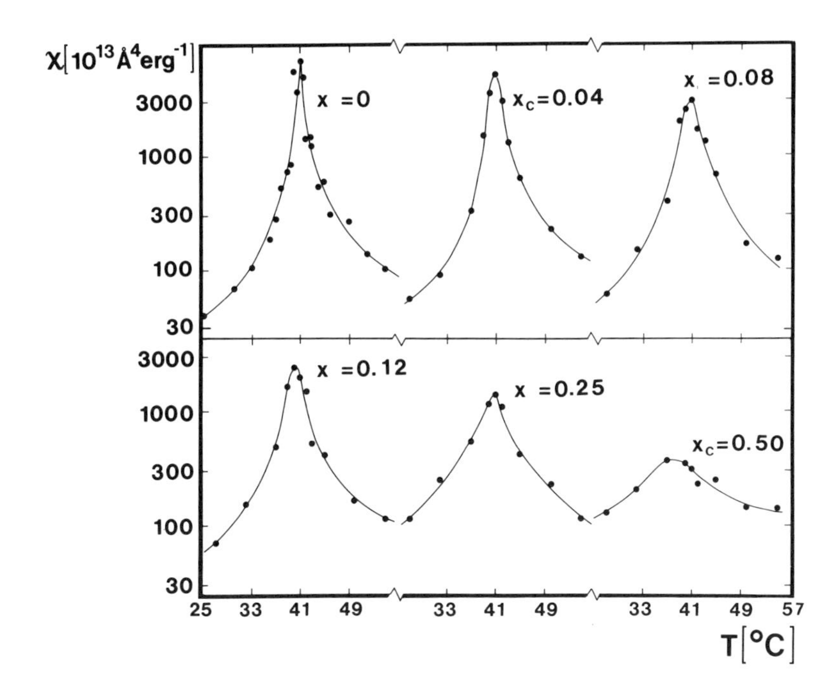

FIGURE 11. Semilogarithmic plot of the isothermal lateral compressibility, $\chi(T)$, for a DPPC bilayer with different concentrations, x, of cholesterol. The data are obtained from Monte Carlo simulations on the ten-state Pink model including lipid-cholesterol interactions.

$\chi(T)$ can be associated with the transition temperature. The compressibility has been measured experimentally by micromechanics (Needham et al., 1988) for DMPC vesicles containing cholesterol. It was found that $\chi(T)$ decreases with x at high concentrations, in agreement with the Monte Carlo results in Figure 11. The experimental data are not able to discern the enhancement in the wings of $\chi(T)$ at low concentrations. The suppression of the intensity of $\chi(T)$ at the transition point is intimately related to the model assumption of cholesterol being mobile in the membrane. For a stationary dilution of cholesterol, the phenomenological model study by Jähnig (1981) shows that the peak intensity is increased by cholesterol (driven towards criticality), in contradiction with experiment (Needham et al., 1988).

B. LATERAL DENSITY FLUCTUATIONS AND LIPID-DOMAIN FORMATION

Cruzeiro-Hansson et al. (1989) have extended their earlier study (Cruzeiro-Hansson and Mouritsen, 1988) of the heterogeneous dynamic membrane structure formed in pure lipid bilayers near the main transition to bilayers containing cholesterol. Snapshots obtained from Monte Carlo calculations on the ten-state Pink model showing the influence of cholesterol are presented in Figure 12 for low cholesterol concentrations. Only the lipid domain boundaries and the lateral distribution of cholesterol molecules are indicated on this figure. It is seen that the presence of cholesterol leads to larger and more ramified domains and even induces a clustering among the domains themselves.

A quantitative analysis of the domain distributions is given in Figure 13 in terms of the average domain size $\bar{l}\,(T,x)$ (in units of molecules), as calculated from the lipid domain distribution function, $n_l^i(T,x)$, i = gel or fluid. The overall behavior of $\bar{l}$ as a function of T and x is very similar to that of the lateral compressibility in Figure 11. In particular, cholesterol at low concentrations decreases the peak intensity at the transition, but increases $\bar{l}$ away from the transition. The enhancement in the wings persists up to about $x \cong 0.25$, whereas at $x = 0.50$, cholesterol suppresses $\bar{l}$ at all temperatures. This finding is consistent with the presence of a critical point in this model in between these two concentration values.

It is not possible to directly monitor the lipid-domain formation phenomena in bilayers. However, in lipid monolayers it has been observed by epifluorescence microscopy (Weis and McConnell, 1985) that cholesterol leads to a ramification of the solid domain interfaces.

C. CHOLESTEROL AT LIPID-DOMAIN INTERFACES

A closer inspection of the snapshots of microconfigurations in Figure 12 indicates that the cholesterol molecules are not randomly distributed in the plane of the membrane, but exhibit a tendency to accumulate at the lipid-domain interfaces. Cruzeiro-Hansson et al. (1989) performed a detailed analysis along the lines described for the pure lipid bilayer case. The fractional interfacial area is defined as the area of the interfaces, with respect to the total area. An interface is defined as the first layer of acyl chains and cholesterol molecules outside a minority phase cluster. In this analysis the fractional interfacial area, a_{i}, is divided into a lipid (L) and a cholesterol (C) part, $a_{\mathrm{i}} = a_{\mathrm{iL}} + a_{\mathrm{iC}}$. It was found (Cruzeiro-Hansson et al., 1989) that at the transition for low x, there is hardly any change in a_{iL} when cholesterol is introduced, whereas a_{iL} increases with x at temperatures away from the transition. In contrast, a_{iC} increases steadily with x at all temperatures. This effect is most clearly seen by examining the excess fractional interfacial area

$$a_{\mathrm{iL}}^{\mathrm{excess}}(x) = a_{\mathrm{iL}}(x) - a_{\mathrm{iL}}(x=0) \tag{11}$$

which is a measure of the change due to cholesterol in the lipid part of the interfacial area, relative to the pure system. The Monte Carlo data for $a_{\mathrm{iL}}^{\mathrm{excess}}\,(T,x)$ are shown in Figure 14. This figure clearly demonstrates that at low cholesterol concentrations and away from the transition temperature, not only is the total interfacial area increased, but also the lipid part of this area is increased. Hence, the interface is not simply increased by an amount corresponding to the accumulated cholesterol. This is a highly nontrivial result that is caused by the cooperative behavior of the lipid-cholesterol mixture and leads to some important consequences for the permeability of the cholesterol-containing mem-

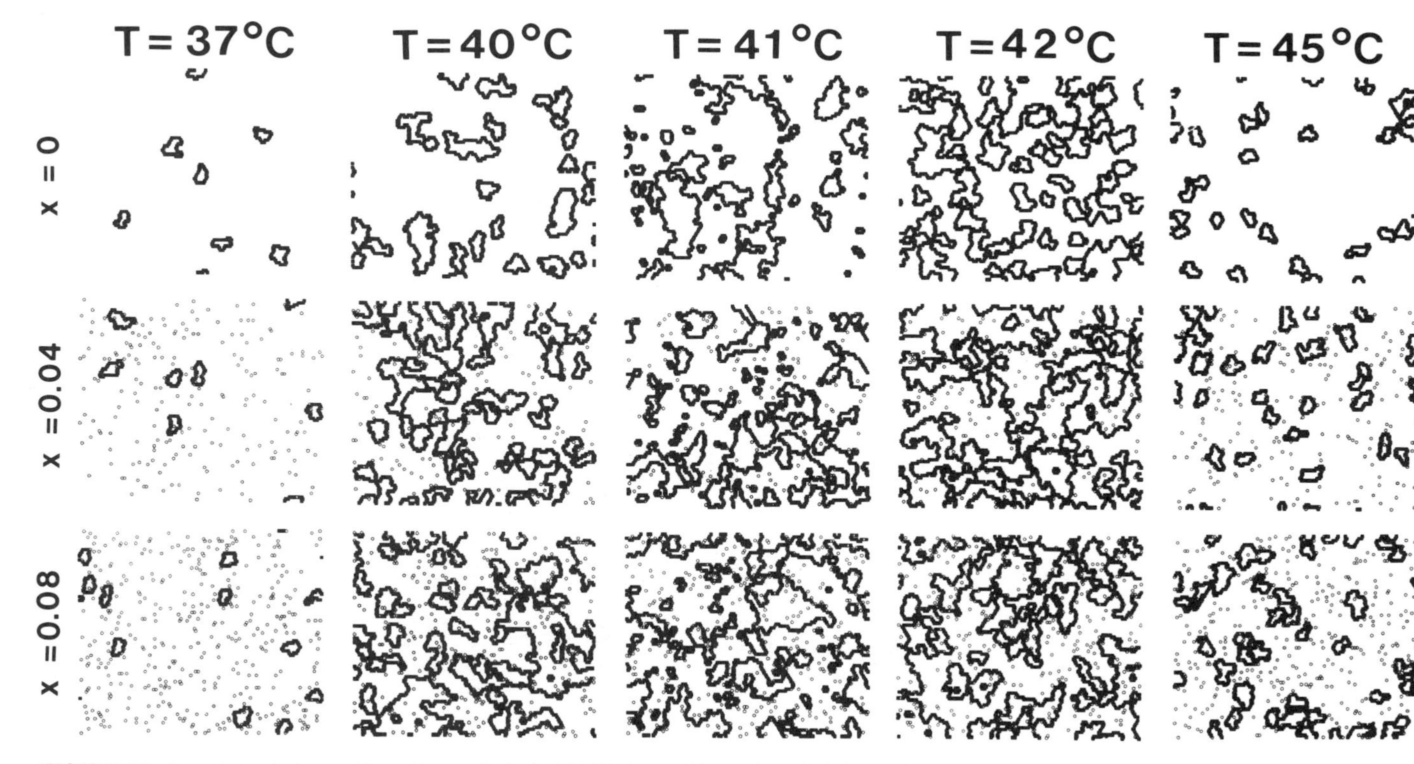

FIGURE 12. Snapshots of microconfigurations typical of a DPPC bilayer with a variety of cholesterol concentrations, x, for different temperatures in the region of the main transition. The data are obtained from Monte Carlo simulations on the ten-state Pink model including lipid-cholesterol interactions. Only the interfacial regions (the solid network) and the cholesterol molecular distribution (o) are shown.

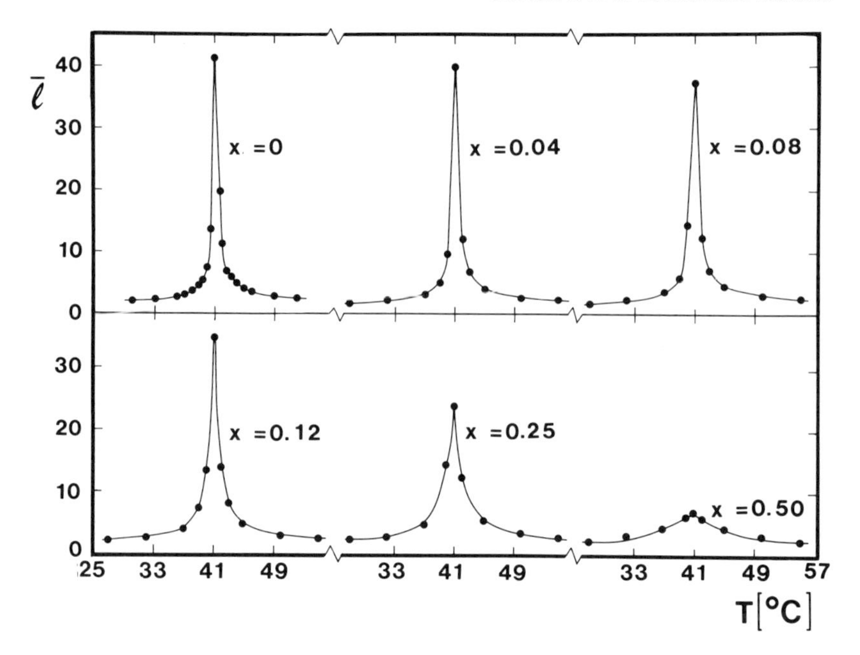

FIGURE 13. Average lipid domain size, $\bar{l}(T, x)$, as a function of temperature and cholesterol concentration, x, in a DPPC bilayer. $\bar{l}$ denotes the number of lipid molecules in the average-sized domain of the minority phase. The results are obtained from Monte Carlo simulations on the ten-state Pink model with lipid-cholesterol interactions.

branes. The subtle temperature dependence of a_{iL}^{excess} in Figure 14 may be rationalized in physical terms by remarking that cholesterol acts to lower the interfacial tension between the lipid domains and the bulk. In general, the interface acts as a sink for impurities. The reason why the effects of cholesterol in low concentrations are so weak at the transition is due to the fact that the interfacial tension at this point is already small due to the proximity of a pseudocritical point (Mouritsen, 1991).

The claimed accumulation of cholesterol at the lipid-domain interfaces is made quantitative by the data presented in Figure 15, which show that the level of cholesterol in the interfacial region is more than twice that of cholesterol in the bulk, for low global cholesterol concentrations, *x*. This accumulating effect is found to be sharply localized in the first interfacial layer (Cruzeiro-Hansson et al., 1989), and the decay to the bulk distribution is very rapid.

Cruzeiro-Hansen et al. (1989) showed that the the finding from the Monte Carlo simulations, of a cholesterol-induced increase in the lipid part of the interfacial area away from the transition, has the consequence that the passive ion permeability (as predicted from a simple model of passive permeability) varies with *x* and *T*, as seen in Figure 16. The simple model (Cruzeiro-Hansson and Mouritsen, 1988) on which these results are based assumes that the main temperature dependence of the passive transmembrane permeability is described by the temperature dependence of the interfacial regions bounding the

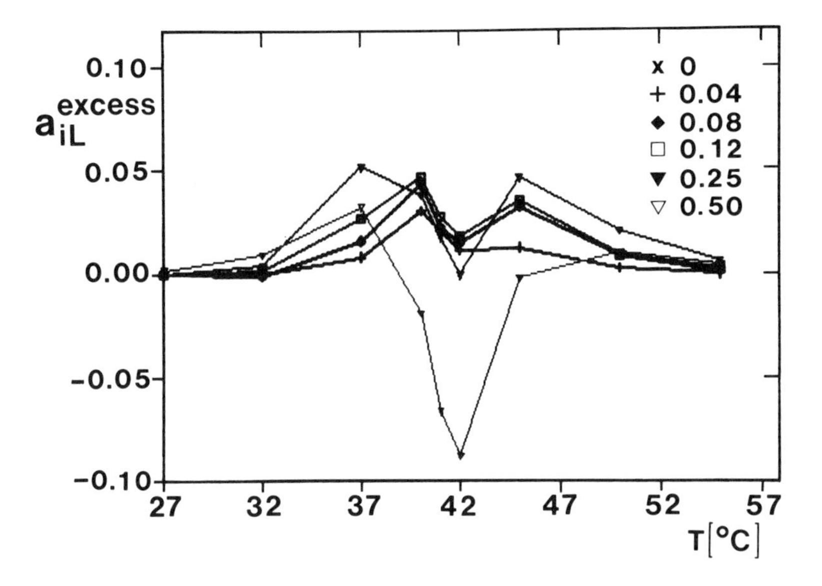

FIGURE 14. Excess fractional interfacial area, a_{iL}^{excess}, of the lipid part of the interfacial area as a function of temperature for a DPPC bilayer containing different concentrations of cholesterol, x. The results are derived from Monte Carlo simulations on the ten-state Pink model including lipid-cholesterol interactions.

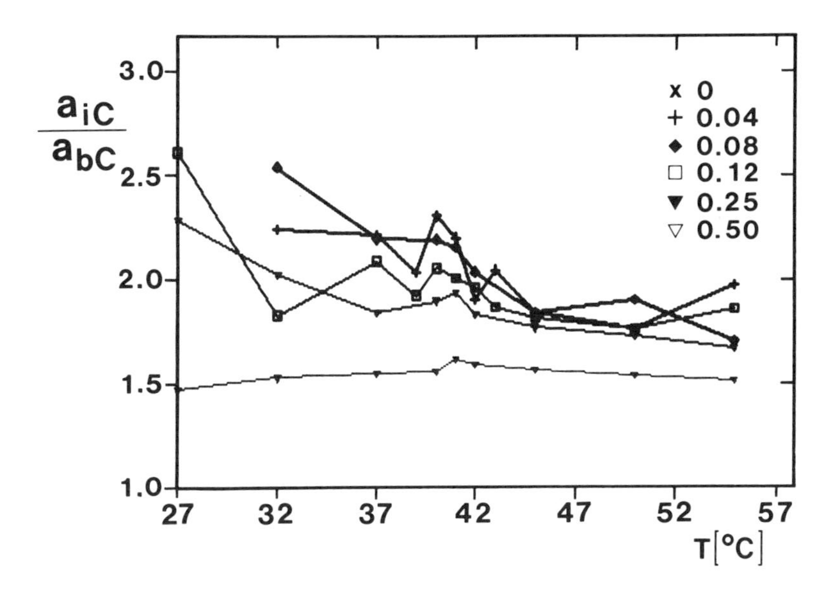

FIGURE 15. Ratio, a_{iC}/a_{bC}, between the cholesterol concentration in the interfaces and in the bulk for different cholesterol concentrations, x, in a DPPC bilayer. The data are obtained from Monte Carlo simulations on the ten-state Pink model with lipid-cholesterol interactions.

bulk and the lipid domains, cf. Figure 12. In the simple model the interfacial region is ascribed a transmission coefficient that is much larger that those describing the bulk and the lipid domains. Figure 16 shows that the permeability is almost constant and increases slightly with cholesterol concentration at low x, but has decreased considerably for high x. The rather scarce experimental data available (Corvera et al., 1992) in the low concentration region shows that the permeability increases considerably with increasing x in this region. However, this is a rather controversial result, since it has been observed in micropipette experiments that the vesicles often break in the transition region: an effect that may lead to an anomalous permeability. A recent micromechanical measurement (E. Evans, unpublished data) has shown that in the case of water permeability, cholesterol leads to a suppression of the permeability even at low cholesterol contents. For high cholesterol concentrations the experimental measurements (Papahadjopoulos et al., 1973) show, in agreement with Figure 16, that the permeability is lowered in the presence of 50-mol% cholesterol. As pointed out by Cruzeiro-Hansson et al. (1989), the influence of cholesterol on the interfacial regions in membranes may also be to inhibit permeation of other molecular species. The experimental finding of a lowered water permeation in membranes containing large amounts of cholesterol (Carruthers and Melchior, 1983; Blok et al., 1977) is in accordance with this suggestion. Moreover, the quantitative result in Figure 16 of a strong suppression of the permeability at high cholesterol concentrations is in line with the general idea of cholesterol being a molecular agent that assures high mechanical coherence and low leakiness of membranes (Bloom et al., 1991; Needham et al., 1988).

Experimental measurements of the kinetics of the phase transition in lipid bilayers containing cholesterol have been interpreted in terms of cluster-formation phenomena (Blume and Hillmann, 1986; Genz et al., 1986). For example, it has been found that for $x = 0.075$ the kinetic processes associated with domain formation are suppressed at the transition, but enhanced away from the transition, in agreement with Figure 13. At higher cholesterol concentrations the kinetics associated with the cooperative domain-formation processes are fully suppressed (Genz et al., 1986). Furthermore, extrapolation of the interpretation of specific heat anomalies for pure lipid bilayers in terms of domain distributions (Freire and Biltonen, 1978) to the specific heat measured for cholesterol-containing membranes (Mabrey et al., 1978; Estep et al., 1978) bears further testimony to the model results in Figure 13.

Finally, it should be pointed out that the finding of the ability of cholesterol to modulate the interfacial regions of membranes is of marked interest in relation to studies of interfacially active enzyme processes (Sandermann, 1978; Gheriani-Gruszka, 1988). The specific example of pancreatic phospholipase A_2 is particularly interesting in this context. In the presence of cholate, which has a structure similar to that of cholesterol, it was found (Gheriani-Gruszka, 1988) that small amounts of cholate increase the rate of activation of the phospholipase away from the phase transition. Since it has been suggested (Menashe et al., 1986) that the rate of activation is closely linked to the lipid-membrane fluctuations and the

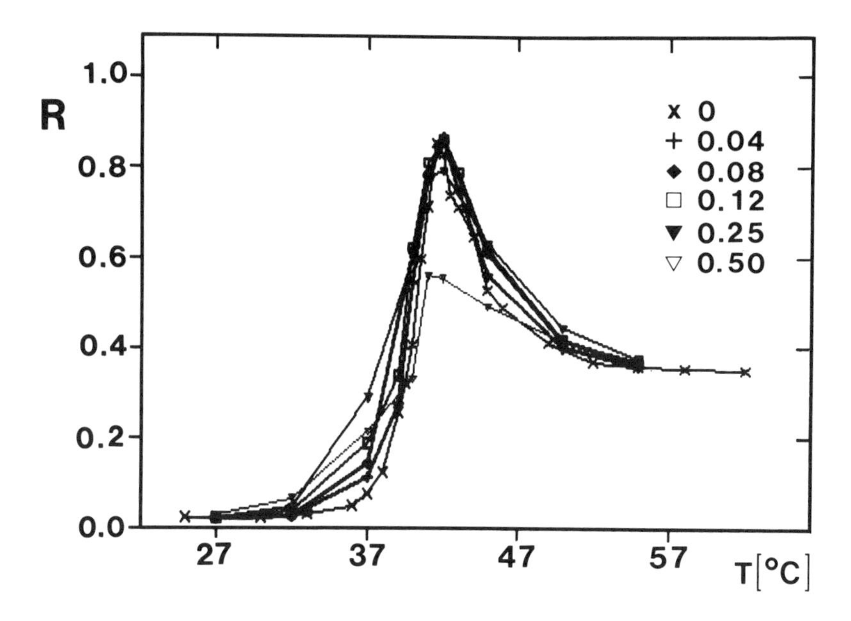

FIGURE 16. Reduced permeability, $R(T)$, of Na^+-ions in DPPC liposomes as a function of temperature and cholesterol concentration, x. $R(T)$ is given in arbitrary units. The data are obtained from Monte Carlo simulations on the ten-state Pink model with lipid-cholesterol interactions.

formation of a particular interfacial environment that supports the active conformation of the enzyme, the effect of cholate is in good agreement with the results of the computer simulation on the simple model of cooperative phenomena in membranes with small intrinsic molecules.

D. CHAIN-LENGTH DEPENDENCE

Corvera et al. (1992) used the model of Cruzeiro-Hansson et al. (1989) for lipid-cholesterol bilayers at low cholesterol concentrations to predict the thermodynamic properties and the passive ion permeability of these systems as a function of acyl-chain length and cholesterol concentration. Numerical simulations based on the Monte Carlo method were again used to simulate the equilibrium state of the system near the main gel-fluid phase transition. The theoretical results can be summarized as follows:

1. The effect of cholesterol on DMPC, DPPC, and DSPC bilayers is to broaden the main phase transition.
2. In all three systems it was shown, by calculating the response functions (specific heat and lateral compressibility), that the effect of cholesterol is to cause the thermal fluctuations to decrease at the main phase transition, but to increase in the thermal wings of the transition. In addition, it was found that, for all cholesterol concentrations studied, the longer the chain length, the larger the fluctuations at the transition temperature. In the

wings of the transition, however, an enhancement is observed in the response functions for decreasing chain length.

3. It was shown that the permeability of systems with low cholesterol content increased substantially as a function of cholesterol concentration for all temperatures in the transition region. This result is presented in Figure 17. Furthermore, the model predicts that the relative permeability increases with decreasing chain length at a given value of the reduced temperature for all the cholesterol concentrations studied. This is in contrast to the case of lipid bilayers containing high cholesterol concentrations, where the cholesterol strongly supresses the permeability.
4. The model predicts that for DMPC-cholesterol and DPPC-cholesterol bilayers, the peak in the relative permeability occurs slightly above T_m for all cholesterol concentrations studied. This is in agreement with the experimental results of Corvera et al. (1992). The prediction is due to the aggregation of lipid domains very close to T_m which causes the fractional area of the interfaces to decrease.

The experimental results for C15PC- and C16PC (DPPC)-cholesterol bilayers (Corvera et al., 1992) demonstrate that there is an increase in the permeability between the pure bilayers and bilayers containing 10% cholesterol. Furthermore, the maximum in the permeability is found to decrease with increasing chain length, confirming the predictions given in (3) above. It was also found experimentally that for values of the concentration greater than 20 mol%, the permeability decreases with increasing cholesterol concentration. This is in agreement with the theoretical results shown in Figure 17. It should be remarked at this point, that these permeability measurements may suffer from artifacts due to breaking of the vesicles as the membrane system is taken through the transition region.

VI. SUMMARY AND DISCUSSION

We have in this paper shown that the unusual phase equilibria, cf. Figures 1 and 2, of the DPPC-cholesterol bilayer system give rise to a set of specific heat thermal profiles with distinct anomalies, which reflect the topology of the underlying phase diagram. A number of specific heat peaks can be discerned, depending on the thermal resolution and the composition of the membrane, cf. Figures 3 and 4. Our results are based on a theoretical calculation on the microscopic interaction model for lipid bilayers containing cholesterol, presented in Sections III and IV, and hence, the specific heat is basically derived from first principles, with no special mechanisms having been assumed. It has been pointed out that it is possible to rationalize the experimental phase diagram if proper account is taken of both the conformational and translational degrees of freedom of the lipid molecules. The key to understanding the lipid-cholesterol phase behavior lies in anticipating the very different ways cholesterol interacts with these two types of degree of freedom. We therefore argue that an interpre-

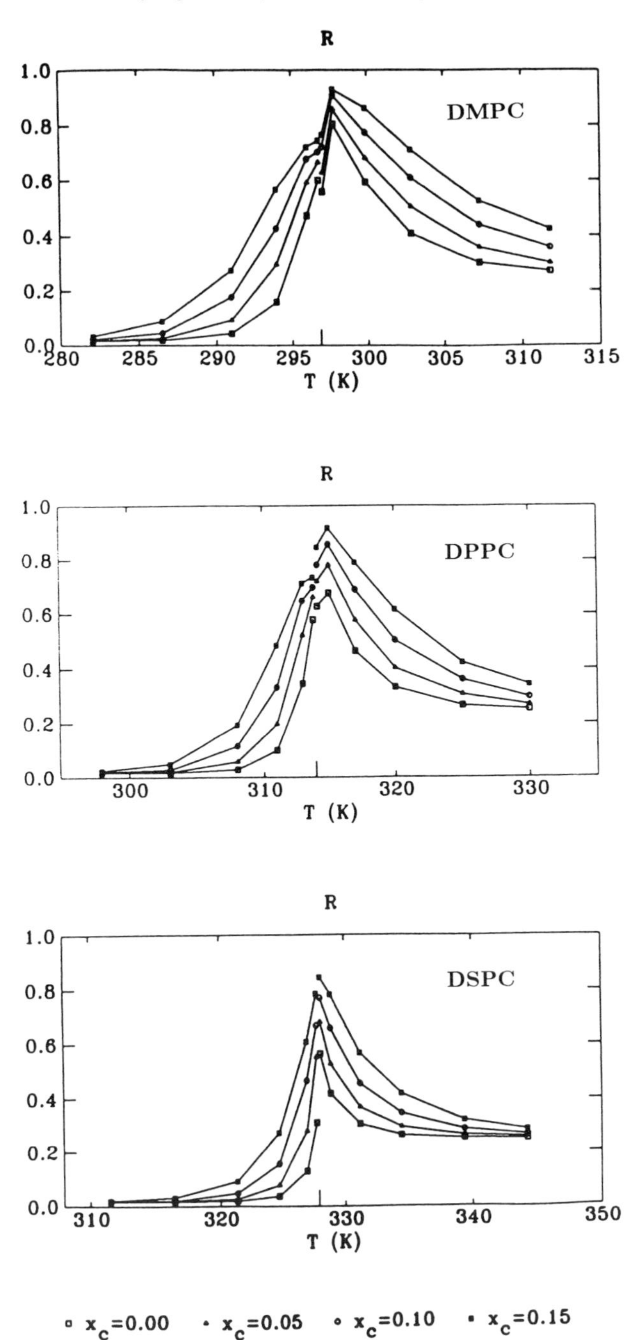

FIGURE 17. Numerical results for the relative permeability, R, of DMPC-cholesterol, DPPC-cholesterol and DSPC-cholesterol bilayers. The units of R are arbitrary. The data above and below T_m appear disconnected in order to show more clearly the behavior at both sides of T_m.

tation of the experimentally observed specific heat anomalies does not require specific molecular mechanisms to be operative, such as long-lived lipid-cholesterol complexes or the formation of special interfacial regions (Estep et al., 1978; Imaizumi and Hatta, 1984). The absence of hydrogen bonding in such complexes and, given our interpretation, the absence of the complexes themselves in lipid-cholesterol systems can be inferred from recent calorimetric measurements by Lai et al. (1985) on DPPC-cholesteryl hemisuccinate multibilayers.

As pointed out by Ipsen et al. (1987), the theoretical phase diagram, as derived from the microscopic model, is in close agreement with the most reliable experimental measurements of the phase equilibria in the DPPC-cholesterol system (see Figure 1). It should be mentioned, however, that there are still some experimental controversies regarding the low-temperature structure of the phase diagram and the interference with the ripple phase (Mortensen et al., 1988). It is notoriously difficult to determine accurately the equilibrium phase diagram for mixtures using bulk techniques, such as calorimetry, or local probe techniques, such as magnetic resonance (for a recent discussion, see Ipsen and Mouritsen, 1988). In the case of calorimetry, this is due to the difficulty of relating the shape of the specific heat scan to the phase boundaries, given a finite thermal resolution. Often not all of the anomalies can be resolved, as is clearly demonstrated by the present work, cf. Figures 3 and 4. In the case of local probe techniques, the probe molecules may report on local density fluctuations rather than macroscopic equilibrium phases and, hence, lead to a misinterpretation of the phase lines. Our results in Figure 3 suggest that a most careful calorimetric study, with very slow scan rates (and consequently high resolution) at low cholesterol concentration, may reveal new features in the specific heat, which should prove useful in determining accurately the extent of the (*s***o**-*l***d**) coexistence region and the location of the low-x terminus of the three phase line.

The specific heat traces shown in Figures 3 and 4, and their thermal anomalies are very similar, even quantitatively, to experimental calorimetry measurements of several experimental groups (Mabrey et al., 1978; Estep et al., 1978; Imaizumi and Hatta, 1984; Vist and Davis, 1990). In particular, the calculations closely reproduce the second broad component rather accurately. Moreover, the separation of the total heat content into contributions attributed to the two peaks (see Figure 5) is in quantitative accordance with experimental findings (Estep et al., 1978; see also Chapter 5 by Finegold and Singer in this volume).

Mechanical properties of the lipid-cholesterol bilayers were presented in Figure 6–8, which showed the disordering and expansion effect of cholesterol at low temperatures and the ordering and condensation effect at high temperatures. Comcomitant with these effects is a membrane-thinning effect that sets in when the temperature is raised.

The results shown in Figures 2 to 5 were obtained for a set of model parameters pertinent to DPPC bilayers. However, our general findings for the specific heat anomalies will also apply to other lipid membranes containing cholesterol if they have a phase diagram of the type shown in Figure 2. Specifically, we expect the results to apply to DMPC membranes for which it is found experimentally (Mabrey et al., 1978; Imaizumi and Hatta, 1984) that two sharp specific heat peaks can be discerned around $x \approx 0.15$. Within our model this can be rationalized by a phase diagram similar to that in Figure 2, where the *l***d**-(*l***d**-*l***o**) phase line is more horizontal (and the (*l***d**-*l***o**) coexistence region more narrow) and therefore gives rise to a sharper and more intense second component.

It is important to point out that the theoretical calculations, which led to the results shown in Figures 2 to 5, were performed in a mean-field approximation to the statistical mechanical problem posed by the microscopic interaction model. This approximation suppresses lateral density fluctuations, which may have some effect on the phase equilibria, in particular close to $x = 0$ where the main phase transition in the pure system is known to be strongly influenced by fluctuations (Mouritsen, 1991). Consequently, mean-field theory neglects effects due to inhomogeneous lateral distributions of cholesterol and domain formation near the phase lines. Such effects will lead to wings in the specific heat peaks, cf. Figures 3 and 4, but are not expected to change quantitatively the overall behavior of the specific heat as a function of temperature and cholesterol content.

We have shown that by taking the theoretical description beyond the mean-field approximation, using computer simulation of microscopic, statistical, mechanical models, it is possible to calculate the effects of density fluctuations as these manifest themselves in macroscopic properties, cf. Figures 9 and 11, and and in the microscopic lateral organization of the membrane, cf. Figures 10, 12, and 13. It was found that the presence of cholesterol in low amounts led to increased dynamic membrane heterogeneity, cf. Figures 12, 14, and 15, and enhanced transmembrane permeability, cf. Figures 16 and 17. We propose that cholesterol acts as an interfacially active agent that can be used to control membrane organization in the transition region. At high cholesterol contents the membrane fluctuations are strongly suppressed, leading to a lowered permeability. For these high cholesterol contents, the membrane is in the liquid-ordered phase, which is characterized by high acyl-chain conformational order and fluid-like viscosity. The liquid-ordered phase is the physiologically important membrane phase, and it has very exceptional material characteristics (Bloom et al., 1991).

Finally, we would like to emphasize that the models of lipid-cholesterol interactions presented in this paper are two-dimensional models and therefore do not include detailed three-dimensional interactions between the lipid and cholesterol molecules. The reader is referred to the paper of Scott (Chapter 8 in this volume) for details of full three-dimensional modeling.

In summary, the use of theoretical models in studying the phase behavior of lipid bilayers containing cholesterol has greatly helped promoting our understanding of the basic physical principles that control the phase behavior. In particular, the modeling has helped in the interpretation of a variety of experimental data and, furthermore, served as a useful guide in proposing more well-focused experiments. The modeling is still, however, at a rather primitive stage, and much more detailed and accurate modeling is required to make substantial progress in relation to the grand problem of the relationship between microscopic lipid-cholesterol interactions and the exceptional macroscopic material characteristics of lipid membranes containing cholesterol (Bloom et al., 1991). It is to be anticipated that modern computer-simulation techniques (Mouritsen, 1990) will continue to provide a powerful tool to elucidate the cooperative phenomena in models of lipid membranes and how such phenomena are influenced by cholesterol.

ACKNOWLEDGMENTS

This work was supported by the Danish Natural Science Research Council under grants J.nr. 11-7785 and 11-7488, by The Danish Research Academy under grant J.nr. S910015, by NSERC of Canada, and by Le FCAR du Quebec under a center grant and a team grant. MJZ wishes, in particular, to express his gratitude to Dennis Chapman, Sebastian Doniach, Martin Grant, Alexandros Georgallas, David Pink, Erich Sackmann, and Michael Singer for many stimulating discussions over the past years.

REFERENCES

Alecio, M. R., D. E. Golan, W. R. Veatch, and R. R. Rando. 1982. *Proc. Natl. Acad. Sci. U.S.A.*, 79:5171.

Blok, M. C., L. L. M. van Deenen, and J. de Gier. 1977. *Biochim. Biophys. Acta*, 464:509.

Bloom, M., E. Evans, and O. G. Mouritsen. 1991. *Q. Rev. Biophys.*, 24:293.

Blume, A. and M. Hillmann. 1986. *Eur. Biophys. J.*, 13:342.

Caillé, A., D. A. Pink, F. de Verteuil, and M. J. Zuckermann. 1980. *Can. J. Phys.*, 58:581.

Carruthers, A. and D. L. Melchior. 1983. *Biochemistry*, 22:5759.

Chapman, D., R. M. Williams, and B. D. Ladbrooke. 1968. *Chem. Phys. Lipids*, 1:44.

Copeland, B. R. and H. M. McConnell. 1980. *Biochim. Biophys. Acta*, 599:95.

Corvera, E., O. G. Mouritsen, M. A. Singer, and M. J. Zuckermann. 1992. *Biochim. Biophys. Acta*, 1107:261.

Cruzeiro-Hansson, L. and O. G. Mouritsen. 1988. *Biochim. Biophys. Acta*, 944:63.

Cruzeiro-Hansson, L., J. H. Ipsen, and O. G. Mouritsen. 1989. *Biochem. Biophys. Acta*, 974:166.

Davis, J. H. 1988. In *Proc. Int. School Phys. 'Enrico Fermi' on Medical Applications of NMR*, Varenna, Maraviglia, B., Ed., North Holland, Amsterdam, p. 302.

Demel, R. A. and B. De Kruyff. 1976. *Biochim. Biophys. Acta*, 457:109.

Doniach, S. 1978. *J. Chem. Phys.*, 68:4912.

Dufoure, E. J., E. J. Parish, S. Chitrakorn, and I. C. P. Smith. 1984. *Biochemistry*, 23:6062.

Engelman, D. M. and J. E. Rothman. 1972. *J. Biol. Chem.*, 247:3694; Huang, C.-H. 1976. *Lipids*, 12:348.

Estep, T. N., D. B. Mountcastle, R. L. Biltonen, and T. E. Thompson. 1978. *Biochemistry*, 17:1984.

Evans, E. and D. Needham. 1986. *Faraday Discuss. Chem. Soc.*, 81:267.

Freire, E. and R. L. Biltonen. 1978. *Biochim. Biophys. Acta,* 514:54.

Genz, A., J. F. Holzwarth, and T. Y. Tsong. 1986. *Biophys. J.*, 50:1043.

Gheriani-Gruszka, N., S. Almog, R. L. Biltonen, and D. Lichtenberg. 1988. *J. Biol. Chem.*, 263:11808.

H. L. Scott. 1992. Lipid-cholesterol phase diagrams: theroretical and numerical aspects. In *Cholesterol in Membrane Models*, Finegold, L., Ed., CRC Press, Boca Raton, FL.

Hicks, A., M. Dinda, and M. Singer. 1987. *Biochim. Biophys. Acta,* 903:177.

Hui, S. W. and N.-B. He. 1983. *Biochemistry*, 22:1159.

Imaizumi, S. and I. Hatta. 1984. *J. Phys. Soc. Jpn.*, 53:4476.

Ipsen, J. H. and O. G. Mouritsen. 1988. *Biochim. Biophys. Acta*, 944:121.

Ipsen, J. H., G. Karlström, O. G. Mouritsen, H. Wennerström, and M. J. Zuckermann. 1987. *Biochim. Biophys. Acta,* 905:162.

Ipsen, J. H., O. G. Mouritsen, and M. Bloom. 1990. *Biophys. J.*, 57:405.

Ipsen, J. H., O. G. Mouritsen, and M. J. Zuckermann. 1989. *J. Chem. Phys.*, 91:1855.

Ipsen, J. H., O. G. Mouritsen, and M. J. Zuckermann. 1990. *Biophys. J.*, 56:661.

Jackson, M. B. 1976. *Biochemistry*, 15:299.

Jähnig, F. 1981. *Biophys. J.*, 36:329–347.

Kawato, S., K. Kinosita, and A. Ikegami. 1978. *Biochemistry*, 17:5026.

Knoll, W., G. Schmidt, K. Ibel, and E. Sackmann. 1985. *Biochemistry*, 24:5240.

Kusumi, A., W. K. Subczynski, M. Pasenkiewicz-Gierula, J. S. Hyde, and H. Merkle. 1986. *Biochim. Biophys. Acta,* 854:307.

Ladbrooke, B. D., R. M. Williams, and D. Chapman. 1968. *Biochim. Biophys. Acta,* 150:333.

Lai, M.-Z., N. Düzgünes, and F. C. Szoka. 1985. *Biochemistry*, 24:1646.

Lentz, B. R., D. A. Barrow, and M. Hoechli. 1980. *Biochemistry*, 19:1943.

Levin. I. W., E. Keihn, and W. C. Harris. 1985. *Biochim. Biophys. Acta,* 820:40.

Lindblom, G., L. B. A. Johansson, and G. Arvidson. 1981. *Biochemistry*, 20:2204.

Lippert, J. L. and W. L. Peticolas. 1971. *Proc. Natl. Acad. Sci. U.S.A.*, 68:1572.

Mabrey, S., P. L. Mateo, and J. M. Sturtevant. 1978. *Biochemistry,* 17:2464.

Martin, R. B. and P. L. Yeagle. 1978. *Lipids*, 13:594.

Marčelja, S. 1976. *Biochim. Biophys. Acta,* 455:1.

McIntosh, T. J. 1978. *Biochim. Biophys. Acta,* 513:43.

Melchior, D. L., F. J. Scavitto, and J. M. Steim. 1980. *Biochemistry*, 19:4828.

Menashe, M., G. Romero, R. L. Biltonen, and D. Lichtenberg. 1986. *J. Biol. Chem.*, 261:5328.

Mortensen, K., W. Pfeiffer, E. Sackmann, and W. Knoll. 1988. *Biochim. Biophys. Acta,* 945:221.

Mouritsen, O. G. 1990. In *Molecular Description of Biological Membrane Components by Computer Aided Conformational Analysis*, vol. I, Brasseur, R., Ed., CRC Press, Boca Raton, FL, pp. 8–83.

Mouritsen, O. G. 1991. *Chem. Phys. Lipids,* 57:179.

Mouritsen, O. G. and M. Bloom. 1984. *Biophys. J.*, 46:141.

Mouritsen, O. G. and M. J. Zuckermann. 1987. *Chem. Phys. Lett.,* 135:294.

Mouritsen, O. G., A. Boothroyd, R. Harris, N. Jan, L. MacDonald, D. A. Pink, and M. J. Zuckermann. 1983. *J. Chem. Phys.,* 79:2027.

Needham, D., T. J. McIntosh, and E. Evans. 1988. *Biochemistry*, 27:4668.

Nezil, F. and M. Bloom. 1991. *Biophys. J.*, submitted.

O'Leary, T. 1983. *Biochim. Biophys. Acta,* 731:47.

O'Leary, T. J. and I. W. Levin. 1986. *Biochim. Biophys. Acta*, 854:321.

Owicki J. G., and H. M. McConnell. 1979. *Proc. Natl. Acad. Sci. U.S.A.*, 76:4750.

Owicki, J. G. and H. M. McConnell. 1980. *Biophys. J.*, 30:383; Slater, G. and A. Caillé. 1982. *Biochim. Biophys. Acta,* 686:249.

Papahadjopoulos, D., K. Jacobsen, S. Nir, and T. Isac. 1973. *Biochim. Biophys. Acta,* 311:330.

Pink, D. A. and C. E. Carroll. 1978. *Phys. Lett.*, 66A:157.

Pink, D. A. and L. MacDonald. 1988. *Biochim. Biophys. Acta,* 937:417.

Pink, D. A., T. J. Green, and D. Chapman. 1981. *Biochemistry*, 20:6692.

Presti, F. T. 1985. In *Membrane Fluidity in Biology*, Aloia, R. C. and J. M. Boggs, Eds., Academic Press, New York, pp. 97–146.

Presti, F. T. and S. I. Chan. 1982. *Biochemistry*, 21:3821.

Presti, F. T., R. J. Pace, and S. I. Chan. 1982. *Biochemistry*, 21:3831

Recktenwald, D. J. and H. M. McConnell. 1981. *Biochemistry*, 20:4505; Shimshick, E. J. and H. M. McConnell. 1973. *Biochem. Biophys. Res. Commun.*, 53:446.

Rooney, M., W. Tamura-Lis, L. J. Lis, S. Yachnin, O. Kucuk, and J. W. Kauffman. 1986. *Chem. Phys. Lipids*, 41:81.

Rubenstein, J. L. R., B. A. Smith, and H. M. McConnell. 1979. *Proc. Natl. Acad. Sci. U.S.A.*, 76:15.

Sandermann, H. 1978. *Biochim. Biophys. Acta,* 515:209.

Slater, G. and A. Caillé. 1981. *Phys. Lett.*, 86A:256.

Smutzer, G. and P. L. Yeagle. 1985. *Biochim. Biophys. Acta,* 814:274.

Snyder, B. and E. Freire. 1980. *Proc. Natl. Acad. Sci. U.S.A.*, 77:405.

Stockton, G. W. and I. C. P. Smith. 1976. *Chem. Phys. Lipids,* 17:251; Jacobs, R. and E. G. Oldfield. 1979. *Biochemistry*, 18:3280.

Tessier-Lavigne, M., A. Boothroyd, M. J. Zuckermann, and D. A. Pink. 1982. *J. Chem. Phys.*, 76:4587.

Ulmius, J., H. Wennerström, G. Lindblom, and G. Arvidson. 1975. *Biochim. Biophys. Acta,* 389:197.

Van Ginkel, G., L. J. Korstanje, H. Van Langen, and Y. K. Levine. 1986. *Faraday Discuss. Chem., Soc.*, 81:49.

Vist, M. R. and J. H. Davis. 1990. *Biochemistry*, 29:451.

Weis, R. M. and H. M. McConnell. 1985. *J. Chem. Phys.*, 89:4453.

Yeagle, P. L. 1985. *Biochim. Biophys. Acta*, 822:267.

Zuckermann, M. J. and O. G. Mouritsen. 1987. *Eur. Biophys. J.*, 15:77.

AUTHOR INDEX

A

B

C

L

M

S

T

SUBJECT INDEX

D

K

L

M

N

O

P

Q

R

S

T

V

W

X

Y

Z